T0140556

Fast Track Land Occupations in Zimbabwe

Kirk Helliker · Sandra Bhatasara ·
Manase Kudzai Chiweshe

Fast Track Land Occupations in Zimbabwe

In the Context of the Zvimurenga

Springer

Kirk Helliker
Department of Sociology
Rhodes University
Makhanda, South Africa

Sandra Bhatasara
Department of Sociology
University of Zimbabwe
Harare, Zimbabwe

Manase Kudzai Chiweshe
Department of Sociology
University of Zimbabwe
Harare, Zimbabwe

ISBN 978-3-030-66350-6   ISBN 978-3-030-66348-3   (eBook)
https://doi.org/10.1007/978-3-030-66348-3

© The Editor(s) (if applicable) and The Author(s), under exclusive license to Springer Nature Switzerland AG 2021
This work is subject to copyright. All rights are solely and exclusively licensed by the Publisher, whether the whole or part of the material is concerned, specifically the rights of translation, reprinting, reuse of illustrations, recitation, broadcasting, reproduction on microfilms or in any other physical way, and transmission or information storage and retrieval, electronic adaptation, computer software, or by similar or dissimilar methodology now known or hereafter developed.
The use of general descriptive names, registered names, trademarks, service marks, etc. in this publication does not imply, even in the absence of a specific statement, that such names are exempt from the relevant protective laws and regulations and therefore free for general use.
The publisher, the authors and the editors are safe to assume that the advice and information in this book are believed to be true and accurate at the date of publication. Neither the publisher nor the authors or the editors give a warranty, expressed or implied, with respect to the material contained herein or for any errors or omissions that may have been made. The publisher remains neutral with regard to jurisdictional claims in published maps and institutional affiliations.

This Springer imprint is published by the registered company Springer Nature Switzerland AG
The registered company address is: Gewerbestrasse 11, 6330 Cham, Switzerland

Preface

This book is an initiative of the Unit of Zimbabwean Studies in the Department of Sociology at Rhodes University in South Africa, of which the lead author (Kirk Helliker) is the Head. The Unit focuses on the supervision of Master's and Ph.D. students working on a diverse range of topics on Zimbabwean history and society, as well as on research projects and publications led by the Head. The other two authors (Dr. Sandra Bhatasara and Dr. Manase Kudzai Chiweshe) are former Ph.D. students of Professor Helliker, and they both lecture currently in the Department of Sociology at the University of Zimbabwe. Together, we edited a recent book titled *The Political Economy of Livelihoods in Contemporary Zimbabwe* (Routledge Press, 2018) which consisted of chapters written by current or former Master's and Ph.D. students of Professor Helliker.

This book emerged out of research undertaken in Shamva and Bindura Districts by Professor Helliker and Dr. Bhatasara in 2015 and 2016 on the nation-wide land occupations from the year 2000, with these occupations leading to the Zimbabwean state's Fast Track Land Reform Programme. Two journal articles were published in 2018 based on this research, in the *Journal of Asian and African Studies* and the *African Studies Quarterly Journal*. Based on the findings from this research, and a reading of the prevailing scholarly literature on the broader occupations throughout the Zimbabwean countryside, it seemed that there were important parallels between these occupations and two earlier major episodes of struggle around land in Zimbabwean history, namely, the anti-colonial revolt of 1896–97 and the war waged by guerrillas during the 1970s. The ruling party labels each of these three episodes as a war of liberation, or *chimurenga* (with *zvimurenga* as the plural).

The main purpose of this book is to understand and examine the land occupations from the year 2000, which we label as the "fast track" occupations. In doing so, we provide a comparative and historical analysis of the fast track occupations in the context of the two earlier heightened periods of struggle. This entails a comprehensive overview of the literature on the *zvimurenga* alongside the presentation of original research. Despite the deep significance of these three sets of struggles to the configuring of past and present Zimbabwe, no historical-comparative analysis exists.

We hope that this book will be of value to both established and young scholars in
Zimbabwe.

Makhanda, South Africa   Kirk Helliker
Harare, Zimbabwe   Sandra Bhatasara
Harare, Zimbabwe   Manase Kudzai Chiweshe

Contents

About the Authors

Kirk Helliker is a Research Professor and Head of the Unit of Zimbabwean Studies in the Department of Sociology at Rhodes University, South Africa. In 2018, he co-edited two books by Routledge titled *Politics at a Distance from the State: Radical and African Perspectives* and *The Political Economy of Livelihoods in Contemporary Zimbabwe*. He supervises a large number of Zimbabwean Ph.D. students, and writes primarily on livelihoods, land struggles, civil society and democratisation in Zimbabwe.

Sandra Bhatasara has a Ph.D. in Sociology from Rhodes University, South Africa. She is a senior lecturer in the Sociology Department at the University of Zimbabwe and is also a Research Associate in the Department of Sociology at Rhodes University. Her research focuses on intersectional studies of gender, land and agrarian issues, environment and social dimensions of climate change in Zimbabwe, and southern Africa. She co-edited *The Political Economy of Livelihoods in Contemporary Zimbabwe*, a book published by Routledge in 2018.

Manase Kudzai Chiweshe has a Ph.D. in Sociology from Rhodes University, South Africa. He is a Senior Lecturer in the Sociology Department, University of Zimbabwe and a Research Associate in the Department of Sociology, Rhodes University. He is the winner of the 2015 Gerti Hessling Award for the best paper in African studies. His work revolves around the sociology of everyday life in African spaces with special focus on promoting African ways of knowing with specific reference to agrarian studies, livelihoods, gender and football studies.

Chapter 1
Zvimurenga Reflections

Abstract To label the initial anti-colonial revolt (from the 1890s), the war of liberation during the 1970s and the fast track land occupations as *zvimurenga* is not without controversy. To portray them as such, without raising any objections, would be to accept the ruling party's rendition of the history of anti-colonial struggle, with this struggle—according to the party—being reactivated dramatically during the fast track occupations. The existence of white-owned agricultural landholdings twenty years into independence signified the ongoing presence of the colonial condition in Zimbabwe, and fast track restructuring undercut this agrarian condition in certain ways. However, recognising this does not entail an acceptance of politicised history-telling as articulated by the ruling party. Simultaneously, no clear analytical understanding of the occupations has been forthcoming in the Zimbabwean literature. This arises because scholarly analysis is ruled out from the start by the a priori conclusion that the occupations were initiated and organised by the ruling party. This chapter unpacks and analyses critically the ruling party's *chimurenga* discourse about the fast track land occupations, as well as the inadequate explanations of the land occupations offered in the scholarly literature.

Keywords Zimbabwe · *Chimurenga* · Patriotic history · Fast track · Nationalist historiography · Historiography of nationalism

We knew and still know that land was the prime goal of King Lobengula as he fought British encroachment in 1893; we knew and still know that land was the principal grievance for our heroes of the First Chimurenga led by [spirit mediums] Nehanda and Kaguvi. We knew and still know it to be the fundamental premise of the Second Chimurenga ... Indeed we know it to be the core issue of the Third Chimurenga which you and me are fighting, and for which we continue to make such enormous sacrifices (Mugabe 2001: 93).

© The Author(s), under exclusive license to Springer Nature Switzerland AG 2021
K. Helliker et al., *Fast Track Land Occupations in Zimbabwe*,
https://doi.org/10.1007/978-3-030-66348-3_1

1.1 Introduction

The thoughts by the then-president of the ruling party and state president of Zimbabwe, Robert Mugabe, as outlined in the epigraph, were articulated during the height of the occupations of white commercial farms and other landholdings which arose in the early months of the year 2000 and soon flowed across almost the entire countryside. Though these extensive land occupations tapered off in the year 2001, a significant number of sporadic and isolated occupations have taken place since then. In glowing terms, these occupations were spoken about by Mugabe as a crucial component of the "third *chimurenga*"[1] (or third war of liberation) as they involved a strident undoing of the colonial legacy of land alienation and an unfolding process of land expropriation and reclamation bolstered by the Zimbabwean state's Fast Track Land Reform Programme (or simply fast track) launched in mid-2000.

Mugabe did not coin the idea of another (third) *chimurenga* at the turn of this century. Nevertheless, the argument that the land occupations alone or in combination with the state's fast track programme entailed a *chimurenga* is now common currency within the ruling party (the Zimbabwe African National Union-Patriotic Front, ZANU-PF). This is part of the grand ruling party narrative of successive but interrupted *zvimurenga* (plural for "*chimurenga*") dating back over a century, with the first *chimurenga* referring to the original anti-colonial revolt in the 1890s and the second *chimurenga* to the nationalist-backed guerrilla war during the 1970s which ended Rhodesian rule in 1980.

In this context, the main aim of this book is to provide an examination of the land occupations which led to the state's fast track land reform. In pursuing this, we designate these occupations (i.e. from early 2000) as "fast track land occupations". In addition, in our focused investigation of the fast track occupations, we seek to identify their specificities more clearly and fully through a comparative analysis of the land occupations on the one hand, and the first and second *zvimurenga* on the other hand. In this respect, and unlike many other Zimbabwean scholars, we draw a clear distinction between the land occupations and fast track reform. In doing so, and for the purposes of our analysis, we apply the notion of the third *chimurenga* specifically and exclusively to the land occupations and not to fast track land reform as well.

In examining the land occupations (as constituting the third *chimurenga*) with reference to the two earlier *zvimurenga*, the book also seeks to offer scholars and students of Zimbabwean history and society a detailed overview of the dynamics of the three *zvimurenga*. To date, only limited attempts to investigate the "fast track" occupations in relation to one or two of the earlier *zvimurenga* have appeared in the Zimbabwean literature, with the work of Sadomba (2008, 2013) being particularly important with respect to the second and third *zvimurenga*. Other scholarly works have offered short reviews of literature on two of more *zvimurenga*, but without

---

[1]The term "*chimurenga*" is developed from the name Murenga Soro Renzou, who was a chief of the Munhumutapa dynasty. The meaning of "*murenga*" remains unclear though "*soro renzou*" relates to the head of an elephant (Pfukwa 2007). See also Chiwome (1990) and Chigwedere (1991).

offering any useful comparative analysis (Raftopolous and Mlambo 2009; Ndlovu-Gatsheni 2009). In drawing upon and further contributing to the prevailing scholarly literature on Zimbabwe, the book therefore offers comprehensive summaries and insightful analyses of all three *zvimurenga*.

While we argue quite strongly against the ruling party's exclusionary discourse about the three *zvimurenga* (as many Zimbabwean scholars likewise do), we nevertheless claim that the three periods of heightened struggle identified by the ruling party's *zvimurenga* discourse refer to key moments in the colonial and postcolonial history of Zimbabwe. If only because of this, they are worthy of investigation under the rubric *zvimurenga* and on a comparative basis. Once these three moments of struggle are stripped bare of the ruling party's highly partisan discourse, which bears down heavily upon any capacity to understand them, it is possible and pertinent to offer a proper analytical account of the complex and convoluted intricacies of the *zvimurenga* and specifically the third *chimurenga*.

The challenge and necessity of undertaking this comparative work is particularly important because, soon after they began in the year 2000, the land occupations became the subject of intense political scrutiny and debate from the diverging perspectives of the main party-political antagonists, namely, the ruling ZANU-PF party and the new opposition party (Movement for Democratic Change, or MDC) which emerged in 1999. These political discourses only served to obfuscate in-depth analysis and, even more troublesome, most scholarly commentaries on the occupations were not based on any in-depth empirical accounts or case studies so that, in effect, they simply dressed up political discourses in intellectual garb.

At the helm of the fast track land occupations were significant numbers of ex-guerrillas (known as war veterans) involved in the war of liberation waged against white Rhodesian rule during the 1970s (the second *chimurenga*), who were aggrieved by their marginalised status and the slow pace—and minimal extent—of land reform in postcolonial Zimbabwe since independence in 1980. The fast track land reform programme, as promulgated and implemented from mid-2000, was the main policy response of the ruling party to the occupations. As we will see, many Zimbabwean scholars deeply critical of ZANU-PF argue that the ruling party (and the state) initiated and directed the land occupations and, because of this, there is a prevailing tendency to conceptualise the occupations and fast track reform as intertwined temporal phases in a seamless political process.

Ironically, this seamless transition—from land occupations to fast track—is also central to the ruling party's own narrative about the third *chimurenga* (as understood broadly), with the party presenting itself—ex post facto—as immediately and unreservedly defending and legitimising the occupations through fast track in common cause with the occupiers and against ongoing colonial land dispossession. In the case of critics of the third *chimurenga*, the seamless process is seen as regressive whereas, for the ruling party, the process is seen as progressive; the analyses are the same, and only the moral connotations differ. We argue against, as noted, any kind of conflation between land occupations and fast track, thereby justifying an examination of the occupations in terms of their own sets of modalities, rhythms and dynamics in a way irreducible to ruling party and state machinations.

There is now significant scholarly work on the fast track land programme, including full-length manuscripts based on original case-study research (Scoones et al. 2010; Matondi 2012). Key themes in this burgeoning literature include post-occupation developments on farms redistributed and resettled under fast track (such as the character of agricultural production and livelihoods), and the ways in which the land reform programme entails significant agrarian restructuring along class and racial lines. The implications of fast track for the performance of the agrarian and national economies have also led to important controversies within Zimbabwean studies and, as well, debates exist about fast track around the themes of citizenship, nation-building and statecraft in relation to agrarian restructuring in postcolonial Zimbabwe.

While the fast track literature expands, and competing arguments about fast track continue (though now in a less intense manner), the land occupations have received significantly far less scholarly attention and empirical study. Even with regard to the major studies on fast track, the occupations are typically given only a cursory and fleeting glance as political and historical background to fast track restructuring. To date, focused and localised case studies of the land occupations hardly exist.

Undoubtedly, the most ambitious study is by Laurie (2016). Though his study provides useful insights into land occupation dynamics, sweeping national claims are made based in large part on scattered evidence across the three Mashonaland provinces, thus going contrary to our central recognition, discussed below, of spatial diversity across the countryside in terms of the tempo and character of the land occupations. Further, Laurie's analysis of the occupations rests ultimately on the claim that the occupations were planned, organised and undertaken in some form, and to some degree, by the ruling party and the state's coercive agencies (specifically, the Central Intelligence Organisation), a perspective we explicitly dispute. He makes this argument while also admitting that farm-level occupations were "highly spontaneous, organic, and complex, varying considerably even within the same district" (Laurie 2016: 159).

In the end, there is simply no existing manuscript which provides a thematic overview of the nation-wide occupations as well as a meaningful analytical under-standing of them. This book contributes to filling these lacunae in the prevailing liter-ature, including by means of a thematic dissection, empirical fieldwork (in Mashona-land Central Province) and analytical input. But it goes beyond a mere contemporary account of the fast track land occupations by also examining them historically and comparatively by way of earlier *zvimurenga*.

In this respect, we select and use a particular angle into studying the land occupa-tions. In detailing the contingencies and complexities of the occupations, our main point of entry is the relationship between the occupiers (including war veterans) on the one hand and the ruling party and state on the other. In arguing against the claim that ZANU-PF led, directed, organised and coordinated the occupations, we point to considerable evidence which suggests that the occupations took on a more horizontal and decentralised form. This entry point facilitates our examination of other crucial dimensions of the occupations, including questions around the memory, motiva-tion, make-up and mobilisation of occupiers and, for us, the crucial theme of deeply localised dynamics entailing spatially and historically embedded variations across the occupations. It also allows for illuminating and investigating parallels (and divergences) between the fast track occupations and the two earlier *zvimurenga*, including intriguing debates about the character of all three *zvimurenga*.

In terms of the first *chimurenga*, controversies exist for example around the extent to which the revolts in Matabeleland and Mashonaland in 1896 and 1897 were in some way centrally organised or merely coordinated through local networks, and about the role of kingly, chiefly and spiritual authorities in the revolt. In the case of the second *chimurenga*, the complex interfaces between the nationalist movements, guerrilla armies and rural people, as well as the involvement of Christian-aligned churches, spirit mediums and ancestral cults in the war, are regularly brought to the fore and debated in the literature.

The main nationalist movements from the 1960s and 1970s now constitute, in a deeply reconfigured manner, the ruling party and state (or party-state), and the ex-guerrillas demanded and acquired the status of war veterans in the post-1980 period. The shifting relationships between these (ex-liberation movement) political forces were played out during the fast track land occupations, and in the context of fluid and contested guerrilla—party-state dynamics during the 1980s and 1990s. In understanding more clearly the form of the third *chimurenga* occupations and the specificities of this form, we thus link this *chimurenga* to previous ones through a historical-comparative analytical lens.

The book thus provides an examination of the fast track occupations in the light of the two pre-1980 *zvimurenga*, as well as with reference to relevant dimensions of colonial and postcolonial restructuring in Zimbabwe (as outlined briefly in the last section of this chapter).[2] The occupations continued beyond the year 2000 but the intensity and breadth of the occupations were particularly prevalent during 2000. We focus primarily on the years 2000 and 2001 without denying the existence of subsequent occupations and the possible shifting dynamics of these more sporadic occupations. In concluding the book, we also consider the implications of our arguments for post-fast track Zimbabwe up until and beyond the unseating of Mugabe as both ruling party and state president in November 2017.

---

[2]In this volume, we do not make extensive use of literature which simply seeks to justify the repressive actions taken against those who waged the first and second *zvimurenga*, or literature which serves to romanticise the ruling party's involvement in the third *chimurenga*. In the case of the second *chimurenga*, a significant amount of this literature exists, such as a study of the special forces Selous Scouts by Lt. Col. Ron Reid Daly (as told by Peter Stiff) (Reid-Daly and Stiff 1982) and of the Rhodesian SAS (Special Air Service) (Cole 1984). Also, there are a number of personal accounts of the three *zvimurenga* as told by those who sided against the *zvimurenga*. These include, from first through to third *chimurenga*, the works by Baden-Powell (1901), Lemon (2006) and Buckle (2001). We also make no significant attempt to draw upon these. More broadly, we focus our attention on scholarly literature.

1.2   Critical Thoughts on *Chimurenga* Historiography

To label the initial anti-colonial revolt, the war of liberation during the 1970s and the fast track land occupations as *zvimurenga* is not without controversy. To portray them as such, without raising any queries or objections, would be to accept the ruling party's particular rendition of the history of anti-colonial struggle, with this struggle—according to the party—being reactivated dramatically during the fast track occupations. The pervasive existence of white-owned agricultural landholdings twenty years into independence no doubt signified the ongoing presence of the colonial condition in Zimbabwe, and fast track restructuring did undercut this agrarian condition in certain ways.

However, recognising this does not necessarily entail an acceptance of history-telling as articulated by the ruling party. Despite lengthy periods marked by the absence of large-scale struggles and upheavals over land, ZANU-PF's narrative speaks about—in a teleological manner—a seamless (though disrupted) progression of land struggles from the first to the second through to the third and, apparently, final *chimurenga*. Further, these three sets of struggles somehow became crystallised and embodied in their entirety in the party and its historical legacy, and ultimately in Robert Mugabe himself (as specifically ordained as the leader of the Zimbabwean nation by the spiritual-ancestral realm) (Manzira 2018).

Concerns about this *chimurenga* historiography (or historical narrative) were correctly put forward by for instance Terence Ranger, who wrote path-breaking works on both the first and second *zvimurenga*. Although Ranger never wrote extensively about the third *chimurenga*, he reflected upon the ruling party's *chimurenga* historiography in the context of his own studies of the previous *zvimurenga*. In an article in 2004, Ranger offered a critical analysis of "a new variety of historiography" (Ranger 2004: 218), namely, the "patriotic history" (i.e. *chimurenga* historiography) increasingly propagated since the year 2000 by the ruling party and its public intellectuals. Undoubtedly, this is a monolithic history based on a narrow, exclusive and authoritarian nationalism which ignores and indeed erases alternative histories (including urban-based trade union struggles) and portrays the ruling party as the very epitome of any and all revolutionary truth. It is this historical story which led to the depiction of the land occupations as the third *chimurenga* or part thereof. Ranger distinguishes this history from both "nationalist historiography" and the "history of nationalism".

1.2.1 *Nationalist Historiography*

Nationalist historiography presents nationalism, as it existed historically in Zimbabwe, as inherently "inclusive" and "emancipatory" (Ranger 2005: 8) with Ranger citing his early works (including *Revolt in Southern Rhodesia, 1986–1987* on the first *chimurenga*—Ranger 1967) as exemplifying this type of historiography. Ranger also recognises that his key text—*Peasant Consciousness and Guerrilla*

War in Zimbabwe (1985)—on the second *chimurenga* also falls within the celebratory nationalist historiography. This partial, biased and romanticised historical
understanding of anti-colonial struggles is—according to Ranger—different from
patriotic history, with the latter arising more recently to justify the idea that the land
occupations (as a land-centred revolution) entail a third *chimurenga* (Ranger 2002).

Patriotic history, for Ranger, somehow transcends nationalist historiography and
goes to the extreme in the logic of excluding alternative understandings of the past
and present: "It is a doctrine of 'permanent revolution' leaping from *Chimurenga*
to *Chimurenga*. It has no time for questions or alternatives. It is a doctrine of
violence because it sees itself as a doctrine of revolution" (Ranger 2005: 8). As
Bull-Christiansen (2004: 107) notes,

[T]he discursive appropriation of the first Chimurenga in the third Chimurenga discourse
works by articulating the history of the first Chimurenga and the liberation war [second
chimurenga] into a temporal schema that equates the Chimurengas with an ongoing struggle
against colonial and neo-colonial forces.

Ranger goes on to argue that "[m]ost ... people, including those who, like myself,
were sympathetic to the 'second Chimurenga' of national emancipation, believe this
anachronistic re-rerun [or 'third *chimurenga*'] threatens to plunge the country into
chaos" (Ranger 2002: 25). In effect, Ranger claims that patriotic history, and it alone,
is equivalent to the ruling party's post-2000 rejuvenated *chimurenga* historiography
and that a nationalist historiography falls outside it, a point we challenge below.

As a partisan history which masquerades as undisputable truth (or as a false
universal), the form and content of patriotic history has been unpacked quite extensively by Tendi (2008, 2010), with Ranger writing a foreword in Tendi's book (2010)
on the subject. As Tendi (2010: 2) argues: "Patriotic History has become politically
significant because it plays a key role in legitimising ZANU PF's authoritarianism
and the party's hold on power. It has severely curtailed the development of an alternative view of Zimbabwe's history". In this way, the ruling party presents itself as "the
ordained guardian of Zimbabwe's political past, present and future" (Tendi 2010: 5),
with the third *chimurenga* becoming "the teleological culmination of the first two
Chimurengas" (Tendi 2010: 94).

This entails "a conspiracy of silence" (Tendi 2010: 24) with reference for instance
to the relevance of human rights and democracy (and not of just land) to the liberation
struggle, as well as to major episodes in the ruling party's chequered history (such
as the massacres in Matabeleland in the 1980s) which clearly blemish and undercut
any genuine claim to revolutionary status on the part of the party. But also intrinsic
to patriotic history is the "discursive motion of ... remembering" (Bull-Christiansen
2004: 69) certain events, or privileging these events and sometimes even fabricating
them, including perhaps even the centrality of Mugabe to the liberation struggle. In
the end, as Raftopoulos and Mlambo (2011: 4) note, the ruling party is "claiming
for itself the sole right to script the nation's past" as the one-and-only possible
embodiment of present struggles.

As argued by Bull-Christiansen (2004: 109), "articulations of the relation between
the past, present and future" which challenge patriotic history can "blur, fragment

and distort the imagined relation between the first, second and third Chimurenga". In this respect, the identification, critical appraisal and outright condemnation of patriotic history-based *chimurenga* historiography is very pervasive in the Zimbabwean literature, including fictional work (Hove 2011). Ndlovu-Gatsheni (2011) for example has written extensively about this historiography. He refers to it as the "*chimurenga* monologue" involving a "hegemonic and monologic narrative of the nation" (Ndlovu-Gatsheni 2011: 2). By way of a linear and unitary mode of historiography, the history of Zimbabwean nationalism becomes "nothing other than a catalogue of *Chimurengas*" (Ndlovu-Gatsheni 2014: 184).

In a similar vein, Nyambi and Mangena (2015: 141) refer to the ruling party's "homogenisation of the three liberation struggles". Writing from the perspective of land and agrarian history, Alexander (2007: 183) highlights that—for patriotic history—"[r]epossessing the land in the name of the nation has been cast as the singular, unwavering goal of Zimbabwe's three 'chimurengas'". In her major work with Ranger and McGregor on Matabeleland history, patriotic history is likewise labelled as "the master narrative of official Zimbabwean nationalism" (Alexander et al. 2000: 4). As Alexander stresses as well, land "cannot be reduced to a static role in a single narrative" (2007: 183).

In addition to this, Alexander brings to the fore patriotic history's "love of dichotomies and dualities" (2007: 193), such as between revolutionaries and sell-outs. These dichotomies are also used in relation to factional battles within ZANU-PF, whereby once liberation heroes such as Joice Mujuru become villains and sell-outs, which implies that "recognition of … [revolutionary] heroes is more of the present than past historical participation in the liberation struggle" (Masiya and Maringira 2017: 7). This dualistic portrait of the past and present fails to appreciate, and is deeply dismissive of, the sheer plurality of memories, motivations and movements which animated the anti-colonial struggle and which are crudely silenced in the *chimurenga* monologue. A deeply militaristic and violent tone emanates from this monologue in which enemies of the revolution are to be eliminated in one way or another.

But also evident is a fundamentally patriarchal dimension in patriotic history's three wars of liberation, involving a "masculinisation of nationalist history" (Muchemwa and Muponde 2008: xiii). Parpart (2008) thus claims that the war of liberation (in the 1970s) was, at least in part, a means by which to regain an African manhood subjugated by colonial oppression, and to feel and act like real (hyper-masculine) men in order to "penetrate" and win back the country. In his historical examination of the patriarchal underpinning of Zimbabwe's liberation struggle, Campbell refers to the "androcentrism in the current liberation discourse" (Campbell 2003: 269) and Bull-Christiansen (2004) offers an analysis of "male dominated narratives of Zimbabwean history" (2004: 38). For this reason, post-2000, a feminist activist in Zimbabwe—in a play of words—highlights the need for another *chimurenga*, namely, a "shemurenga" (Essof 2013).

Returning to Ranger's argument about the distinction between nationalist historiography and patriotic (or *chimurenga*) historiography, it seems that the difference is overstretched. It is not clear why the latter is not simply seen and understood as the logical outcome of the former under post-2000 conditions in Zimbabwe. Certainly, many of the criticisms of patriotic history as a mode of historiography apply equally to nationalist historiography. Some scholars note this, but it is not often foregrounded in their analyses. Tendi (2010) argues though that there are antecedents to patriotic history prior to the third *chimurenga* such that patriotic history is not solely a post-2000 discourse. In fact, in the early 1990s, Ibbo Mandaza (1994: 257), who is identified by Tendi as one of the main public intellectuals of patriotic history, spoke about the ways in which the ruling party portrayed itself discursively as the "sole and authentic liberation movement". Going back even further, as Ndlovu-Gatsheni (2012) argues, the actual term *chimurenga* began to be used widely by ZANU and its guerrilla forces during the 1970s to depict the war of liberation. As he notes elsewhere:

> ZANU (before it became ZANU-PF in 1980) appropriated the history of African history to construct the ideology of *Chimurenga* and to eventually claim to be the divinely ordained heir to the nationalist revolutionary spirit running from primary resistance of the 1890s to mass nationalism of the 1960s and armed liberation struggle of the 1970s. (Ndlovu-Gatsheni 2011: 2)

Bull-Christiansen (2004: 52) also notes that, during the liberation war of the 1970s, "the notion of the first Chimurenga began to gain mythological status" and spurred on the second (and now third) *chimurenga*. This pre-2000, and even pre-1980, *chimurenga* history, what Ranger would label as nationalist historiography, is not qualitatively different from patriotic history.

In fact, many scholars argue that Ranger's book on the 1896–97 anti-colonial revolt, written in the 1960s, contributed directly to ZANU's *chimurenga* narrative of the 1970s. It was part of the "triumphalist nationalist historiography of an earlier period" (Raftopoulos 1999: 128) and, for us, an earlier version of patriotic history. Ranger depicted the first *chimurenga* as the initial unfolding of the nationalist struggle which was reinvigorated in the 1950s and 1960s through mass nationalism. In his criticisms of Ranger's first *chimurenga* work, Cobbing (1976: 6) argues against Ranger's attempt to link the 1986–1987 risings to the "Zimbabwean nationalist movement of the 1950s" and, by extension, to the second *chimurenga*. For Cobbing, the Ndebele rising (in 1896) in particular should be viewed as "the last act of the independent Ndebele state, and not as the first of act of 'Zimbabwean' nationalism" (Cobbing 1976: 442).

As Raftopoulos (1999: 117) argues more specifically, Ranger's 1967 book "helped to feed the nationalist invention of a continuous thread of anti-colonial struggle", thereby deepening "nationalist nostalgia and mythology" (Raftopoulos 1999: 120). In a similar way, Ranger's nationalist historiography on the war of liberation in the 1970s (Ranger 1985), as well as other second *chimurenga* works coming out during that decade (Martin and Johnson 1981; Lan 1985), further emboldened the *chimurenga* discourse by speaking about the "unifying capacity of nationalism" (Raftopoulos 1999: 134). Combined, Ranger's studies of the first and second *zvimurenga* therefore were marked by "teleological implications" (Smith 1998: 294).

By the late 1990s (that is, before the fast track land occupations), "authoritarian notions of unity" were being widely "peddled by nationalist politicians" (Raftopoulos

1999: 134) in line with the kind of nationalist historiography articulated by Ranger. In her work on post-1980 contestations within the ruling party amongst political and military stalwarts of the second *chimurenga* around the discursive construction of "war veteran", Kriger (2006) identifies clear continuities between nationalist historiography and patriotic history. She concludes that patriotic history draws upon "patriotic versions of history [patriotic memories] advanced by the military and political veterans in the 1990s" (Kriger 2006: 1194) and that "[i]mportantly, 'patriotic history' is deeply rooted in post-independence discourses and practices [pre-2000] … in the entire nationalist struggle" (Kriger 2006: 1165). Post-1999, Kriger adds that this pre-existing historical discourse was merely drawn and elaborated upon by the ruling party in legitimising its reclamation of land through fast track land reform.

Central to patriotic history has been the significance of spirit mediums. As Mugabe says: "Nehanda … appears in our war annals of postcolonial Zimbabwe as the first heroine and martyr" (Mugabe 1983: 73). This grand claim, as we show in the next chapter, is deeply problematic. Nevertheless, in the light of such claims, Bull-Christiansen (2004: 72) argues that any external interference from the former British colonial masters during the third *chimurenga* (such as sanctions) is seen (from the perspective of patriotic history) "as not just a reversal to colonial times; it is also a re-enactment of the first colonial occupation"—given the central role of the British South African Company in expropriating land from the late 1800s. As Tendi (2010: 96) argues: "The role of spirit mediums such as Nehanda is played up in Patriotic History in order to conscript her martyrdom and spiritual attachment to the lost lands for the legitimisation and mobilisation of support for the Third *Chimurenga*". We would argue, however, that the nationalist historiography of the first and second *zvimurenga* (as least in the case of the early work by Ranger and Lan) embodies the same thread of spirituality in drawing and tying them together.

At the same time, we would go further and suggest that the story told by patriotic history has a significant degree of popular resonance and that it is not simply and wholly reducible to a top-down party-state discursive imposition. This claim in fact becomes central to our overall analysis of the third *chimurenga* land occupations. Other scholars note this, including Tendi (2010: 22) in positing that patriotic history "plays on real grievances", notably around land, and that it is "not without some historical foundation" in relation to land dispossession and demands. In this context, Alexander (2007: 192) also refers to the "the rootedness of patriotic history in genuine injustice".

This is a point which Mujere et al. (2017: 87) focus on quite extensively in exploring "what 'patriotic history', in its different guises, afforded … often long-muted localised agendas" and "how it gained traction [or saliency] in specific material and imaginative contexts". They examine this in specific relation to monumentalising two massacre sites (in which the dead included guerrillas and local villagers) dating back to the late 1970s in Gutu (Masvingo Province), and the building of shrines in the late 2000s and early 2010s as led by war veterans and local communities. The underpinning motive for this was "a desire for national recognition" (Mujere et al. 2017: 101) which, for war veterans, meant public acknowledgement of the role played by Gutu district during Zimbabwe's liberation war and, for the community, this was tied to demands for compensation and development:

> The renewed drive to commemorate Gutu's massacres sites …. [were] closely associated with FTLR [fast track land reform]. This was explicit. The same people were involved in both processes in Gutu, which were linked by the rhetoric of patriotic history in which they were imbricated. Both FTLR and the commemorative efforts involved a material and imaginative remaking of the district's landscapes. (Mujere et al. 2017: 101)

Overall, the commemorative efforts chimed in some way with the remaking of the land as provoked by the occupations and then fast track and, crucially, this did not involve mere "cynical efforts to access resources": "Patriotic history, and the broader contexts in which it played out, offered Gutu's massacre survivors and bereaved relatives a renewed opportunity to remake futures that the violence of war had constrained and denied" (Mujere et al. 2017: 106, 107). On this basis, the survivors and relatives sought to reinterpret, translate and exert some "ownership" over patriotic history, though in a manner involving a "negotiated" tension with the official version of history. For us, this implies that the fast track occupations should not be understood and analysed from the top-down, as a mere manifestation of ZANU-PF's political intrigues and impositions. Instead, the memories, motivations and mobilisation of occupiers need to be brought to the fore.

This point has been made more broadly by James Scott in his attempt to decode agrarian politics (Scott 2013). He shows through a number of case studies that a discursive distance often exists between elites (in our case, nationalist elites) and non-elites (in the countryside), despite the holding of seemingly common ideas. He thus claims that "[i]deological principles [such as nationalism] are replaced [at local levels] most often by an identification … with concrete social groups and by political reasoning from … immediate experience" (Scott 2013: 11). Hence, the memories, meanings and perspectives of rural people (including during the third *chimurenga* occupations) are of great significance in understanding the basis for mobilisation during the occupations. Perhaps even more important, Scott goes on to argue that "one cannot necessarily infer a consensus of beliefs among elites and non-elites from their participation in the same [agrarian] movement" (Scott 2013: 11). This is of relevance certainly for both the second and third *zvimurenga*, as the claims and aspirations of rural people which become crystallised in processes of intense struggle are deeply experiential and situational.

1.2.2 Historiography of Nationalism

This brings us to the question of the historiography of nationalism, as identified by Ranger. For Ranger, the history of nationalism—when compared to nationalist historiography and even more so to patriotic history—offers a critical and pluralist account of nationalism (including contestations within it), with the idea of histories of nationalism*s* being particularly important in this regard. Ranger here cites his work on Matabeleland (Ranger 1999; Alexander et al. 2000) as illustrations of this. With reference to the historiography of nationalism, Bull-Christiansen (2004: 103) thus speaks of a "polyphony of articulations [rather than a unified articulation] of the [Zimbabwean] nation's history" consistent with the importance of rewriting Zimbabwean history, given "the monopolisation of the discursive space of national history" through patriotic history by the ruling party. As already indicated, in his study of the first *chimurenga*, Ranger sought to demonstrate a unity of purpose between the Shona and Ndebele in the risings as a prelude to modern-day mass nationalism (Raftopoulos 1999). Cobbing (1977) and Beach (1971) paint a (somewhat) more complex, less romanticised, version of the first *chimurenga*, possibly in some way contributing to a historiography of nationalisms.

Heterogeneity is also highlighted in many post-Ranger books on the second *chimurenga*, including by Kriger (1992). Further, Raftopoulos (1999) identifies a range of works (particularly in the 1990s) which focus on the significance of nationalisms in urban spaces and the relationship between nationalism and labour, amounting in effect to urban social histories which recognise the relevance of democracy and not just land to anti-colonial struggles. For Raftopoulos (1999: 124, 130), this involved bringing to the fore "an uneven and differentiated picture of nationalist struggles" and the importance of a "variegated analysis of nationalism in Zimbabwe". Likewise, there is now literature which considers localised belongings and rural nationalisms of smaller ethnic groups, such as Tonga (McGregor 2009), Hlengwe (Chisi 2019) and Basotho (Mujere 2012).

This is part of rectifying a broader discursive "ethnic cleansing" about the second *chimurenga*. ZANU-PF's official representation of the history of the guerrilla struggle during the 1970s, and initially national historiography as well, is a distortion of the past with reference to downplaying the contributions of the armed wing of the Zimbabwe African People's Union (ZAPU) (Alexander and McGregor 2017). In his edited work on Joshua Nkomo, Ndlovu-Gatsheni (2017: 3) notes that the book's chapters are "part of the emerging literature that writes back into the nation those who have been excluded and at the same time subverting the official [ZANU-PF] historical narratives". After ZAPU's loss in the independence elections in 1980, Nkomo lost his liberation credentials as well and became "disqualified from speaking in the name of the nation" (Ndlovu-Gatsheni 2017: 5). However, as Ndlovu-Gatsheni admits, this correction of history can itself become unbalanced and partisan, as many of the chapters in his edited collection offer a deeply romanticised and celebratory image of Nkomo and ZAPU.

Nevertheless, we argue that the historiography of nationalism is crucial to an analysis of the third *chimurenga* and it is consistent with our claim about the ways in which the land occupations cannot be read from the manipulations and machinations of the party-state (as per the arguments made by critics of fast track) or seen as leading seamlessly to fast track land reform (as per ZANU-PF's *chimurenga* monologue).

All of the major studies on the first *chimurenga* took place decades ago, and there are many details about these risings which we will likely never know. Despite the important differences between the main protagonists (Ranger, Beach and Cobbing) in the debate about this *chimurenga*, most notably about the relative influence of political and spiritual factors, no detailed localised studies of the risings exist. However, a more decentralised understanding, in which national coordination and direction of the revolt is questioned, has arisen (Beach 1979). This attempt at a history of nationalism, in relation to the first *chimurenga*, becomes a crucial point of reference for our examination of the third *chimurenga* land occupations.

Compared to the first and third *zvimurenga*, there is a burgeoning literature on the second *chimurenga*. Two initial trends in this literature are quite apparent. First of all, as Bhebe and Ranger (1995: 2) wrote in the mid-1990s:

> The major studies of the war from an African perspective do not deal with armies or military tactics or the experience of fighting men and women. Instead they deal with the impact of the war on Zimbabwe's peasantries; with the war experience of African women; with ideology and religion; and with the need for healing after the war.

Since then, there has been a greater emphasis on the role, activities and experiences of guerrillas (including women guerrillas) as well as on nationalist-guerrilla relationships during the war of liberation. Related to this have been attempts at correcting the initial, almost exclusive focus on the activities of the guerrilla forces aligned to Mugabe's ZANU nationalist movement (as in the work of Ranger, Lan and Kriger) by considering the forces attached to Nkomo's ZAPU as well. Secondly, and perhaps in part as an indirect consequence of the early limited focus on the internal dynamics of the guerrilla-nationalist movements, most of the early scholarly literature was highly celebratory of the nationalist and guerrilla movements in a manner consistent with nationalist historiography. More recent literature on the second *chimurenga* is firmly rooted in the history of nationalisms discourse.

Unlike the second *chimurenga*, as noted, no significant body of literature has yet to emerge on the third *chimurenga* occupations, at least literature grounded firmly in empirical research. From the start, analyses of the third *chimurenga* (to reiterate, specifically the fast track land occupations) have been highly critical of the war veteran-led land occupations. The abiding argument has been that the occupations were instigated and led by the ruling party, in centralised fashion, as a basis for ZANU-PF's political survival. This has, therefore, involved interpreting the occupations in line with a broad and sweeping critique of state authoritarianism in post-2000 Zimbabwe, with the specific claim that the occupations were marked first and foremost by coercion and violence (Raftopoulos 2006; Raftopoulos and Phimister 2004). In this regard, patriotic history (and, by extension, nationalist historiography) came under direct attack.

No clear alternative analytical understanding of the occupations has been forthcoming from these critics though, by process of elimination, we would imagine that it would be based on a history of nationalism approach. However, analysis is in effect ruled out from the start by these critics, as an a priori conclusion (or mere assertion) has been reached about the political force behind the occupations. Additionally, as we have suggested (and show later in the book), this analysis—but not its moral evaluation—is in fact quite consistent with patriotic history. A few scholars (notably Sam Moyo) have provided a more sympathetic interpretation of the occupations and their work has been criticised as standing on the edge of a nationalist historiography or patriotic history. We seek to show however that the work of scholars like Moyo in fact draws extensively—and analytically—upon a historiography of nationalism.

In a sense, the literature on the third *chimurenga* has taken a different temporal course compared to the second *chimurenga* literature in that the former has been dominated throughout by a critical stance, or one seemingly unsympathetic to patriotic history. Intriguingly, though, it is the (apparently) nationalist historiography of sympathetic critics like Moyo which has highlighted the decentralised character of the third *chimurenga* occupations in a manner aligned to a history of nationalism. This is ironic, as critics claiming sensitivity to a history of nationalism would be expected to unpack the localised complexities of postcolonial struggles in Zimbabwe.

1.3   Overview of Chapters

In the chapters that follow, we develop these arguments by examining, in turn, the first, second and third *zvimurenga*. The next chapter (Chapter 2) focuses on the first *chimurenga*, while two chapters (Chapters 4 and 5) discuss the second *chimurenga* and three chapters (Chapters 7–9) consider the third *chimurenga*. Other chapters (namely Chapters 3 and 6) set out the historical contexts for the second and third *zvimurenga* respectively by detailing state programmes and political struggles which, combined, set the conditions for the emergence of the ensuing *zvimurenga*. An understanding of these intervening periods, between the first and second *zvimurenga* and between the second and third *zvimurenga*, is crucial for identifying any historical continuities between the three major episodes of land struggles (or *zvimurenga*) in Zimbabwean history. In the concluding chapter, we offer a critical analysis of the third *chimurenga* and speak about post-third *chimurenga* developments through our *zvimurenga* lens.

References

Alexander J (2007) The historiograhy of land in Zimbabwe: strengths, sciences, and questions. Safundi: J S Afr Am Stud 8(2):183–198

Alexander J, McGregor J (2017) African soldiers in the USSR: oral histories of ZAPU intelligence cadres' Soviet training, 1964–1979. J South Afr Stud 43(1):49–66

Alexander J, McGregor J, Ranger T (2000) Violence and memory: one hundred years in the 'dark forests' of Matebeleland. James Currey, Oxford

Baden-Powell R (1901) The Matabele campaign 1896. Methuen & Co, London

Beach DN (1971) The rising in South-Western Mashonaland, 1896–97. Unpublished PhD Thesis, University College of Rhodesia, Rhodesia

Beach DN (1979) Chimurenga': the Shona rising of 1896–97. J Afr Hist 20(3):395–420

Bhebe N, Ranger T (1995) General introduction. In: Bhebe N, Ranger T (eds) Society in Zimbabwe's liberation war (Volume Two). University of Zimbabwe Publications, Harare, pp 1–5

Buckle C (2001) African tears: the Zimbabwe land invasions. Covos Day, Johannesburg

Bull-Christiansen L (2004) Tales of the nation: feminist nationalism or patriotic history? Defining national history and identity in Zimbabwe. Nordiska Afrikainstitutet, Uppsala

Campbell H (2003) Reclaiming Zimbabwe: the exhaustion of the patriarchal model of liberation. Africa World Press, Trenton

Chigwedere AS (1991) The forgotten heroes of Chimurenga I. Mercury Press, Harare

Chisi TH (2019) Hlengwe memories of the Zimbabwean liberation struggle, 1975–1979. Oral Hist J S Afr 7(1):1–13

Chiwome E (1990) The role of oral traditions in the war of national liberation in Zimbabwe: preliminary observations. J Folk Res 27(3):241–247

Cobbing J (1976) The Ndebele under the Khumalos, 1820–1896. Unpublished PhD Thesis, University of Lancaster, UK

Cobbing J (1977) The absent priesthood: another look at the Rhodesian risings of 1896–1897. J Afr Hist 18(1):61–84

Cole B (1984) The elite: the story of the Rhodesian special air service. Three Knight Publishing, Durban

Essof S (2013) Shemurenga: the Zimbabwe women's movement, 1995–2000. Weaver Press, Harare

Hove M (2011) Strugglers and stragglers: imagining the "war veteran" from the 1890s to the present in Zimbabwean literary discourse. J Lit Stud 27(2):38–57

Kriger N (1992) Zimbabwe's guerrilla war: peasant voices. Cambridge University Press, Cambridge

Kriger N (2006) From patriotic memories to 'patriotic history' in Zimbabwe, 1990–2005. Third World Q 27(6):1151–1169

Lan D (1985) Guns and rain: guerillas and spirit mediums in Zimbabwe. James Currey, London

Laurie C (2016) The land reform deception: political opportunism in Zimbabwe's land seizure era. Oxford University Press, New York

Lemon D (2006) Never quite a soldier: a Rhodesian Policeman's war 1971–1982. Galago Publishing, Alberton

Mandaza I (1994) The state and democracy in Southern Africa: towards a conceptual framework. In: Osaghae EE (ed) Between state and civil society in Africa: perspectives on development. CODESRIA, Dakar, pp 249–271

Manzira R (2018) The power of mysticism—understanding political support for President Robert Mugabe in Zimbabwe. Unpublished Master's Thesis, Rhodes University, South Africa

Martin D, Johnson P (1981) The struggle for Zimbabwe: the Chimurenga war. Ravan Press, Johannesburg

Masiya T, Maringira G (2017) The use of heroism in the Zimbabwe African National Union-Patriotic Front (ZANU-PF) intra-party factional dynamics. Strat Rev South Afr 39(2):3–24

Matondi P (2012) Zimbabwe's fast track and reform. Zed Books, London

McGregor J (2009) Crossing the Zambezi: the politics of landscape on a central African Frontier. James Currey, Suffolk

Muchemwa KZ, Muponde R (2008) Introduction: manning the nation. In: Muchemwa KZ, Muponde R (eds) Manning the nation: father figures in Zimbabwean literature and society, Weaver Press, Harare

Mugabe R (1983) Our war of liberation: speeches, articles, interviews 1976–1979. Mambo Press, Gweru

Mugabe RG (2001) Inside the Third Chimurenga. Department of Information and Publicity, Harare

Mujere J (2012) Autochthons, strangers, modernising educationists, and progressive farmers: Basotho struggles for belonging in Zimbabwe 1930s–2008. PhD Thesis, University of Edinburgh, UK

Mujere J, Sagiya ME, Fontein J (2017) 'Those who are not known, should be known by the country': patriotic history and the politics of recognition in Southern Zimbabwe. J East Afr Stud 11(1):86–114

Ndlovu-Gatsheni S (2009) Do 'Zimbabweans' exist? Trajectories of nationalism, national identity formation and crisis in postcolonial state. Peter Lang, Bern

Ndlovu-Gatsheni S (2011) The construction and decline of Chimurenga Monologues in Zimbabwe: a study in resilience of ideology and limits of alternatives. Paper presented at the 4th European Conference on African Studies, The Nordic Africa Institute Uppsala, Sweden

Ndlovu-Gatsheni S (2012) Rethinking Chimurenga and Gukurahundi in Zimbabwe: a critique of partisan national history. Afr Stud Rev 55(3):1–26

Ndlovu-Gatsheni S (2014) Zimbabwe and the crisis of Chimurenga nationalism. In: Ndlovu-Gatsheni S (ed) Coloniality of power in postcolonial Africa: myths of decolonisation. Council for the Development of Social Science Research in Africa (CODESRIA), Dakar, pp 179–236

Ndlovu-Gatsheni S (ed) (2017) Joshua Mqabuko Nkomo of Zimbabwe: politics, power, and memory. Palgrave Macmillan, London

Nyambi O, Mangena T (2015) The past is the present and future: ambivalent names and naming patterns in post-2000. African Language Association of Southern Africa and the Foundation for Education, Science, and Technology

Parpart JL (2008) Masculinities, race and violence in the making of Zimbabwe. In: Muchemwa KZ, Muponde R (eds) Manning the nation: father figures in Zimbabwean literature and society, Weaver Press, Harare

Pfukwa C (2007) The function and significance of war names in the Zimbabwean Armed Conflict (1966–1979). Unpublished PhD Thesis, University of South Africa, South Africa

Raftopoulos B (1999) Problematising nationalism in Zimbabwe: a historiographical review. Zambezia 26(2):115–134

Raftopoulos B (2006) The Zimbabwean crisis and the challenges for the left. J South Afr Stud 32(2):203–219

Raftopolous B, Mlambo A (eds) (2009) Becoming Zimbabwe: a history from the pre-colonial period to 2008. Weaver Press, Harare

Raftopoulos B, Mlambo A (2011) Outside the Third Chimurenga: the challenges of writing a national history of Zimbabwe. Crit Afr Stud 4(6):2–14

Raftopoulos B, Phimister I (2004) Zimbabwe now: the political economy of crisis and coercion. Hist Mater 12(4):355–382

Ranger T (1967) Revolt in southern Rhodesia 1896–7: a study in African resistance. Heinemann Educational Books, London

Ranger T (1985) Peasant consciousness and Guerrilla war in Zimbabwe: a comparative study. Zimbabwe Publishing House, Harare

Ranger T (1999) Voices from the rocks: nature, culture and history in the Matopos Hills of Zimbabwe. Indiana University Press, Bloomington

Ranger T (2002) Cultural revolution. The World Today 58(2):23–25

Ranger T (2004) Nationalist historiography, patriotic history and the history of the nation; the struggle over the past in Zimbabwe. J South Afr Stud 30(2):215–234

Ranger T (2005) The uses and abuses of history in Zimbabwe. In: Palmberg M, Primorac R (eds) Skinning the skunk—facing Zimbabwean futures. Nordic Africa Institute, Uppsala, pp 7–15

Reid-Daly R, Stiff P (1982) Selous Scouts: top secret war. Galago Publishing, Johannesburg

Sadomba WZ (2008) War veterans in Zimbabwe's land occupations: complexities of a liberation movement in an African post-colonial settler society. Unpublished PhD Thesis, Wageningen University, The Netherlands

Sadomba WZ (2013) A decade of Zimbabwe's land revolution: the politics of the war veteran Vanguard. In: Moyo S, Chambati W (eds) Land and agrarian reform in Zimbabwe: beyond white-settler capitalism. African Institute for Agrarian Studies, Harare, pp 79–122

Scoones I, Marongwe N, Mavedzenge B, Mahenehene J, Murimbarimba F, Sukume C (2010) Zimbabwe's land reform: myths and realities. Weaver Press, Harare

Scott J (2013) Decoding subaltern politics. Routledge Press, New York

Smith N (1998) Theorising discourses of Zimbabwe, 1860–1900: a foucauldian analysis of colonial narratives. Unpublished PhD Thesis, University of Natal, South Africa

Tendi BM (2008) Patriotic history and public intellectuals critical of power. J South Afr Stud 34(2):379–396

Tendi BM (2010) Making history in Mugabe's Zimbabwe: politics, intellectuals, and the media. Peter Lang AG, International Academic Publishers, Oxford

Chapter 2
The First *Chimurenga*

Abstract This chapter discusses the first major anti-colonial revolt in colonial Zimbabwe (in 1896 and 1897), which began to be labelled retrospectively as the first *chimurenga* from the late 1950s as a nationalist movement arose, leading a decade later to the guerrilla-based war of liberation or second *chimurenga* in the 1970s. It considers the overall character and trajectory of the first *chimurenga*, in particular its modalities of organisation and mobilisation, as the book seeks to draw parallels between the first anti-colonial revolt and the second and, notably, the third *chimurenga*. In addition to understanding the general dynamics of the first *chimurenga*, the chapter also highlights certain themes, such as gender and spirituality, which are pertinent to our latter analyses of the other *zvimurenga*. At the same time, it must be highlighted that there are very few full-length analyses of the first *chimurenga*, and no real insights about local dynamics, which means that any examination of this revolt remains somewhat thin and shallow.

Keywords First *chimurenga* · Terence ranger · Nehanda · Spirit mediums · Mashonaland · Matabeleland

2.1 Introduction

This chapter discusses the first major anti-colonial revolt in colonial Zimbabwe (in 1896 and 1897), which was to be labelled retrospectively as the first *chimurenga* from the late 1950s as a nationalist movement arose, leading a decade later to the guerrilla-based war of liberation or second *chimurenga* in the 1970s. We consider the overall character and trajectory of the first *chimurenga*, in particular its modalities of organisation and mobilisation, as the book seeks to draw parallels between the first anti-colonial revolt and the second and, notably, the third *chimurenga*. In addition to understanding the general dynamics of the first *chimurenga*, we also highlight certain themes, such as gender and spirituality, which are pertinent to our latter analyses of the other *zvimurenga*. At the same time, it must be highlighted that there are very few full-length analyses of the first *chimurenga*, and no significant insights about local dynamics, which means that any examination of this revolt is somewhat thin and shallow. Besides this, it is also readily noticeable, and both unfortunate and

© The Author(s), under exclusive license to Springer Nature Switzerland AG 2021   19
K. Helliker et al., *Fast Track Land Occupations in Zimbabwe*,
https://doi.org/10.1007/978-3-030-66348-3_2

regrettable, that no new pronounced archival research (at least published work) has taken place on the first *chimurenga* since the mid-1980s.

In this context, we turn primarily to academic works from the 1960s to the 1980s, mainly of those historians who entered into a vigorous scholarly debate about the character of the 1896–97 revolts in Matabeleland and Mashonaland. In this regard, we focus on the analyses of Terence Ranger, David Beach and Julian Cobbing, but also on the largely unknown and unpublished thesis of Mark Horn. Combined, the work of these scholars brings to the fore a range of crucial themes relevant to our comparative analysis of the three *zvimurenga*. After discussing the work of Ranger, Beach, Cobbing and Horn, we lay out these themes more explicitly and, in drawing upon other academic literature which discusses at least in passing the first *chimurenga*, highlight the importance of some of these themes for our later analysis of the third *chimurenga*.

2.2   Ranger's *Revolt in Southern Rhodesia*

Based on significant archival research, Terence Ranger (in 1967) wrote what is generally considered as the first full account of the first *chimurenga*, or the anti-colonial revolt of both Shona and Ndebele peoples in 1896 and 1897. As with his account of the war of liberation during the 1970s (see Chapter 4), Ranger's depiction of the revolt of the late 1890s has sparked considerable controversy. One key point of contention, which is particularly pertinent to our comparative analysis of the three *zvimurenga*, relates to the organisation, mobilisation and coordination of the revolt, notably the extent to which it was centrally directed and nationally coordinated. Directly related to this is the existence of historically constituted spatial variations in the countryside (for instance, between Matabeleland and Mashonaland) and thus the ways in which this first anti-colonial revolt was shaped by subnational and local interests, factors and dynamics. Also, of great importance in Ranger's work on the first *chimurenga* is the role of spirituality with specific reference to the so-called *Mwari* cult and spirit mediums, as well as the reasoning and motivations behind Shona and Ndebele people's involvement in the revolt. In this light, we first detail Ranger's key arguments in his 1967 book and then consider criticisms of his claims by other scholars who have thoughtfully examined the first *chimurenga*, in particular Cobbing and Beach, both of whom had primarily a subnational focus.

Ranger argues that, prior to the entry of the Pioneer Column from the south, Ndebele society was caste-based and unified around an "authoritarian monarch" involving "a highly centralised military system" (Ranger 1967: 32–33) with regiments commanded by *izinduna*. Meanwhile, throughout most of Mashonaland, a more decentralised tribal system existed which incorporated scattered and localised paramount chiefs (and other chiefs) who were in conflict with each other and indeed "extremely disunited" (Ranger 1967: 32). Some people from Shona subgroups and other ethnic groups (such as the Tonga) were assimilated into the Ndebele socio-political-military system as lower caste members (*Holi*), while some Shona and other chiefdoms existed as political tributaries of the Ndebele state (particularly in western Mashonaland). Additionally, others were subjected to occasional Ndebele raids and demands. In the case of the tributary areas, "[t]he Ndebele made no attempt to re-model Shona political or religious institutions …, but in many cases entered into relations of varying kinds with Shona paramounts and [spirit] mediums" (Ranger 1967: 28).

In terms of spiritual institutions, the Shona had one deity (*Mwari*), who was approached either through spirit mediums (or *mhondoro*) when possessed by an ancestral spirit (in eastern Mashonaland), or by way of the *Mwari* cult (in western Mashonaland). Senior spirit mediums tended to minimise the unbridled power of chiefs and "acted as a centralising and stabilising factor" (Ranger 1967: 20). The *Mwari* (or *Mlilo*) cult though was particularly pronounced amongst the Ndebele and it was well-organised by way of shrines, priests and officers in Matabeleland including in the Matopo hills. By 1893, Matabeleland was (at least partially) conquered and Mashonaland began to be occupied increasingly by white settlers, though on a very limited basis. Both regions were under the emerging (but still weak and haphazard) rulership of the British South Africa Company (simply, Company) by the time of the first *chimurenga*.

From 1893 to 1896, Ndebele land and cattle were forcibly extracted by the white settlers, and the Ndebele and their tributary subjects were compelled to enter wage employment (more so than in Mashonaland). Native Reserves were soon established in Matabeleland and these were "profoundly unsuitable" (Ranger 1967: 104) as a means of subsistence, with Ndebele people refusing to move onto them and "squatting" rather on white or crown land. The *Matabeleland Order in Council* of 1894 soon led to the setting up of a Native administration which was concerned primarily with seizing Ndebele cattle: "Native Commissioners were kept busy registering, branding and distributing cattle almost right up to the [1896–97] rising itself" (Ranger 1967: 116). African police often played a central role in enforcing key aspects of Company rule, including recruiting labour, and hence they were despised by the people (including in Mashonaland). The conquering of Matabeleland (in 1893) entailed an attempt at undercutting the centralised Ndebele monarch including its militarised regimental kraals. Ndebele people were deeply angered by the increasing colonial pressures on their precolonial way of life, and they wanted to re-establish their lost kingdom. However, they failed to enter into widespread (at least armed) resistance against cattle confiscation and labour compulsion.

In the case of Mashonaland, Company rule did not seek to dismantle the chiefly system such that the decentralised paramounts were in large part left intact. But Shona people began to experience increasingly the detrimental effects of British-based colonial rule, including through the loss of well-established trading relationships with the Portuguese, granting of mining concessions to white settlers, press-ganging of labour, hut tax (and cattle being seized in lieu of not paying tax) and loss of land. Unrestrained by any effective centralised colonial administration, farmers, mining prospectors and other settlers were also able to abuse and mistreat Shona people almost at will. Unlike in Matabeleland, the Shona paramounts and their subjects in Mashonaland engaged in significant clashes and resistance prior to the first *chimurenga*, in large part because they considered themselves as unconquered and still independent of Company authority. In this context, there is "ample evidence of widespread grievances which in themselves would account for a general disposition to rebel" (Ranger 1967: 81) in Mashonaland, but in Matabeleland as well.

In the light of the deep and multifaceted challenges faced by indigenous people in both Matabeleland and Mashonaland, the rebellion of 1896–97 emerged. Thus, the revolt cannot be explained "merely in terms of a generalised will to resist, but .... also in terms of the peculiar pressure of Company rule" (1967: 48). It began in early 1896 in Matabeleland. In this respect, from the time the Company's administrator Jameson led an abortive attack on Johannesburg with white police officers in January 1896, it was clear that "a rising was decided upon. In February meetings of *izinduna* took place to prepare plans" (Ranger 1967: 124). Ndebele emissaries were sent apparently to Mashonaland villages to seek Shona support. Meanwhile, if only below the surface, there was the "mass determination of the Shona to rid themselves of the white presence", leading the paramount chiefs to "plan open and general armed resistance" (Ranger 1967: 87) later in 1896. However, to start and sustain the revolt in Matabeleland and Mashonaland, an effective form of coordination and organisation was necessary; and Ranger searches for and identifies this organisational arrangement. This is based on the claim, or even assumption, that the first *chimurenga* was in fact coordinated and organised across the entire countryside.

Ranger argues that, despite the conquering of Matabeleland in 1893, the Ndebele regimental system had not been completely broken by 1896. As a result, "most of the military effectiveness of the rebellion came from a revival of the old regimental system" (Ranger 1967: 135), augmented by—Ranger adds—considerable if uneven support from the Native (or African) police. At the same time, this did not guarantee the overall coherence and coordination of the revolt, as the Ndebele king as supreme commander had been overthrown. Divisions had arisen between the senior *izinduna* (and their regiments) around attempts to proclaim particular candidates as the new king in order to restore the Ndebele monarchy. Likewise, the Ndebele aristocracy was split on the matter of kingship. These divisions indeed continued during the 1986 rebellion, such that the regiments "never did act according to a plan conceived by a single Ndebele military authority" (Ranger 1967: 139), with some *izinduna* even standing aloof from the revolt at least at first.

But the high priest of the *Mwari* cult "symbolised unity of the Ndebele nation" (Ranger 1967: 136) and was "at the centre of the planning which preceded the rising and after its outbreak he was a focus of loyalty" (Ranger 1967: 139) for all *izinduna*. This spiritual dimension meant that the Matabeleland rising was "not merely a chaos of independent action" (Ranger 1967: 139). Though there was no one single cult centre, the *Mwari* cult offered "considerable possibilities of coordination" (Ranger 1967: 144) in part because of its "high degree of centralisation" (Ranger 1967: 146) and the tight links between the main cult centres. In fact, "most of the leading officers of the *Mwari* cult lent ... their organisational apparatus to the preparations for the rebellion" (Ranger 1967: 148) and the "Ndebele [aristocratic] leaders were in contact with leading priestly families in the early months of 1896" (Ranger 1967: 147) in plotting the revolt.

The lower caste people (*Holi*) and particularly the (now) ex-tributary Shona chiefdoms also became involved in the revolt. This was not because of any loyalty to the Ndebele aristocracy (and hence so as to restore the kingdom), nor was it due to threats and intimidation by the *izinduna*. Rather, it arose because of their own self-interest considering that, for instance, white settlers were confiscating the cattle of the former tributaries and compelling their subjects to labour. Again, it was the *Mwari* cult and not the Ndebele aristocracy which facilitated their involvement. This entailed a significant coordinating role for the *Mwari* cult officers, as "allies of the Ndebele aristocracy on the one hand and as [spiritual] authorities amongst the ex-tributary Shona on the other" (Ranger 1967: 142). The *Mwari* cult hence encouraged the ex-tributary Shona to enter the revolt, to the extent that they were in effect incorporated directly into the Ndebele rising. As well, the *Mwari* cult priest Mkwati became "deeply involved" (Ranger 1967: 190) in planning the later Shona rising in Mashonaland.

Overall, "[t]he rising in Matabeleland was an alliance of convenience" or "a coalition of different, and even hostile, groups combined in the common interest of overthrowing whites" (Ranger 1967: 160), with *Mwari* cult officers having an unquestioned moral authority and regimental *izinduna* becoming military commanders during the uprising. Additionally, the various actions, including attacks, undertaken across the Matabeleland countryside were "roughly synchronised" due to the "continued authority and efficiency of Ndebele institutions" and the "existence of a widely influential religious organisation which was at one and the same time in touch with the Ndebele leadership and with men of influence ... in the tributary Shona areas" (Ranger 1967: 160). However, more tentatively, Ranger notes (if only in passing) that it is "hard to say whether there was a combined strategy jointly devised between the various rebel factions; probably there was not" (Ranger 1967: 175). Indeed, just as some Ndebele groups and the south-western Kalanga failed to join the revolt, there were also certain divisions between the *izinduna* regiments involved in the uprising.

The armed revolt in Mashonaland, which took the white settlers by greater surprise than the one in Matabeleland, arose in June 1896. The Shona rising started only two months before the Company's negotiations with the Ndebele began, and it was only subdued forcefully near the end of 1897. It started notably in the Hartley District, escalated rapidly in western Mashonaland and then spread to central Mashonaland before engulfing virtually the whole of Mashonaland. Though Ndebele rebels were central to the initial armed outbreaks in Hartley and western Mashonaland more broadly, there was "little direct Ndebele influence in the greater part of Mashonaland" (Ranger 1967: 196).

As with the Ndebele, the Shona showed signs of disunity before the revolt arose; but, like in the case of Matabeleland, the Mashonaland revolt showed signs of significant organisation. Pre-revolt tensions did in fact exist between a number of Shona paramount chiefs and these were expressed during the revolt, with some chiefs expressing loyalty to the white settlers (or remaining neutral) simply because their long-standing chiefly competitors joined the revolt. All paramount chiefs joining the revolt, though, were supported by a substantial majority of their subject peoples.

Amongst these paramount chiefs, the existence of "coordination ... lay behind the almost simultaneous outbreak of the rising itself; ... and there was constant coordination of military matters" during the uprising (Ranger 1967: 200) including joint ambushes.

As in Matabeleland, coordination (across paramounts and lesser chiefs in the case of Mashonaland) was achieved mainly because of the key involvement of spiritual authorities—the presence of these authorities helps to explain the sheer commitment of Shona people to the uprising at the paramountcy level. Both the *Mwari* cult and spirit mediums participated in the Shona rising, though there were regional variations in this regard. A fully organised *Mwari* cult system (with shrines and priests) existed only in western Mashonaland; however, the prestige of the cult extended beyond this area. Thus, Mkwati and the *Mwari* cult stirred up discontent in western Mashonaland and, in this way, were able to "link the rising there to the Ndebele rising, to help coordinate the rising within the area itself, and to maintain morale once the rising had begun" (Ranger 1967: 205). In other parts of Mashonaland, spirit mediums became more critical to coordinating the revolt, notably the Chaminuka-Nehanda hierarchy of spirit mediums of central and western Mashonaland and the Mutota-Dzivaguru hierarchy in the north-eastern and eastern parts of the region. However, "[n]o one senior medium enjoyed a determining influence over the whole area of the Shona rising nor did the mediums as a whole unanimously agree to support it" (Ranger 1967: 208).

Nevertheless, in emphasising the link between revolt and mediums, Ranger argues that "the influence of senior mediums was even in normal times more extensive than that of any secular authority and could bring together for ritual purposes rival, and even warring, paramounts" (Ranger 1967: 208). It becomes clear then that the "principle of unity" for the Shona revolt was "embodied in the senior mediums" (Ranger 1967: 209), such as the Nehanda medium in the Mazoe area. As well, the fact that both the *Mwari* cult and spirit mediums existed in western Mashonaland (such as in the Hartley and Charter districts) explains why the Shona rising arose there. More generally, in understanding the seemingly intimate relationship between the Matabeleland and Mashonaland risings, it is notable that there existed "a working alliance between the *Mwari* officers, the mediums of the Chaminuka circle and the mediums of the Nehanda circle" (Ranger 1967: 212), with the *Mwari* cult priests and officers central to both the Matabeleland and Mashonaland revolts.

In addition, the Kaguvi medium (as a "national" medium) emerged during the Shona revolt and became the equivalent of Mkwati in Matabeleland, as the medium "provided coherence to the Shona rising" (Ranger 1967: 217). In fact, the medium of Kaguvi held meetings in western Mashonaland with Ndebele emissaries along with representatives of certain paramounts from central Mashonaland. At these meetings, the representatives received assurances of significant Ndebele support for the Shona rising. The Kaguvi medium also constantly urged Shona peoples to participate in the revolt against white settlers, received frequent reports of progress in the fighting, and issued military advice like Mkwati did in Matabeleland. Given the heavy input of the *Mwari* cult and spirit mediums during this anti-colonial revolt, it is not surprising that Christian mission stations, white missionaries and their very few Native converts came under attack in both Matabeleland and Mashonaland.

Some paramount chiefs, as noted, failed to join the revolt (labelled by Ranger as "collaborators"). These were living in areas where the detrimental effects of Company administration were "much lighter" (Ranger 1967: 207) and hence, comparatively speaking, they had not experienced tax collection and cattle seizures to any significant extent. Therefore, as was the situation in Matabeleland, "the risings [in Mashonaland] were more or less co-extensive with white settlement" (Ranger 1967: 348) and "the area which rebelled was that under the heaviest economic pressure" (Ranger 1967: 349). The extent and character of the presence (or absence) of the Rhodesian state in different parts of the countryside was likewise pertinent during the war of liberation in conditioning the guerrilla struggle. Ultimately, during 1897, the Shona paramounts one by one surrendered unconditionally while, in Matabeleland, negotiations with the Ndebele senior *izinduna* eventually ended the bulk of the rising.

2.3 Responses to Ranger: Cobbing, Beach and Horn

In this context, we now discuss the criticisms of Ranger's work by Beach and Cobbing, and then Horn's overall critique of all three scholars. It is clear that Ranger, in the main, understood the first *chimurenga* as both well-organised and well-coordinated (by way of spiritual authorities in particular) and characterised by a centralised thrust. But, at times, he intimates the possibility of localised specificities if only in terms of the broad distinction between "rebels" and "collaborators". Despite the significant revisions which both Beach and Cobbing offer, they do not provide a full reworking of Ranger's overall claims, except in the case of Beach's later writings.

2.3.1 Cobbing

Cobbing is particularly concerned with Matabeleland, though he speaks about the revolt more broadly. He entered the controversy around the first *chimurenga* through his PhD thesis in 1976, a scholarly work which was primarily focused on revising the then-current understanding of the Ndebele kingdom pre-1893. Like Ranger, he speaks about the presence of solid organisation animating specifically the Ndebele revolt, but he provides a political (and not a religious) twist to his argument, as Beach does as well. He claims that the political organisation underlying the Ndebele state (involving the centralised monarchy and outlying decentralised chieftaincies) was unaffected substantially by the British conquest in 1893, and that the monarchy coordinated and solidified the revolt. In this regard, in a later article, he argues that "the Ndebele state was in much better shape in early 1896 than has been supposed" (Cobbing 1977: 63). Not all Ndebele settlements took the form of regiments, but the regimental villages were still in large part intact immediately post-1893, though they had been "temporarily shaken" (Cobbing 1976: 366). Thus, no significant demilitarisation took place because of the 1893 conquest. With the military regiments subordinate to the state, "the Ndebele [and its monarchy] began the [1896] war impressively united" (Cobbing 1977: 65).

Simultaneously, Cobbing argues that Ranger over-estimates the level of coordination in the Ndebele rebellion (in the case of Ranger, through the *Mwari* cult), as numerous "deficiencies of co-ordination" (Cobbing 1976: 388) contributed to the loss of the war by the Ndebele. Cobbing (1976: 407) thus speaks of "difficulties of strategic co-ordination of … [the] … various armies". As well, a deterioration in supplies of grain (millet and maize) and arms and ammunition took place during the course of the revolt, and this undercut the capacity to engage militarily over an extended period. Prior to the revolt, the Ndebele had mobilised and stored grain and arms in preparation for a revolt, which speaks to "long term planning and co-ordination" for an impending war (Cobbing 1976: 408). Indeed, one of the many ways by which the Company sought to undercut the rising (in Matabeleland and Mashonaland) involved destroying kraals as well as crops to undercut the food supply of the rebels (Hodder-Williams 1967), as happened during the second *chimurenga*.

Post-1893 and up to 1896, as Cobbing argues, the Company increasingly seized Ndebele cattle and coerced Ndebele people almost indiscriminately into labouring for the emerging settler economy (for the mining claims, for instance). Land had been expropriated but, because it remained in large part unutilised by white settlers, this was not a major pressure upon the lives of Ndebele, who remained living undisturbed on now-white land. Massive cattle seizures from December 1895 to February 1896 in particular though "came as the last straw for the Ndebele and may be taken as the single most important immediate cause of the rising which followed" (Cobbing 1976: 381). Given the intense settler coercion around cattle and labour, "the causes of the Ndebele reaction in March 1896 are self-evident" (Cobbing 1976: 382). Because the Ndebele political and military structures continued in some form after the 1893 conquest, "[i]t was the royalised caste, who together with the lesser chiefs, instigated the first murders, arranged for the distribution of ammunition, … [and] organised the supply routes" (Cobbing 1977: 68) for the Ndebele revolt. The first attacks (for example, in Inyathi) "immediately became general" (Cobbing 1976: 386) and these were led in large part by *izinduna*. In engaging in attacks, the Ndebele also sought to recover their seized cattle.

In this light, "[t]here is no need to look for hidden [mystical] influences" (Cobbing 1977: 68) in explaining the revolt, as Ranger does by referring to supposed *Mwari* cult machinations. For Cobbing, no *Mwari* priests were involved in any significant manner and, by emphasising otherwise, Ranger is simply reproducing the myths propagated by the Company. In the end, Cobbing questions Ranger's entire narrative about the history of the *Mwari* cult, including its alleged connections to the Ndebele (and Rozvi) state, as the Ndebele focused on "(ancestor-spirit) worship and on the Nguni high-God" (Cobbing 1976: 9). At the start of the Ndebele revolt, certainly "the cult had no say in Ndebele military decisions" and there was a "lukewarm attitude of the Ndebele towards the cult" (Cobbing 1977: 75). Thus, "the mechanics of the war, its causes, leadership and strategy, were mundane [secular] rather than mystical [spiritual]" (Cobbing 1976: 396).

As well, the revolt in Matabeleland took place at a time of great divisions within the Ndebele kingdom. A number of Ndebele chiefdoms sided with the Europeans, so that other chiefdoms "directed much of their attention against these collaborators, ... diverting themselves from the main task of defeating the Europeans" (Cobbing 1976: 390–391). This meant that "a war against the Company could easily become an Ndebele civil war, as was in fact to happen in 1896" (Cobbing 1976: 384), and this entailed a continuation of earlier intra-Ndebele conflicts predating 1893. Cobbing (1976: 422) hence speaks of an incipient civil war, varying in extent regionally, within Matabeleland.

The non-Ndebele (including Kalanga, Rozvi/Shona and Venda) chiefs falling under the Ndebele tributary state, as a "motley collection of tribes" (Cobbing 1977: 71), entered the revolt (as Ranger also claims). They did so because of "loyalty to a long-standing [political] alliance" (Cobbing 1977: 70) or according to "long-standing allegiances" (Cobbing 1976: 387) under the Ndebele monarchy, but they also had grievances similar to the core of the Ndebele state. Importantly, though the Ndebele brought their Shona tributaries into the war, "the [Ndebele] kingdom was not a direct co-ordinating factor in the part of the Mashona country which rose in June" 1896 (Cobbing 1977: 77), specifically, Hartley and Charter. Cobbing in fact argues that Ranger's list of grievances for Mashonaland is sufficient to explain the Shona. The Shona decided simply to revolt when the Company was at its weakest, rising "in response to the opportunity provided by European difficulties in Matabeleland" (Cobbing 1976: 417).

Intriguingly, he refers to the unevenness and "patchiness" (Cobbing 1977: 77) of the Shona rising which was devoid of any major acts of coordination across paramounts, an important issue which he does not seek to explain. As well, the influence of the *Mwari* cult by the mid-1890s did not reach into Mashonaland as extensively as Ranger claims. Additionally, spirit mediums such as Kaguvi and Nehanda had a spatially limited sphere of influence and these mediums (in terms of any political function) simply "reflect[ed] the claims of the incumbent [political] chiefly line" (Cobbing 1977: 79). Like Matabeleland, the spiritual factor in Mashonaland simply "provided an additional sanction for, rather than initiated a political uprising" (Cobbing 1976: 417), with mediums rarely acting autonomously from chiefly rule and prerogatives.

Based on these specific claims, Cobbing provides an overall analysis of the first *chimurenga*, as follows:

The clearest implication ... is that with the exception of the Ndebele kingdom's co-ordination of its tributary peoples and the timing of the Shona outbreaks in June, 'the problem of scale' was not even temporarily solved by Africans in Rhodesia in 1896–7, nor did they make any attempt to do so in the supra-tribal sense that Ranger has suggested. The dominant theme of those eighteen months was, on the contrary, disunity: the splits within the Ndebele kingdom which led to their third civil war; divisions between the Shona, some of whom rose – though not in alliance with the Ndebele – and others who collaborated with the Europeans against both the Ndebele and the rebel Shona for a number of sometimes contrasting reasons. The quest for a secret history or leadership is unnecessary. (1977: 82)

Beyond Cobbing's focus on politics rather than spirituality, his concluding commentary on the revolts is a far cry from Ranger's theory of centralised organisation and wide-scale coordination and, when compared to Beach's initial argument (in his PhD thesis of 1971), Cobbing adopts a more decentralised understanding of the revolt. Beach's later work is more consistent (in particular, explicitly) with the notion of decentralised and localised mobilisation.

2.3.2 *Beach*

The analysis by Beach (1971) in his PhD thesis is specifically about the revolt in Mashonaland, though he provides—in particular—a local (though it is more aptly labelled a regional) case study of what of he calls "south-western Mashonaland". His regional study includes parts of Matabeleland where Shona tributaries existed (such as in the Selukwe and Belingwe districts) as well as the Mashonaland revolt's epicentre in notably Hartley district.

Like Ranger, Beach argues that the Mashonaland risings entailed a degree of inter-chiefdom coordination but, as with Cobbing (and unlike Ranger), he focuses specifically on the political underpinnings of the revolt. In relation to the Shona revolt, Beach claims (like Cobbing) that the actual spatial areas in which the *Mwari* cult officers and the Kaguvi and other mediums operated with any substantial influence were considerably more limited than Ranger suggested, and thus their coordinating role likewise needs to be downplayed. Religion was therefore not a unifying force. However, in Beach's PhD analysis, the political basis of the Mashonaland risings refers merely to the political context in which the revolt arose, rather than to any political organisation during the revolt as such.

In this regard, political organisation animating the Matabeleland rising (in which there was a centralised Ndebele state not broken in its entirety in 1893) is easier to identify than in the case of the Mashonaland rising. In Mashonaland, there was simply no clear pan-Shona unity for decades prior to the first *chimurenga*. Thus, in Mashonaland broadly, no "national unity" (Beach 1971: 25) existed as the Shona polities under the Rozvi empire were deeply decentralised, and the subsequent Shona chiefdoms acted "for the most part like so many independent states" by the 1890s (Beach 1971: 26). Detailing the specific arguments in Beach's PhD thesis would simply entail speaking to queries by Beach of particular points found in both Ranger's and Cobbing's work.

In his 1971 PhD thesis, Beach seems to want to resolve the "problem of scale"— that is, to demonstrate that the first *chimurenga* of 1896–97 entailed a degree of regional (subnational) if not national coordination. However, his actual evidence goes against this attempt, as the importance of localised dynamics often come to the fore, which Beach hesitantly admits. This, we suggest, is the same challenge that critics of the third *chimurenga* face when trying to go beyond merely asserting that this *chimurenga* was a centrally organised initiative of the ruling party. However, by the end of the 1970s, Beach came to realise that the "problem of scale" was not a problem after all. In an article in 1979, he started with a position which went contrary to the search in his earlier work. The uprisings had a decentralised structure and, because of this, "the [analytical] need for a 'religious' [Ranger] or 'political' [Beach and Cobbing] overall organisation falls away" (Beach 1979: 401). Put simply, there was "no widespread concerted planning [including between the Ndebele and Shona, or amongst the Shona] of the main raising in advance of the outbreak" (Beach 1979: 404).

No overall coordination was evident across the countryside, with only rare joint attacks on white settlers and, even within Mashonaland, the archival evidence does not justify the claim that a pre-planned simultaneous uprising simply took place: if anything, the uprising rippled across the countryside, with local chiefdoms deciding either to join, oppose or remain neutral (Beach 1979). Beach (1979) therefore shows the local complexities of the Shona rising in particular, including deeply embedded intra-Shona tensions and diverse motivations in joining or not joining the revolt, with reasons for joining not necessarily reducible to an anti-colonial stance. His overall conclusion is revealing:

> [A] complex set of personality and interest-groups in each area reacting with remarkable swiftness and decision to the events and opportunities and pressures of the 1890s, but doing so according to their conception of their own territory as an independent, undefeated entity, rather than as part of a larger organization that solved 'problem of scale'. In short, the history of the 1896 central Shona *Chimurenga* promises to be the history of many local *zvimurenga* with their similarities, differences and connexions – or lack of them. (Beach 1979: 419)

This concluding claim is highly significant and resonates with our later analyses of the latter two *zvimurenga* and the third *chimurenga* in particular.

2.3.3 Horn

Horn's Master's thesis from 1986, involving significant archival research, entails a major revision in interpreting the first *chimurenga* and yet it has not become widely known. He has a number of interrelated arguments which he claims are also contained implicitly in the work Ranger, Cobbing and Beach, but they refuse to draw out the full logical conclusions. Some of these arguments are controversial, but there is evidence certainly in the writings of Ranger, Cobbing and Beach consistent with them. These include the claims that colonial pressure on the Ndebele and Shona was not overbearing and intrusive, and that the uprising was not an uprising at all (but a mere expression of long-standing intra-Ndebele and intra-Shona civil wars). His other main claim is the revolts were not centrally organised and coordinated, rather, they entailed localised conflicts. Horn acknowledges that Beach's later work (from 1979) is important in refusing to acknowledge any overall organisation for the risings, but Beach still speaks of a general Shona revolt based on the vague idea of ripple effects across Mashonaland.

Like the other scholars, Horn speaks of acts of violence (against settlers) taking place in Mashonaland from March to June 1986. These were, though, "local responses to particular grievances" which "cannot be distinguished from the conflicts there-after" (Horn 1986: 184), that is, from June 1896. In this sense, the idea of a distinct Mashonaland revolt was a white settler idea, based on the heightened white insecurity arising from the Ndebele disturbances starting in March. As such, the Mashonaland settlers now interpreted the prevailing local violence, which also entailed intra-chiefdom rivalries, as part of a general rising. There was no overarching anti-colonial revolt in Mashonaland and indeed no revolt at all. Rather, a range of long-standing localised power struggles within and between Shona chiefdoms were taking place in which white settlers became implicated.

In the case of the Ndebele rising, there was likewise an "uncoordinated response" (Horn 1986: 110). The supposed leadership of the Ndebele rising was divided, dispersed, disunited, disorganised and localised. Like in Mashonaland, any semblance of a rising was not planned or organised, and the "local leadership responded to the situation on the basis of their localised perception" (Horn 1986: 140). This led to local decisions around either 'collaborating' with the British or not, based on political considerations vis-à-vis broader Matabeleland politics.: "The internal divisions prevented any planned 'Rebellion' and retarded any movement towards a cohesive and united determination to expel the European settlers" (Horn 1986: 141).

In Horn's revisionist account, the first *chimurenga* was not a *chimurenga* but, rather, "a series of local conflicts precipitated to a large degree by the actions of Europeans" (1986: 246). The idea of the first *chimurenga* is a "powerful myth" (Horn 1986: 284), in the sense of a centralised and coordinated revolt. While Horn's claims about the absence of any significant anti-colonial impulse during 1896–97 seems deeply problematic, his argument about the localised character of anti-colonial struggle, and tensions within local polities and communities, resonates with—but does not replicate fully—the conclusion reached in particular by Beach about multiple *zvimurenga* during the rising. Of course, Horn comes across as an apologist for white settlerism by arguing that intra-African disputes were vastly more important than any agitation by Africans against white settlers.

The notion of a *chimurenga* myth reappears during the second and third *zvimurenga*. During the war of liberation, the Rhodesian regime claimed that communist-inspired terrorists were agitating peace-loving and satisfied Africans against the government: the second *chimurenga* would not have existed otherwise. Based on arguments presented in Chapter 1, it would appear that many (but not all) scholars deeply critical of the national land occupations may have constructed their own chimurenga myth-making. From their perspective, what took place was at best a hollowed-out version of a *chimurenga* orchestrated and imposed by the ruling party and, at worst, no *chimurenga* at all. Occupiers were manipulated and used by the ruling party for its own political gain, an argument which is inconsistent with the historiography of nationalism which soon became embedded in the second *chimurenga* literature and needs to be more central to the third *chimurenga* literature, as we argue in later chapters.

In this context, we now turn to a consideration of some key and pertinent themes in the first *chimurenga* literature.

2.4 First *Chimurenga* Themes

Knowledge about the details and complexities of the first *chimurenga* remains exceedingly limited and will likely remain so. The almost complete reliance on the colonial archives for reconstructing the rising is problematic, leading scholars at times to gaze upon the revolt from a broadly colonial perspective. Ranger, instance, has regularly been accused of the "uncritical, use of the evidence of colonial officials" (Smith 1998: 213).

Despite these limitations, and drawing upon other literature which indirectly or in part addresses the anti-colonial revolts of 1896–97, we identify three interrelated themes about the risings in Matabeleland and Mashonaland which are of some significance for our comparative analysis of the three *zvimurenga*.

First of all, a central theme for our examination, as already brought out, is the organisational form and dynamics underpinning the revolt (including questions around the degree of centralised coordination and mobilisation). The two other interconnected themes relate to spirituality and gender. Incorporated in these themes are issues pertaining to the spatial and temporal variation of the revolt and the diverse motivations behind involvement in the rising.

In the preface to the second edition of his *Revolt*, Ranger (1979) accepts the criticisms raised by Cobbing and in particular Beach about overprivileging the centralised character of the organisation, mobilisation and coordination of the rising. With a specific focus on Mazoe and Victoria areas of Mashonaland, Tsomondo (1977) retains Ranger's original perspective on the centralised character of the rising in Mashonaland, what he labels as a "national liberation war" (Tsomondo 1977: 21). Inter-paramount rivalries did play out, such that chief Mutasa for instance aligned with the British while chief Makoni did not (see also Barnes 1976). These rivalries, though, were largely overcome because of the coordinating role of supra-ethnic mediums like Kaguvi and Nehanda. For Tsomondo, the Mashonaland revolt should not be read from the Matabeleland revolt as the two revolts had different logics, though this is not clearly articulated. Nevertheless, while the Shona fought a proto-nationalist struggle, the Ndebele engaged in "a war against some specific injustices" (Tsomondo 1977: 27). At the same time, significant cooperation existed between the Ndebele and Shona in the first *chimurenga*.

In an article in 2011, and reminiscent of Beach's later formulation, Dawson argues that the first *chimurenga* was "a complex set of [local] struggles over land, cattle and taxes", "[r]ather than a planned, cohesive movement intended to overthrow the whites" (Dawson 2011: 144). The largely uncoordinated form of the revolt, at least in Mashonaland, had been emphasised for some time, even at the time of Ranger's original work. Thus, Hodder-Williams (1967), in examining the area around Marandellas in Mashonaland, speaks about the "extremely haphazard nature" (Hodder-Williams 1967: 40) of the Mashonaland rising, highlighting that it "was not a well-organised movement nor was the strategy coordinated" (Hodder-Williams 1967: 39). In examining the third *chimurenga*, and even the second *chimurenga*, debates about the layers and levels of organisation and mobilisation become very important.

In the case of spirituality, Ranger (also in his 1979 preface) acknowledges that, in his earlier analysis, he likely overprivileged the role of the *Mwari* cult and spirit mediums in explaining the mass mobilisation and coordination which took place during the rising. But he still spoke back to Beach and Cobbing by claiming that "it is not true to assert that mediums were always limited in influence" and that the *Mwari* cult was not "without influence in the Ndebele state" (Ranger 1979: xiv). The significance of the spirit mediums for the first *chimurenga* is deeply entrenched in the scholarly literature. For example, Garlake (1966) refers to Kaguvi literally giving instructions for the rising, east of Salisbury.

In 1972, based on oral history from his home area (falling at the time under Chief Mashonganyika) but also borne along by Ranger's work, Vambe argued that mediums and the *Mwari* cult "provided the yeast which fermented the whole movement and enabled the [Ndebele and Shona] participants to co-ordinate their policies and tactics" (Vambe 1972: 115). The Ndebele-Shona "war policy was worked out" at the place of the paramount chief Mashayamombe in western Mashonaland where Mkwati and Kaguvi met and "put the final seal to the Shona-Ndebele pact" (Vambe 1972: 119). For Vambe (1972), the Nehanda medium had spatially limited following. Likewise, in his 1976 book on the Zezuru in the Chiota Tribal Trust Lands in the Marandellas area, Fry (1976) brings out the significance of spirit mediums during the 1896–97 rising, though he disputes Ranger's notion of a rigid Nehanda-Chaminuka spirit hierarchy existing in central Mashonaland. The centrality of spirit mediums to the first *chimurenga* in whatever form remains central to current analyses (see Kaoma 2016 for instance).

Spirit mediums are crucial to the literature on the second *chimurenga*, and less so the third chimurenga literature. Ranger (1979: xviii) in fact notes that there are "tantalising echoes and re-enactments" of the 1896–97 revolt "in the practices of the guerrilla war itself" during the war of liberation in the 1970s. The relevance of spirit mediums to nationalist struggle arose alongside the birth of mass nationalism in the late 1950s. According to Fry (1976), this relates to the fact that, by the 1960s, chiefs had acquired a coercive authority in and through indirect rule, in contrast to the more consensual authority of spirit mediums.

Christian missionaries entered Matabeleland and Mashonaland before the risings, and some came with the Pioneer Column in 1890 or soon thereafter, with mission stations established on land granted by the Company and largely supporting the colonial intrusions (McLaughlin 1996). The Salvation Army, like other early missionary outreaches, "proved to be impressively imperial" (Murdoch 2015: 7). The fact that Christian churches joined settlers and Company officials in appropriating Shona and Ndebele "property, labor and cattle, power, and treasured sacred landscapes" (Murdoch 2015: 31) contributed to local suspicions of white missionaries' spiritual project.

Typically, as Vambe (1972) notes, those white settlers "who had been noted for their arrogance and harshness towards black people were the first objects of attack" (1972: 125) in the late 1890s, a point relevant to the second *zvimurenga*. In the case of the killing of a Salvation Army captain in 1896, this relates to the fact that the Army's Mazoe mission station was, at the time, home to the Company's local Native

Commissioner and a post for the Company's African police (Pollet 1957; Howland 1963; Murdoch 2015). In the case of Chishawasha Mission, which came under attack, mission cattle were taken by mission tenants including the few Native converts (Rea 1975). This occurred in the context of a preceding rinderpest plague, during which the Jesuit brothers at Chishawasha had aided the Company in slaughtering African cattle. In certain instances, Africans converted to Christianity were murdered during the first *chimurenga* (Hodder-Williams 1967).

Because mission stations and the spread of Christianity only became entrenched in the Rhodesian countryside during the twentieth century, the issue of Christian churches in the context of *zvimurenga* arose most forcefully during the second *chimurenga*, including the variegated character of the relationship between guerrillas and missions.

In relation to the third theme, very limited knowledge is available about the position and role of women in the 1896–97 rising. In reflecting on the first *chimurenga* literature in the early 1990s, Schmidt (1992) thus noted: "Although the causes, organisation, and consequences of the rising have been debated at length, none of the historical accounts has been concerned with gender. What did the rising mean for women? In what ways did they participate? How did the outcome affect them?" (Schmidt 1992: 36). In her history of rural Shona women as peasants and traders in the Goromonzi area (up until the 1930s), Schmidt is regrettably unable to provide any substantial or detailed answer to the question of women's participation in the Mashonaland revolt, in large part because of the sheer absence of any archival record which privileges women in history. Over two decades later, in writing about historical memory pertaining to the Nehanda medium in the 1890s, Charumbira (2015) makes a similar point in arguing that "women and gender dynamics … have not been fully addressed in the historiography of the 1896–97 wars" (2015: 54). In fact, discussions about the involvement of women have been reduced to the role of the Nehanda spirit medium during the Mashonaland revolt.

In an article from 1998, Beach revisited the role of Charwe, the medium of Nehanda, during the Mashonaland revolt. By the 1960s, an oral tradition had emerged about the medium's early anti-colonial role (particularly within the burgeoning nationalist movement), with Ranger contributing to making this part of nationalist gospel through his 1967 book. The fact that Charwe did not recant from "pagan" beliefs and convert to Christianity at her death, whereas Kaguvi reportedly did, likely made her a "symbol of intransigence" (Beach 1998: 52). However, the nearest archival evidence to the events suggests that any link between Charwe and the actual organisation of the revolt (even in Mazoe, where Charwe resided) is "thin and insubstantial" (Beach 1998: 45). In suggesting some kind of gendered analysis, Beach claims that the medium of Nehanda was an innocent victim of both British and Native men, who falsely accused her of wrongdoing at her trial.

A decade later, in 2008, Charumbira directly responds to (and criticises) Beach's reinterpretation of Charwe's historical relevance from what she considers to be a more fully grounded feminist perspective. This entailed examining the Nehanda medium from a gendered analysis of power dynamics with reference to white colonisers and Native colonised, as well as to Native men and Native women. Contrary to Beach,

Nehanda-Charwe was "very much involved in the rallying of rebels and in urging people to participate in the rebellions" (Charumbira 2008: 108). In trying to restore the agency of Charwe and women more generally in the rising, Charumbira argues that the medium of Nehanda was a woman locked into the power struggles of her times and, insofar as she was a heroine, there were also other, common, heroines beyond the Nehanda medium.

Schmidt (1992) sought, no matter how unsuccessfully, to incorporate women generally into an analysis of the first *chimurenga*, as both agents and victims. She notes for instance that, in the case of the mission cattle at Chishawasha, a large number of women (perhaps Charumbira's other "heroines") were seen driving away the cattle and hence they directly participated in this "military operation" (Schmidt 1992: 39). This was significant because men were normally the warriors and raided for cattle and, further, cattle were considered culturally as an exclusively male domain. This episode, therefore, involved "a major violation of gender roles" (Schmidt 1992: 39). However, Schmidt comes to a sweeping, unfounded, conclusion on the basis of this one instance alone. More specifically, she argues unjustifiably that "the Shona-Ndebele rising of 1986–97 witnessed the active involvement of women in ways that shattered gender norms" (Schmidt 1992: 42).

However, she adds that "most women experienced vulnerabilities [as victims] similar to those in previous armed conflicts" (Schmidt 1992: 42). Women, as before the risings, were pawns in the revolt. For example, they were taken hostage by white settlers in order to pressurise rebelling men to surrender, with women forming the majority of captives imprisoned at Chishawasha's police camp during the rising. Some of these captive women were married off to the Company's African police or messengers, so that "the missionaries and colonial officials were simply perpetuating a new version of an old practice – capturing women at war and making them wives of the victors and their clients" (Schmidt 1992: 40). Captive women attached to those chiefs who had surrendered were released to appease the vanquished chiefs, such that women became "pawns" in disputes between men. The position of women, including their agency, is revisited when discussing the second and third *zvimurenga*.

2.5 Conclusion

This chapter reviewed the key texts which provide, based mainly on archival research, an analysis of the first *chimurenga*. In addition, it identified certain themes within the first *chimurenga* literature including a key concern of ours pertaining to the organisation, mobilisation and coordination of the risings, as well as religion and gender. Colonial intrusion and pressure, including the loss of autonomy and the undercutting of the precolonial way of life, were significant in almost compelling both the Ndebele and Shona to revolt. As well, pre-existing tensions within the Ndebele and Shona polities played themselves out during the rising, though the major contradiction remained between the coloniser and colonised. At the same time, it is clear that there was significant local variation in the character of the rising, leading to what Beach referred to as the existence of *zvimurenga*. As we will see, many of these points resonate with both the second and third *zvimurenga*.

Of course, it is tempting to interpret the first *chimurenga* from the perspective of both the second and third *zvimurenga*, which we refrain from doing. For instance, writing at a time when the war of liberation in the 1970s was taking off, Vambe (1972) argued: "Unlike the Ndebele whose military tactics took the form of massive charges and therefore provided excellent shooting targets, the Shona operated as guerrillas, striking only when it was to their advantage and disappearing into natural cover before the enemy had time to recover his balance" (1972: 134). This comes across as reading into the first anti-colonial revolt, the presence of alleged differences in guerrilla tactics during the 1970s, respectively a difference between the guerrilla army attached to ZAPU and the guerrilla army attached to ZANU. The use of a "guerrilla type of warfare" (Burke 1971: 26), involving hit-and-run attacks, is though commonly attributed to the rebels in Mashonaland.

Before discussing the second *chimurenga*, and the parallels between it and the first one, we examine—in the next chapter—peasant struggles and nationalism post-1896/97 until the 1960s. This sets the context for the two chapters (Chapters 4 and 5) on the war of liberation in the 1970s.

References

Barnes J (1976) The 1896 rebellion in Manicaland. Rhodesiana 34:1–7
Beach DN (1971) The rising in South-Western Mashonaland, 1896–97. Unpublished PhD Thesis, University College of Rhodesia, Rhodesia
Beach DN (1979) Chimurenga': the Shona rising of 1896–97. J Afr Hist 20(3):395–420
Beach DN (1998) An innocent woman, unjustly accused? Charwe, medium of the Nehanda Mhondoro spirit, and the 1896–97 Central Shona rising in Zimbabwe. History in Africa 25:27–54
Burke E (1971) Mazoe and the MaShona Rebelion. Rhodesiana 25:1–34
Charumbira R (2008) Nehanda and gender victimhood in the Central Mashonaland 1896–97 rebellions: revisiting the evidence. Hist Afr 35:103–131
Charumbira R (2015) Imagining a nation: history and memory in making Zimbabwe. University of Virgina Press, Charlottesville
Cobbing J (1976) The Ndebele under the Khumalos, 1820–1896. PhD Thesis, University of Lancaster, UK
Cobbing J (1977) The absent priesthood: another look at the Rhodesian risings of 1896–1897. J Afr Hist 18(1):61–84
Dawson S (2011) The first Chimurenga: 1896–1897 uprising in Matabeleland and Mashonaland and continued conflicts in academia. Constellations 2(2):144–153
Fry P (1976) Spirits of protest: spirit mediums and the articulation of consensus among the Zezuru of Southern Rhodesia (Zimbabwe). Cambridge University Press, Cambridge
Garlake P (1966) The maShona rebellion east of Salisbury. Rhodesiana 14:viii–11
Hodder-Williams R (1967) Marandellas and the Mashona rebellion. Rhodesiana 16.27–54
Horn M (1986) "Chimurenga" 1896–1897: a revisionist study. Unpublished MA Thesis, Rhodes University, South Africa
Howland R (1963) The Mazoe patrol. Rhodesiana 8:16–33
Kaoma K (2016) African religion and colonial rebellion: the contestation of power in colonial Zimbabwe's Chimurenga of 1896–1897. J Study Relig 29(1):57–84

McLaughlin J (1996) On the frontline: catholic missions in Zimbabwe's liberation war. Baobab Books, Basel

Murdoch N (2015) Christian warfare in Rhodesia-Zimbabwe: The Salvation Army and African Liberation, 1801–1991. The Lutterworth Press, Cambridge

Pollet H (1957) The Mazoe patrol. Rhodesiana 2:9–38

Ranger T (1967) Revolt in Southern Rhodesia 1896–7: a study in African resistance. Heinemann Educational Books, London

Ranger T (1979) Revolt in Southern Rhodesia 1896–7: a study in African resistance, 2nd edn. Heinemann Educational Books, London

Rea S (1975) Chishawasha mission and the 1896 rebellion. Rhodesiana 33:32–35

Schmidt E (1992) Peasants, traders and wives: Shona women in the history of Zimbabwe, 1870–1939. Baobab Books, Harare

Smith N (1998) Theorising discourses of Zimbabwe, 1860–1900: a foucauldian analysis of colonial narratives. Unpublished PhD Thesis, University of Natal, South Africa

Tsomondo MS (1977) Shona reaction and resistance to the European colonization of Zimbabwe, 1890–1898: a case against colonial and revisionist historiography. J South Afr Aff 2(1):11–32

Vambe L (1972) An ill-fated people. Heinemann, London

Chapter 3
Land Alienation, Land Struggles and the Rise of Nationalism in Rhodesia

Abstract This chapter acts as a prelude which frames the examination of the second *chimurenga* in Chapters 4 and 5. In discussing land alienation, land struggles and the rise of nationalism in Rhodesia in the intervening period between the first and second *zvimurenga*, the chapter brings to the fore the deep grievances around land under colonial subjugation, which resulted in localised resistance and struggles in the Reserves, later Tribal Trust Lands. Grievances and struggles were firmly embedded in the historical memories of rural people as they engaged with guerrilla armies during the second *chimurenga*. The chapter shows social differentiation within the reserves and the tensions which sometimes arose because of this differentiation. In the case of both the pre-nationalist days and the days of emerging mass nationalism from the mid-1950s, the chapter stresses the ways in which Africans drew upon their localised experiences and grievances when confronting the colonial order, including the agrarian and land reconfiguration of the reserves. As well, the chapter has a specific focus on women, as they struggled not only against a colonial order but also a patriarchal order.

Keywords Mass nationalism · Native reserves · Land alienation · Rhodesia · Women's resistance · Cultural nationalism

3.1 Introduction

This chapter acts as a kind of prelude which frames the examination of the second *chimurenga* in Chapters 4 and 5. In discussing land alienation, land struggles and the rise of nationalism in Rhodesia in the intervening period between the first and second *zvimurenga*, we do not seek to claim the presence of an unbroken and linear process of intensifying struggles around land which led invariably to the war of liberation in the 1970s. However, we do wish to bring to the fore the deep grievances and challenges around land and related dimensions of colonial subjugation, which at times resulted in localised resistance and struggles in the Native Reserves, later Tribal Trust Lands. In this light, James Scott (2013: 4) argues that, across the globe, "the rural population is generally treated as the more-or-less passive recipient of projects hatched and implemented from above", only seemingly appearing on the

© The Author(s), under exclusive license to Springer Nature Switzerland AG 2021
K. Helliker et al., *Fast Track Land Occupations in Zimbabwe*,
https://doi.org/10.1007/978-3-030-66348-3_3

historical stage during times of major revolt. The Zimbabwean literature, though, has increasingly brought to the fore the political agency of rural villagers in Native Reserves outside of heightened periods of open revolt. Grievances and struggles were firmly etched and indeed embedded in the historical memories of rural people as they engaged with guerrilla armies during the second *chimurenga*. At the same time, we show social differentiation within the reserves and the tensions and conflicts which sometimes arose because of this differentiation. In the case of both the pre-nationalist days and the days of emerging mass nationalism from the mid-1950s, we stress the ways in which Africans in reserves/Tribal Trust Lands ultimately drew upon their own localised experiences and aspirations when confronting the colonial order, including in relation to agrarian and land reconfiguration. We also have a specific on women, as they struggled not only against a colonial order but also a patriarchal order.

3.2   Land Alienation in Colonial Zimbabwe

Alienating African or black people from their landholdings was the cornerstone of policy in colonial Zimbabwe up until the time of the emergence of the war of liberation in the 1970s. Palmer (1977) thus rightly points out that racial domination in the country, involving a range of material, legal, spatial and discursive dimensions and barriers, was founded upon white settler land alienation and evictions of African people.

In terms of legal instruments, the dispossession of black people actually began before the first *chimurenga*. The Lippert Concession of 1889 was introduced even preceding the occupation of the country in 1890 to allow would-be settlers to acquire land rights from then-African territory. As well, the 1894 Order in Council came into force to pave way for land occupation by the Company (BSAC) and evictions of the Ndebele people. The Ndebele were pushed into dry and less fertile lands such as Gwayi and Shangani Native Reserves in 1894 or, at least, this was the intention. In this respect, Moyana (2004) notes that the Order made it legally obligatory for the Company to assign sufficient land for Africans. As spelt out in the Order, the reserves would guarantee ongoing protection and "comfort" for the rural African population. Though the official justification for the reserves comes across as benign, it is certainly the case that reserve land was merely land left over after the settlers had undertaken their initial land grab (Ndlela 1981). By 1896, reserves in Mashonaland had been demarcated also—though the location and size of reserves in both Matabeleland and Mashonaland were reconfigured a number of times during later decades. Overall, the creation of Native reserves marked the beginning of spatial, social and political segregation between Europeans and Africans.

Post the first *chimurenga*, the 1898 Native Reserves Order in Council Act legally buttressed the creation of Native reserves for Africans. The reserves were also legit-imised by the 1902 Executive Council. As such, by 1902, Africans had been dispos-sessed of more than three-quarters of their land, although this did not mean their immediate eviction and occupation by white settlers. Land was demarcated into reserves for exclusive African occupation, while other land was alienated to mines and farmers (either occupied or in the hands of absentee landlords). Additionally, unalienated lands were owned by the Company (until 1918, when it was converted into Crown land). By 1908, many Africans had been evicted from European lands but others remained and were converted into tenants or squatters on European lands.

Through the Private Locations Ordinance of 1908, Africans were expected to pay rent to a white landlord in return for residence. But, because of high tenant rents and grazing fees (for instance, introduced by white ranchers in Matabeleland in 1912), massive movements of Africans into reserves followed. By 1914, when the Native Reserves Commission was appointed, most of the land in Rhodesia had been appropriated by Europeans. Then, by way of an Order in Council in 1920, the initial formalisation of Native reserves was at last finalised. Later, the 1923 Southern Rhodesian Constitution, which accorded self-rule to the settlers (and ended Company rule) further entrenched the Native reserves (Ndlela 1981). The Native reserves were mostly in the middle or low veld where agricultural production was problematic, and often far from away from railroad lines and other infrastructure as well. All in all, the creation of reserves was the foundation of Rhodesia's "dual economy" which disadvantaged Africans significantly.

Socio-legal segregation along racial lines continued to dominate settler politics in the 1920s. At the recommendation of the Morris Carter Commission of 1925 of the need to avoid "inevitable racial conflict", the Land Apportionment Act (LAA) of 1930 was enacted. Regarded as the most controversial part of colonial agrarian policy by Drinkwater (1991), the Act formalised permanent racial separation, by law, of land between blacks and whites. This further consolidated the dual structure of land settlement and agrarian development in the colonial period. Under this Act, land was demarcated as follows: 22.4% for Native reserves, 50.8% as European areas, 7.7% for Native Purchase Areas (NPAs) and 18.4% as unassigned areas. In the context of the LAA and earlier legislation, evictions of Africans into Native reserves proceeded apace. For example, the five chiefdoms of the western Mazoe District—Chiweshe, Hwata, Makope, Negomo and Nyachuru—were pushed into the Chiweshe Reserve. The Hwata people were moved off the Mazoe Citrus Estates in 1922 and a group of Makope people were forced to move into a reserve from the Umvukwes Ranch in 1925 (Bessant 1992).

Land for Europeans was also privatised (involving freehold title) while that for Africans in reserves was held under traditional/customary tenure. In what was considered as a measure of compensation by the colonial authorities, rights to purchase land under freehold tenure was made possible through the formation of NPAs—as recommended by the Carter Commission (Shutt 1997). These areas were meant to cushion the political impact of the prohibition of African land purchases in European areas (Palmer 1977) and, further, they were often spatially located so as to act as buffer zones between reserves and white settler lands. The establishment of the purchase areas technically and legally preserved the principle of land ownership by Africans, which remarkably became the defence for future land alienation by whites.

Farms in the NPAs were purchased in the main by elite male members of rural African society who were expected to demonstrate their agricultural potential by possession of a Master Farmer certificate issued by colonial district administrators—without, however, competing with white settler farmers. Duggan (1980: 235) argues that "throughout the 1940s, 1950s and 1960s, the NPAs acted as a political safety valve for the frustrations of the better-off". As well, the NPAs were further extended in the 1950s, and given more significant administrative attention after years of neglect, as a political response to growing nationalist mobilisation and activities. This was meant to prevent—through co-optation—the educated elite from joining these activities. In fact, by that time, there was evidence of farmers in the NPAs leaning towards nationalism (Cheater 1984); and in the mid-1960s, NPAs were considered by the Rhodesian government as places of dissent and nationalist mobilisation (Scoones et al. 2018). Interestingly, the NPAs, referred to as *matenganyika* ("buying land") in Shona, poignantly revealed the bitter sentiment and resentment amongst Africans, who were first dispossessed forcefully from their land, then evicted, and then made to buy land. The legacies of the NPAs continue to exist today in areas such as Mutoko, Msengezi and Mushagashe.

The detrimental effects of the LAA and related legislation on Africans were considerable, including in relation to the capacity to engage in agricultural production. In this light, and as noted by Moyana (2004), by 1943 there were as many as forty-three reserves which were overstocked in terms of cattle holdings, and reserves such as in Seke experienced acute landlessness. In reserves in Matabeleland alone during the 1950s, there was an excess population of one hundred thousand people; and overpopulation nationally was 12% on average, with the greatest overpopulation in Victoria province. In Chiweshe reserve in Mashonaland, colonial administrators recognised that the area was overpopulated by almost 100%. Ultimately, Africans were trapped in cycles of low agricultural productivity due to degraded soils, and poverty was rife. Duggan (1980) highlights that the impoverishment of the reserves nation-wide continued throughout the 1920s, 1930s, 1940s and indeed beyond.

There is no doubt that the promulgation of the LAA, and landlessness more broadly, was made with considerations of impoverishing rural Africans in order to ensure a supply of cheap African labour for the settler economy. The process of transforming farmers into labourers has been referred to, respectively, as stimulation and strangulation by Palmer and Parsons (1977) and Arrighi (1970). Besides landlessness, forced taxation (such as poll, hut, cattle, wife and dog taxes) was also used on an ongoing basis to "stimulate" the entry of Africans into the colonial labour force (Arrighi and Saul 1973; van Onselen 1976). Thus, an intended outcome of the LAA was some degree and form of proletarianisation. Many Africans who were victims of the LAA through landlessness became engaged in the colonial labour market by means of *chibaro* (forced labour). *Chibaro* was one of the boldest manifestations of the coercion and compulsion inherent in Rhodesian colonialism (Bessant 1992).

The idea of placing maximum land pressure on Africans under the LAA was also to prevent them from becoming competitors to European farmers. Indeed, initially, many African farmers had responded positively to the colonial commodity market and supplied agricultural produce to towns and mines. Further undercutting of African agricultural production was induced by agricultural marketing policies such as the 1931 Maize Control Act which prohibited Africans from selling their maize directly to the Maize Control Board. Rather, they had to sell through white middlemen at lower prices. The 1934 Cattle Levy Act and the associated Cattle Levy and Beef Export Bounty likewise underpinned differential pricing for African- and white-owned cattle. These and other legal instruments pertaining to agricultural marketing were protected white farmers from economic competition.

This was consistent with the "white agricultural policy" (Ndlela 1981) which was formulated under Company rule in 1908. Compared to African farmers, white settler farmers were to be prioritised and supported by the colonial authority including through grants, loans, access to commodity and labour markets and low land prices. Alongside this emerging white commercial agricultural sector, a stagnant, unsup- ported peasant agricultural sector in the reserves came into existence, and in the context of ongoing evictions and inaccessible commodity markets (even if produce was available for sale). To illustrate the neglect of Africans in reserves, Arrighi (1970: 215) explains that "before 1940 public expenditures on agricultural development were negligible; small sums were made available for road building, bridge construc- tion, and other minor works such as cattle dipping pens and small dams". After all, the objective of so-called Native development was merely, or at least partially, to enable the reserves to accommodate more Africans expelled from land expropriated for settler occupation (Machingaidze 1991).

Alexander (1993) emphasises that colonial policies led to extensive and coercive interventions into the way Africans lived and farmed. But, around the late 1920s, government's approach towards the reserves underwent a fundamental shift. Previ- ously, the establishment of the reserves was seen as a temporary phenomenon, as colonial administrators had thought reserves would vanish as African farmers (or peasants) would be drawn into the wider capitalist economy as labourers and thereby fully proletarianised (Kramer 1998). This was no longer seen as a suitable or even valid proposition, so some attention was paid to reconfiguring the reserves internally. As noted by Kramer (1998), early efforts at reconfiguration through what became known as centralisation began in 1929 and expanded after the passage of the LAA in 1930. In fact, centralisation measures became the cornerstone of colonial land- use planning. This involved demarcating landholdings into residential, grazing and arable lands, with homesteads centralised in linear settlements and permanent plots of arable land allocated to men by colonial agriculturalists.

Under the centralisation policy officially espoused in 1929, the idea was to "develop" the reserves through intensive agriculture. The policy was implemented through the Native Agriculturalist Emery Alvord. Alvordism, as the initiative came to be known, was based on the supposed supremacy of white modernity as it sought to dismantle what were labelled as archaic and inefficient African ways of farming especially shifting cultivation. In Selukwe (now Shurugwi) reserves, for instance, rotational cultivation was viewed as primitive, wasteful and low yielding (Kramer 1998). Moore (2005. 81) also notes that Alvord thought "primitive practices produced a package of poverty, inscribing buildings, bodies and beasts". So, colonial modernity for Africans was to be achieved through sedentarism, linear settlement and fields, and limited commodity market orientation.

As with the labour market, *chibaro* was to become a standard method of implementing the restructuring of African agriculture in the reserves. For instance, in Chiweshe, *chibaro* gangs began soil conservation work during 1940. Boundaries were drawn which were not to be encroached by Africans for farming purposes (such as vlei land, steeply gradient land and land near rivers and streams), thereby deepening landlessness (Bessant 1994). Africans were also expected to practice mono-cropping and use hybrid seeds. Further, the Natural Resources Act of 1941 legislated compulsory stocking rates, mandated contour ridges (paying attention to depth and spacing of furrows) and legally buttressed the centralisation project by the colonial state. The Act further concentrated the power of Native Commissioners in the reserves to intervene decisively in matters of natural resource management, including the forced destocking of cattle (Worby 2001).

However, around the mid-1940s, it was clear that the practice of centralisation was failing. As expected, Africans were seen as the cause of the failure. The colonial government, based on a number of recommendations from agriculturalists and Native Commissioners, simply concluded that Africans were poor farmers. As Machingaidze (1991: 560) puts it, "Alvord squarely put the blame on Africans for the lack of change and adoption of new farming methods, and concluded that Africans will not change without compulsion and control". Bessant (1992) echoes this view, citing colonial officials (with respect to the Chiweshe reserves) arguing that peasants' apparently wasteful farming methods, if left unchecked, would destroy the soil and turn the reserves into a desert, thereby creating a permanent land shortage.

Subsequently, the centralisation policy became even more coercive and authoritarian as reflected in the passage of the Native Land Husbandry Act (NLHA) of 1951. The NLHA was, amongst other things, designed to: promote reasonable standards of animal husbandry while limiting the number of livestock; protect land and natural resources from further degradation; and introduce modernist agricultural methodologies (Duggan 1980). What the Rhodesian state considered as land conservation measures were compulsory, including agricultural practices such as furrowing and contour ridging. Thus, labour contributions by Africans in reserves in pursuing such measures continued on a compulsory basis. Like under the earlier centralisation policy, land was allocated to men only (with each plot equal in size). Allocation of land by chiefs was not permissible (Duggan 1980). Holdings could not be subdivided or sold, nor could they be used as collateral for loans as the farming permit granted conferred use rights rather than ownership (Thompson 2004).

In the end, for Thompson (2004: 14), the intentions behind the NLHA were "conservation, segregation, agricultural modernisation, and intensified state control". Quite likely, highly interventionist and disciplinary forms of colonial governance reached their apogee with the NLHA, under which the planning and forced reorganisation of settlement and land use in the reserves were placed mainly in the hands of technicians in the Native Affairs Department (Worby 2001). This involved an attempt "to govern agrarian livelihoods through rigid, orderly spatial discipline" (Moore 2005: 84). The reinforcement of the colonial socio-spatial order through the NLHA, and related agrarian and land restructuring during the 1950s, became central to African resistance and the growth of nationalism.

In 1950, as well, amendments were made to the LAA which paved way for compulsory evictions of Africans off white land within five years. This, of course, was part of a longer-term process of eviction. Indeed, landlessness had worsened for Africans after the Second World War, including for labour tenants on European lands as they were evicted after white soldiers as veterans of the war descended on the country with offers of land by the Rhodesian government. In this respect, a 1941 amendment to the LAA had already consolidated the legal position with reference to the eviction of Africans and defined Africans living on white land (while not labouring for the colonial economy) as squatters. For instance, there were massive evictions from large estates in the Midlands and Matabeleland (Worby 2001). The evictions from the Rhodesdale Estates in particular are known to have paved way for the settlement of white ex-servicemen, as between 10,000 and 12,000 "squatters" were evicted from this land in the 1950s and dumped literally in Gokwe and the adjoining Sanyati reserve as now "immigrants" (Nyambara 2001; Worby 2001). This was to make life for Africans in reserves even more unbearable as land in reserves became increasingly degraded due to continued population pressure, overgrazing, overstocking and reduced soil fertility—and despite reorganisation of the reserves internally.

Thus, reorganisation and evictions proceeded apace simultaneously in the 1950s. The centralisation programme was pursued unevenly across the countryside, with some reserves reconsolidated earlier than others—for example, in Chinyika, the reserve in Goromonzi District came under the full weight of the NLHA in 1952 and Chinamora Reserve in 1953. Towards the late 1950s, one of the largest development interventions took place, namely, the construction of the Kariba Dam. Though not related to resettlement in reserves, this development project resulted in the forced displacement, in 1956 and 1957, of tens of thousands of Tonga-speaking people from the banks of the Zambezi and its tributaries (Worby 2001). Ultimately, it was hoped the NLHA would create a contented peasantry, with those made landless becoming fully proletarianised in serving the expanding white colonial economy.

However, the NLHA affected the rights of Africans to land, even consolidating colonial land alienation. Thus Machingaidze (1991: 571) notes that "under the NLH Act, land rights, the major form of social security for the vast majority of Africans, ceased to be a birthright and a new generation would grow up cut off from the possibility of ever owning land". The Act also "denied labour migrants and young men access to arable holdings, especially as many reserves were so overcrowded that men who were not working the land at the time the NLHA was implemented would never receive a holding" (Thompson 2004: 18). As a result, young men's aspirations for marrying and establishing independent families were undermined. In Chiweshe reserves, for instance, only the most fortunate of newly married couples managed to acquire a few acres of hilly, infertile land; in the past, this land would be considered unusable (Bessant 1994). African land rights and usage under the NLHA were also made conditional as security of tenure was inextricably dependent upon "good farming".

Overall, "the NLHA colonised the micro-practices of subjects already incarcerated in territories of conquest" (Moore 2005: 86). Nevertheless, scholars converge on the view that NLHA was a failure (Duggan 1980; Moyana 2004; Moore 2005;

Thompson 2004). In fact, it was never fully implemented in all reserves. Nevertheless, the colonial agrarian policy vis-à-vis the reserves further expanded the landless class in rural areas (Andersson 1999). By 1959, more than one-quarter of reserve families entitled to land were landless (Duggan 1980). By August 1961, the Native Agriculture Department had registered more than 45,000 men across the country who had applied for land but could not receive plots as there was no land in their area (Thompson 2004).

Added to this, agricultural productivity continued to decline. A major weakness in the implementation of the Act was concentration on conservation and infrastructure, which did not lead to a direct and immediate rise in productivity (Machingaidze 1991). Further, class differentiation—which had always existed to some extent—widened as the rural African elite emerged as winners. Chiefs and headmen for instance were given larger pieces of arable and grazing land because of their duties and responsibilities. More so, "reserve entrepreneurs" increased their yields by tilling more and more acreage, crowding out other farmers unable to make proper use of their plots, thus leading to even further land shortages in the reserves (Shutt 1997). This heightened the need to enter the colonial labour force (Machingaidze 1991).

The NLHA alongside other agrarian laws and policies created a gruesome legacy of land and natural resource tensions and conflicts (Nyandoro 2012: 307) in the reserves, which became known as Tribal Trust Lands (TTLs) in terms of the Tribal Trust Lands Act of 1965. Thus, not only were there conflicts between the colonial administration and peasants, and between white settlers and African farmers, as there were also acute social disputes within reserves/TTLs between households including along class and age lines. Rights to use and retain land, and controversies around tenure security, often produced tensions that led at times to violence between people (Chitiyo 2007). As well, people found adjusting to living in nucleated or centralised settlements difficult (Thompson 2004), after years of the scattered homesteads arrangement. Various boundary and field disputes (for instance, between households and even between villages) were commonplace.

Chiefs (some whose local power had been reduced under colonial rule) were often ignored or overruled by bureaucrats in these disputes (Chitiyo 2007). Indeed, chiefs were subject to attacks by frustrated villagers, with some having their lands dug up or their trees stripped; while some were forced to award land grants to locals and so-called "foreigners" or immigrants (Ranger 1985). Besides the rise in land and boundary conflicts, the increased (forced and voluntary) migrations of locals and foreigners into and out of kraals, villages, wards and districts, and the ensuing confusion over land entitlements, added to the growing crisis. The period 1961 to 1968 thus saw an upsurge in land disputes, some brought before chiefs for mediation and others not (Chitiyo 2007). In Gokwe, for example, ethnic conflict arose between the Shangwe and Madheruka (Alexander and McGregor 1997; Maravanyika 2012). These two groups clashed over differences in farming practices and religion, with the former following traditional religion and the latter Christian. In Bulilimamangwe, where the state had imposed Ndebele chiefs above Kalanga chiefs, tension was rife.

Importantly, restrictions and demands of the NLHA ruptured local relationships and social bonds. For example, land restrictions aggravated gendered conflicts within households as women and men argued over production decisions (i.e. who could grow what, how and where). Questions around inheritance posed problems while traditions of polygamy and large families also added to land pressures (Chitiyo 2007). Thompson (2004: 20) aptly captures some of the tensions as follows:

> Disagreements about production methods surfaced along generational lines, especially between fathers and sons. Stock and land restrictions threatened broader social networks. Marriages, generally secured by the payment of eight to ten head of cattle to the woman's family, were complicated by the restrictions on individual holdings. Patronage links were strained. The relatively affluent had used surplus grain and lending cattle to hire labour and to secure support from other community members, a practice that became increasingly difficult with destocking and land limitations.

Competition, conflict and intensifying struggles also occurred in relation to the NPAs. NPA farmers clashed, often dramatically, with chiefs and peasant farmers whose lands had become designated as purchase areas (Shutt 1997). Peasant cultivators who were once resident within designated purchase areas stole survey pegs, drove their cattle onto the farms, cultivated within the boundaries of NPA farms and, at the extreme, attacked NPA farmers and killed or maimed animals. This was part of broader tensions between reserve/TTL farmers and elite NPA farmers. Additionally, disputes existed between white farmers and nearby reserve farmers, including encroachment onto commercial farms for grazing and natural resources extraction.

The massive land dispossession, deterioration of land-based livelihoods in reserves, and the tensions and conflicts within reserves with arose subsequent to the first *chimurenga*, are all significant when to comes to understanding the historical, social and political conditions which facilitated the emergence of the second *chimurenga* in the 1970s and shaped its character. We consider these conditions further with specific reference to the position and practices of women in the context of land in Rhodesia.

3.2.1 African Women and Colonial Land Policies

Settler colonialism was not simply a racial project as it was also a patriarchal project. Land expropriation and alienation by male white settlers reconfigured African women's access to land. The colonial regime combined foreign modernist ideologies, gendered discourses and local customary traditions in a manner that ultimately alienated women from land. Customary practices were reinterpreted and became a basis for women's exclusion from access to land (Cheater 1986; Gaidzanwa 1981). They were incorporated into colonial laws of segregation and this produced a strong hierarchical and authoritarian colonial administration often under the indirect rule of chiefs. The fusion of Victorian ideologies, Roman and Dutch laws and customary law produced a very strong patriarchal ideology which led in practice to restrictions around women's access to land once guaranteed (albeit limited) under lineage arrangements in precolonial times.

Gaidzanwa (1981) observes that, as land for Africans as a whole became scarce because of colonial policies, African women experienced their own particular problems. For instance, married women—who in the past had access to small pieces of land to grow "female crops"—regularly no longer had such access (Peters and Peters 1998; Jacobs 1992). The NLHA regulations prevented people working *dambo* land (wetlands) and riverbanks where water was readily accessible. Both of these areas were generally controlled by women and played a vital role in crop diversity and food security (Thompson 2004). This eroded women's independence, which had been rooted in their control of certain crops and types of arable land. The Native Husbandry Act of 1951 led to individual tenure and farming rights for Shona and Ndebele men, which meant that any land rights for women that were guaranteed under the lineage system were lost because land was now registered solely under the names of male heads of households. Women could only have very limited usufruct rights to the land of their husbands or fathers.

Rights to land amongst peasant wives married polygynously, as well as amongst widows and divorcees with dependents, were curtailed (Gaidzanwa 1981). For men in a polygynous marriage, additional land was granted by colonial legislation, equal to a third of the standard area for each wife beyond the first (Peters and Peters 1998). At best, women's landholdings were one-third the size of those of men. Generally, landholdings of all wives, whether monogamously or polygynously married, were considered and confirmed as secondary—for even though the number of wives entered into the calculation of the size of holding, the holding itself was registered in the name of the husband (Cheater 1986). In the case of the Mangwende Tribal Trust Lands for example, only 16% of the women qualified for land rights under the NLHA. Besides land being held in the name of males, landholdings of male chiefs (as noted earlier) and headmen were larger than others, if only to pacify them.

The Act explicitly defined the farmer as a man despite the fact that women were working and living on the land, while the majority of the men between the ages of 16 and 60 often migrated as wage labourers to urban areas. Even the user rights of all categories of women were not registrable. Married women had to prove desertion or extraterritorial residence of their husbands, and divorced women had to prove that they had custody of their children, before they could be easily granted land. This was in fact difficult to do as bride wealth payments ensure that women forfeit their children to the men's patrilineages in cases of marital dissolution (Gaidzanwa 1994). Widows with dependent children were eligible to receive one-third to one full share of a standard area given to individual males (Duggan 1980). Therefore, deepening shortages of land, ongoing fragmentation of landholdings and land deterioration in the reserves (later TTLs) often led to the loss of usufruct rights for women (Bhatasara 2010). The Land Tenure Act of 1969 based on racial segregation continued to prioritise the allocation of land to males.

One aspect of colonial rule that is often ignored is how land alienation affected women's political authority. The colonial government did not recognise African women's rights to govern. However, the loss of land through colonial dispossession interfered in local governance systems which had provided some space for women's administration before the entrenchment of colonial rule. There were several cases of local women rulers in precolonial times, including in Honde Valley, Manyika, Chiduku and Makoni (Schmidt 1992; Ranger 1981). As African lands disappeared under European rule, women found themselves with no territories to preside over. In the new reserve-based spaces created for Africans, men exclusively were appointed as chiefs and headmen.

There were also some challenges for women emanating from the colonial regime through the semi-proletarianisation of African men as migrant labourers (Bhatasara and Chiweshe 2017). Fundamentally, this transformed the gender division of labour, with women's labour contributions to agricultural production increasing. This might be seen in a positive light in that women automatically became "heads of households" (or de facto heads) and were now in control of land and agricultural production while male migrants (including husbands) worked elsewhere. But, in the end, women remained as unpaid labourers on their husbands' land (Gaidzanwa 1988) by engaging in food and cash crop production. Women had to endure the increased burden of agricultural labour while managing households and caring for children in the periodic absence of their husbands and other male household members (Schmidt 1992). However, relatively wealthy reserve husbands could hire agricultural labour while away; and, for male migrants from reserves close to cities and towns (such as Goromonzi reserves near Salisbury), leave from work was possible to attend to their fields. At the same time, even if certain women could proclaim some degree of independence from their husbands' control in Native reserves, there were intra-household relations that women, especially young co-wives and daughters-in-law, could not escape. Mothers-in-law asserted their authority and controlled the labour of their young daughters-in-law.

Women's spatial mobility outside of the reserves was curtailed by various colonial laws. They became confined to reserves as agricultural labourers and, even when agricultural produce was sold through state marketing channels, forthcoming payments were made out to the male as the farmer. The Master Framer training programme introduced by the colonial government was also directed at men, again as the farmers (Parpart 1986). After all, African women were considered as legal minors, including under the customary law as constructed by the colonial state. Women who did enter urban areas were sometimes removed, especially those regarded as "travelling native prostitutes" (Barnes 1999), which included unmarried women or those living with men but not married under the Native Marriages Ordinance. But the subordination of African women also arose because of the views of local male patriarchs in the reserves, including chiefs.

Control over African women, including their mobility and sexuality, served the interests of both African patriarchs and colonial administrators. All men had a vested interest in preventing women from forging new ("liberated") identities in the spaces opened up by colonial capitalism, and they intended to keep them in rural areas as agricultural labourers, as daughters generating bride wealth, as household managers, and as child-rearing wives (Kesby 1996). There was at times collusion between white settlers and African men which worked against women. In the face of aggressive land appropriation by the colonial authorities, male elders cooperated with the state in the creation and preservation of reserves as a suitable space for their defence of patriarchal systems and practices. Antagonistic as they may have been to other aspects of colonial rule, African chiefs, headmen, and older men in general welcomed the state's efforts to restrict women to the rural areas (Schmidt 1991; Gaidzanwa 1994). As we show with regard to the second *chimurenga*, the large-scale presence of women in the Tribal Trust Lands often meant that they bore the brunt of the excesses of war during the 1970s.

3.3   Post-First *Chimurenga* African Resistance

It is inconceivable to fathom that any other grievance besides land alienation (and all the consequences arising from it) could lead to the growth of rural Africans' political consciousness under colonial rule. Post the first *chimurenga* up until the early 1960s, land conquest and dispossession remained as a foundational basis for ongoing racial domination. Moyana (2004: 107) shows the centrality of land to all Africans because its loss created "discontent among elites; poverty and despair among peasants; and landlessness that pushed Africans into wage labor in white farms and forced them into urban areas where they encountered a lot of strife".

What land, and the loss of land, means is open to different interpretations and indeed contestations. Amongst others, Hammar (2007) speaks of the importance of the spiritual and cultural meanings of land as constitutive forces in not only conditioning the interpretation of reality but in shaping history. With reference to the Tangwena people in the Gaeresi, Chief Rekayi Tangwena argued that Africans have a deep spiritual bond with their traditional land (Moore 1998). He regarded the loss of land as involving the loss of humanhood and the uprooting of a person's very soul and being (Moyana 2004). Land alienation led to dehumanising and humiliating identities for Africans, including by being defined legally as squatters on their own ancestral lands. In her study in Chiweshe, Steen (2011) also highlights that land was a symbol of male identity and expressive of lineage responsibility. Amongst both the Shona and Ndebele, community membership implied the right to a plot of land and access to communal grazing (Thompson 2004). Though land was also a "productive asset" and a source of livelihoods for Africans historically, this was invariably mediated in and through connections to the spiritual realm. In the case of Shona-speaking peoples, spirit mediums were central to this, as they communicated with the ancestors.

Successive colonial regimes in Rhodesia saw this spirituality as fundamental to the fact that Africans were irrational and backward, and thus steeped in tradition or ways of the past which had no place in the modernising world of the colonialists. Because of this, spirit mediums never received any recognition from the colonial authorities. In seeking to erase tradition, the Rhodesian colonial government invoked an authoritarian administrative logic and optic, as all modern states tend to do (Scott 1998). This involved mapping colonially constructed ethnic identities onto fixed territories (the reserves) and reinventing tribal polities governed through the dictates of British indirect rule, overseen by salaried and appointed chiefs (Moore 1998).

This was linked to agrarian and land policies and programmes, such as centralisation, involving what Bessant (1992) labels as coercive development. As Douglas
Moore (2005: 83) put it, "targeting the point of production and critical node of rural
power, centralisation through spatial discipline unsettled Africans in several senses".
The NLHA in particular generated significant anger and resentment amongst Africans
(Moyana 2004) in the two decades prior to the war of liberation, including amongst
those who were deprived of land and those whose larger landholdings were dimin-
ished (van Velsen 1964). Ranger (1985: 159) thus spoke about "embittered peasants
of Weya and Tanda; the returning labor migrants in Chiduku and elsewhere who
found themselves without land or cattle entitlement". Yet, in the early 1960s, it was
estimated that 6.6 million hectares of European land was lying idle. Individual land
tenure in the reserves as well undercut the unalienable right to land by Africans and
intervened directly in lineage-based practices of land access and indigenous agricultural methodologies, and the role of state-appointed agriculturalists in allocating land
interfered with chiefly authority over land allocation. Further, the NLHA imposed
an onerous labour regime (such as constructing contour ridging) that undermined
farmers' production strategies and ecological practices rooted in indigenous knowledge (Thompson 2004). Likewise, destocking (the forced sale of livestock in excess
of the permitted number) loomed large in the lives of reserve farmers.

Combined, and based on everyday lived experiences and encounters, all these
motivated local acts of resistance and eventually led to coordinated political agitation which increasingly took on a nationalist form and involved a rejection of racially
based colonial modernity (Ranger 1985; Thompson 2007). In the attempt to capture
African responses to colonialism and also counter-responses by the colonial administrators, we argue that the African responses were differentiated, including revolt
and acceptance-cum-compliance.

Thus, our first line of argument is framed in terms of revolt. Localised acts of resistance to colonial administrations in Rhodesia are discernible historically in reserves
across the countryside. To give but one example: in Makoni, it was reported by a
Native Commissioner in 1937 that it was common practice for Africans to settle
themselves on any land outside reserves on which no European was visible (Shutt
1997). We focus in particular, however, on African resistance to the NLHA as this
was the immediate context for the emergence of the second *chimurenga*. Within the
Native reserves, the implementation of the NLHA was resisted by a broad section
of Africans. As noted by Thompson (2004), once its implications became clear,
criticism was almost universal—from Africans of all social positions.

A dominant strategy for dealing with the state's authoritarian agrarian intervention was seemingly non-cooperation. This for instance involved tax evasion and an
outright refusal to engage in wage labour on adjacent white farms (McGregor 1991).
Ranger (1985), is his study in Makoni, also highlights the role of passive resistance,
including the refusal to provide information required for land registration. People
refused to give authorities any information about their stock and landholdings or
simply ran away during the initial survey and census phases of the NLHA's implementation (Thompson 2004). In Chiweshe, Bessant (1992) observed that sometimes
resistance took the form of evasion, such as when young men ran away from *chibaro*,

or when their fathers refused to build contour ridges "voluntarily". If *chibaro* was not avoided entirely, then go-slows or sub-standard work would be undertaken. (Chitiyo 2007). The destocking programme was met with passive resistance in the reserves, with the refusal to dip cattle being the chief weapon (Machingaidze 1991). There was significant derision of chiefs and members of the Native Affairs Department, and the unwillingness to communicate with the colonial state's field workers.

At first, Africans had voiced complaints about the NLHA through the structures of the Native Agriculture Department, engaged in passive resistance, and inter- acted with urban-based proto-nationalist formations. But, as resentment grew, some peasant farmers also began to defy officials and white control openly through angry confrontations, violent incidents and acts of sabotage (Thompson 2007). There were physical attacks on agricultural demonstrators and threats to the white land develop- ment officers as well as against those in the reserve who cooperated with the state's reconfiguration of the reserves. Already, in 1947, 17 peasant farmers, including a headman, were fined for assaulting agricultural demonstrators in Bindura District (Palmer 1977). But violence and sabotage directed against white and black Native Agriculture Department employees, chiefs and village headmen increased over time and became particularly common in the early 1960s (Thompson 2004). As an illus- tration, in 1961, the NLHA land allocation markings were burnt in Urungwe; activists started fires at three schools, the chief's grain store and a courtroom in Nkai; and, in Madziwa, activists filled cattle dips with stones and damaged the tobacco crop on white farms (Ranger 1985).

Importantly, a move towards *madiro* (or freedom farming) arose, which entailed disregarding the NLHA allocations and regulations, and seeking to liberate reserves from the logic and machinations of the colonial state. It was founded on an ideology of individual freedom meant to counter the state's technocratic interventions in reserves (Nyambara 2001) and it sought to blunt implementation of the NLHA (Machingaidze 1991).

Cases of Africans ploughing wherever they wished were reported in many reserves, including in Chilimanzi, Umtali and Plumtree in 1961. In Murambinda in Buhera Reserve, the Native Commissioner noted in 1961 that, as soon as the rains came, ploughing commenced and many landholders were extending their lands either into the roads and drain strips or simply taking additional land (Andersson 1999). A similar situation was reported by the Native Commissioner for Hartley District in Mhondoro Reserve in 1961 (Thompson 2007). In Chinamora Reserve, the people began "illegally" cultivating in 1953 while, in Manyene and Sabi North Reserves, people drove some of their cattle onto the underutilised Wiltshire Estate whenever officials came to conduct stock counts (Thompson 2004). In Nata Reserve, commu- nities which had been moved five times to open up land for white settlers simply refused to recognise the boundaries of their NLHA land allocations, and ploughed wherever they wanted in 1960 and 1961. In Mangwende Reserve, in Umtali and Plumtree Districts, thousands of people who had been made landless by the NLHA opened up fields for themselves in 1960 (Thompson 2007). In effect, these (and earlier) local initiatives entailed a colonial version of land occupations.

In this light, Andersson (1999: 564) further highlights that "those people who had not been allocated land rights in 1960 occupied land; they settled in the grazing areas and built houses in their fields in order to prevent cattle from entering the fields". In Madziwa, Thompson (2004) notes that, after land was allocated as per the NHLA regulations, many peasants: expanded their arable holdings by moving the beacons set by state officers, worked the areas designated for conservation works such as contour ridges, or took over land proclaimed as grazing, particularly where it bordered their allocation. By 1961, rural resistance had reached the stage, at least according to nationalist politician Nathan Shamuyarira, where it was assuming the dimensions of "a major revolt" against the NHLA (Phimister 1993: 228). Similarly, Thompson (2004, 2007) notes that rural opposition and unrest threatened state control of the countryside, creating a state of ungovernability in many reserves.

Besides challenging the reserve reconfiguration based on the centralisation programme in the 1950s and beyond, Africans also resisted evictions—at times, these evictions were related to the NHLA but, at other times, they were not. For example, implementation of the NHLA had to be suspended three times in 1959 in one village in Mhondoro Reserve when people refused to move to new fields and residences (Thompson 2004). Another case is that of the Shangwe in Gokwe who resisted eviction (on the basis of the 1948 Forest Act) from Mapfungautsi state forest in 1963. This resistance expressed itself in the form of squatting in prohibited areas, illegal harvesting of forest products, poaching of game, and leasing out land to Madheruka farmers (a group of large-scale agriculturalists who had in 1953 been evicted from Rhodesdale) (Maravanyika 2012). According to Shutt (1997), peasant opposition was a major reason why some NPAs were never settled; for instance, in the Makoni Reserve, ongoing resistance to evictions forestalled settlement of the Tanda Purchase Area for several decades.

Perhaps the case of Chief Rekayi Tangwena and his people's defiance to eviction from their ancestral lands in the Gaeresi area best illustrates the depth of rural resistance to colonial land and agrarian policies. As early as 1905, the Company had grabbed part of Tangwena land and sold it to the Anglo-French Matabeleland Company which later sold the land in 1944 to Gaeresi Ranch. Despite this, the Tangwena people remained on the land. At the height of mass evictions in the late 1950s and 1960s, they were given an eviction order (in 1963) following on from the recommendations of agricultural officers, who claimed that the Tangwena people's presence was responsible for massive land degradation. The Tangwena people though resisted the eviction order, only to be served with eviction orders again in both 1965 and 1966 and ordered to move to Gokwe (Moyana 2004). Still they resisted the move (even refusing a purchase offer for the land) and subsequently faced violence and raids from the Rhodesian government. The Tangwena soon became nationalist symbols for anti-colonial resistance to racialised land policies (Moore 2005).

Analytically, we should avoid homogenising rural resistance. Colonial rural societies were clearly socially differentiated, such that it is important to clarify who exactly was protesting (Phimister 1993). Machingaidze (1991) notes that opposition was particularly stiff amongst victims of post-Second World War evictions who were now being called upon to go through another process of eviction and resettlement.

Resistance incorporated, amongst others, those who had suffered considerable reductions in land and cattle, entrepreneurial peasants who lost land to standardisation, chiefs who lost power to allocate land, and peasants who resented the coercive agricultural commands (Ranger 1985). Holleman (1968) also observed that in certain instances the majority of "agitators" were in fact young male migrant labourers who lost land access and security of land tenure. This was the case for the migrant labourers from the Mangwende area (who worked in Salisbury) for whom the Land Husbandry Act had made them ineligible for farming rights in their own areas because they were away in the towns at the time of land registration (Ranger 1985). These are just some examples of those who resisted.

The more general point is that, in some way and to some degree, it is likely that all reserve groupings had their own localised and even individualised reasons for wanting to counter colonial land policies and programmes. Diverse localised experiences and conditions of existence, along with historical memories of colonial subjugation, therefore were central to rural resistance and opposition. From the perspective of colonial authorities, though, reserve communities were regarded as intrinsically free from political agitation and disobedience (Moore 2005). Hence, when it emerged, the state attributed all rural discontent to outside agitators (Thompson 2004), especially African political organisations. This is not unlike the views of many contemporary scholars who claim that the third *chimurenga* land occupations were stirred up by outside political formations (namely, the ruling party). In the case of colonial resistance during the late 1950s and early 1960s, while it is no doubt valid to argue that emerging nationalist movements influenced rural resistance, rebellion and revolt also occurred in many rural spaces and places where nationalists were not active (Thompson 2004); and, even where they were active, resistance was often already taking place.

Nationalist or proto-nationalist formations (consisting often of African elites) likely played a role in helping Africans overcome their "fear" of Europeans and in moving protest from expressions of anger and rejection to more pointed attacks on colonial structures and authorities (Thompson 2004). In this context, organised opposition to the NLHA was spearheaded quite often by urban-based formations (Machingaidze 1991). For instance, the Reformed Industrial and Commercial Workers Union (RICU) lobbied the Commonwealth secretary to use his veto on legislation relating to the reserves in 1950. More aggressively, the British African Voice Association, led by Benjamin Burombo, spearheaded a campaign against the Land Apportionment Act, evictions, destocking, and the Native Land Husbandry Bill (and the Act that followed). This association appealed against a number of land eviction cases and hired lawyers who exploited the procedural complexities in sections of the relevant legislation to win a number of cases. For instance, Burombo led the resistance to eviction of Madheruka farmers from Rhodesdale and mobilised resources for legal action against the state. As Phimister (1993: 230) notes:

As early as 1955, some parts of Matabeleland had swung behind Benjamin Burombo's African Voice Association in openly defying Native Department officials – for instance at a large meeting in Shangani Reserve at the beginning of the ploughing season that same year, people declared that they were not going to have anything to do with the Land Development Officers, the Demonstrators and other agricultural officers.

In Matabeleland, then, Burombo's African Voice was implicated in promoting rural defiance to colonial authority. Later on, the Southern Rhodesia African National Congress (SRANC) engaged in significant rural mobilisation (between 1957 and 1959) in Sipolilo, Umtali, Mhondoro and other reserves.

However, to reiterate, localised autonomous forms of resistance are evident, with many cases reported for instance of young rural activists mobilising themselves against the state. In his study in Madziwa, Thompson (2004) claims that young men and women, who were most sharply affected by the NLHA, often took leading roles in the protests, thereby inverting the gender and age hierarchies of rural society—in ways similar to what happened years later during the war of liberation in the 1970s. This implies that colonial agrarian policies may have fueled or reignited contestations within reserves between different social groupings. In addition, as we note when discussing the second *chimurenga*, local spirit mediums retained significant legitimacy amongst reserve villagers at a time when chiefs were often compromised by their association with colonial indirect rule. Spirit mediums thus also propelled rural resistance. For instance, Shangwe spirit mediums responded to the colonial onslaught by telling reserve communities that the ancestors, an important factor in Shangwe day-to-day life, were opposed to the cultivation of a "white man's crops" (Maravanyika 2012).

But Africans did not only resist colonial policies. Hence, we espouse, very briefly, our second argument in terms of patterns of acceptance or at least compliance by Africans, depending on context. Thompson (2004) alludes to grudging acceptance (lack of open opposition), which was interpreted by the colonial state as willing consent—yet it was merely meant to avoid confrontation. Thus, acceptance by rural Africans was shaped by the responses of the colonial state to their resistance. In most cases, colonial state authoritarianism and coercion intensified as a response to African strategies of resistance. In Madziwa Reserve, there was a sense of resignation when the implementation of NLHA began. Based on his case study of Chiweshe reserves, Bessant (1992) observes that occasionally families resisted by evading colonial policies but, for the most part, they complied with the letter of the law because they believed they had no viable alternative given the coercive might of the colonial state.

Often, though, acceptance, compliance and revolt were combined in dynamic ways. With regard to Chiweshe, reserve families neither welcomed nor rejected the idea of centralisation *in toto* in the 1930s. They resisted the modern farming methods but jumped at other agricultural opportunities such as the creation of a market for rice (Bessant 1992). As well, what was once rejected may then be accepted. In response to centralisation in Selukwe, one of the first vociferous objections came from Chief Nhema who cited that the reserve was too small and demanded more land. However, "he later realised the benefits" of the policy (Kramer 1998: 94).

3.3.1 The Growth of "Mass" Nationalism

At this juncture, we now turn to mass nationalism which began to emerge in the late 1950s and early 1960s. Nationalism as a mobilising ideology had a powerful presence in the latter period of colonial Zimbabwe (Raftopoulos 1999). While there are numerous debates within the Zimbabwean literature about the basis, character and tempo of nationalism in colonial Zimbabwe, our main concern here is simply to give the reader a sense of the emergence of nationalist movements and parties and the turn to guerrilla war in the 1960s, which is discussed more thoroughly in the following chapter.

It is broadly recognised that, in the face of decades-long frustrations in seeking to enter into a non-racial partnership with white settlers on the basis of assimilation or at least coexistence, African elites (some with a trade union movement background) became radicalised and more militant. These elites, involving collaboration between Shona and Ndebele elites, began to increasingly dominate nationalist politics in the 1950s, including trying to draw in urban workers and rural peasants to form a broader mass nationalism. This was taking place at a time when, as noted, significant forms of localised resistance was taking place on an independent basis in the reserves. During the 1950s, the new nationalist parties also began to revive the memories of the first *chimurenga*, which had hitherto been in large part ignored:

> [T]he character of this new nationalism was profoundly modified by its discovery of the potentialities of the traditions of resistance of the rural masses. The nationalist leaders, hitherto chary of appeals to the tribal past, discovered that it was possible to appeal to the great supra-tribal spirits or to the heroes of the risings. (Ranger 1967: 382)

This entailed a renewed emphasis on rural radicalism and on the *Mwari* priests and spirit mediums, both of which had lost their influence during the 1920s and 1930s.

This rediscovery of history, though, should not be considered as implying an unbroken line of resistance between the first *chimurenga* and the mass nationalism sixty years later (Cobbing 1977). In fact, as noted in the first chapter, Ranger's unbroken and monolithic chain of resistance spanning a century has been replaced analytically, as Raftopoulos (1999: 115) argues, by "a more complex picture of nationalism, reflecting both its continued resonance and its uneven and differential presence". No such permanent nationalist struggle against Rhodesian colonialism ever existed (Ndlovu-Gatsheni 2012). Rather, as we have sought to show, there was a multiplicity of rural struggles (when there was struggle at all) arising from different locations and for different reasons by different social groupings and marked by ebbs and flows (Kriger 1992). Further, insofar as a broad nationalist movement had been formed by the early 1960s, it was characterised by fissures and tensions.

By the 1930s, in fact, the difficulties of creating a national consciousness which captured the new (colonially created) forms of experiences, identities and struggles in varied spatial locations such as mines, farms, Native reserves and urban centres were clear. Africans confronted different social, economic, political and spatial realities which could not be easily submerged under mass nationalism. There were competing claims, with the admittedly small African elite seeking over an extended period for a more respectable civilised status for themselves (and themselves only), while African workers engaged in labour protests around wages in the hope of even the most basic of living conditions (Ndlovu-Gatsheni 2009). Early attempts at forging a national identity and "nationalising the struggles" were also undercut by the pervasiveness of regional and ethnic politics, at least amongst the elite which was in effect a male (and patriarchal) elite (Raftopoulos and Mlambo 2009; Sithole 1980). As Mlambo (2009) argues, building even elite nationalism from the 1940s and into the 1960s had to grapple with overlapping and competing political and identity claims, beyond the differentiation and tension within both urban and rural spaces.

In urban spaces, gender, class, ethnicity and region all created complexity for a mass nationalist movement. Educated elites made use of their missionary education, assets and emulation of European culture to distinguish themselves from the working classes in the urban townships. Class differences certainly meant that intra-racial linkages were not easy to develop let alone consolidate. For instance, the Reformed Industrial and Commercial Union (led by Charles Mzingeli from 1945 and into the 1950s) claimed to represent the African urban poor and working classes. Yet the union had its own internal contestations regarding strategies to use and there were questions about who it actually represented (Scarnecchia 1993). It was also not easy to align specific worker demands with the nationalist rhetoric propounded by some trade union leaders. Yoshikuni (1989) refers as well to the formation of ethnically based mutual aid societies as well as ethnic responses to the repressive urban work regime though, for Msindo (2007), ethnic identities were not inherently antagonistic to nationalism. Msindo (2007) argues that ethnicity provided local expressions of anti-colonial discontent. Furthermore, ethnic groups provided the leaders who became prominent nationalist figures, with the latter articulating the precolonial history, personalities and monuments that sparked the nationalist imagination. In the end, as Phimister (1988) shows, heterogeneity was a hallmark of urban anti-colonial struggles.

In the late 1950s, calls for national unity in what was now perceived as a nationalist movement increasingly grew from the nationalist elites. In practice, this meant all other African political organisations (such as trade unions) were considered as less important and subordinate to the nationalist movement (Raftopoulos and Mlambo 2009), with signs of commandist and authoritarian politics becoming visible (Raftopoulos 1999)—which is certainly contrary to the more benign view of a unifying nationalism as articulated by for example Ranger (1967) and Bhebe (1989a). This became vividly manifested in the early 1960s in the split within Zimbabwe African People's Union (ZAPU) and the formation of the breakaway Zimbabwe African National Union (ZANU) in 1963. The causes of this split remain disputed, though the common narrative that it was an ethnic split pure and simple is deeply problematic—as questions around ideology, tactics and strategy, as well as personal squabbles, also being significant. Whatever the causes, the split led to intra-nationalist violence in urban centres and seemingly entrenched a politics of factionalism within the national movement, with major consequences for the war of liberation in the 1970s (Scarnecchia 2008).

Here though we discuss nationalism in specific relation to the appeal for the return of land alienated by the colonial regime. During the period immediately preceding the liberation war, "*mwana wevhu/umntwana womhlabati*" (child of the soil) became the nationalists' rallying call, as land became the core issue animating the liberation struggle (Nmoma 2008). Land questions, and the agrarian reconfiguration of the reserves, undoubtedly fueled rural African opposition to colonial authority, but it was also expropriated by nationalist leaders to "unite" African people to challenge settler rule. Leaders of nationalist organisations such as the Southern Rhodesian African National Congress (SRANC) sought to link rural and urban grievances through land, in order to widen their support base and politicise "the masses" (Machingaidze 1991; Nmoma 2008). The SRANC Youth League's leaders (namely George Nyandoro, James Chikerema, Edison Sithole and Dunduza Chisiza) made hostile discursive attacks against the NLHA. The SRANC also managed to draw rural support by winning court cases against some evictions. This appealed to rural Africans irrespective of ethnicity or spatial locations (Moyana 2004).

The responses of the Rhodesian state to attempts by the SRANC and other organisations at building a nationalist movement around land illustrate the on-the-ground effects of mass nationalism. For instance, a state of emergency was declared in 1959 which led to its banning and the arrest of its leaders. In 1959 as well, an amendment was made to the NLHA which made it an offence for an African person to refuse to move to a new dwelling site; the Native Affairs Amendment Act of 1959 increased the state's power against anti-NLHA activists; and the Unlawful Organisations Act proscribed all nationalist organisations. Hence Nyandoro could have been right when he proclaimed that NLHA was "the best recruiter Congress ever had" (Machingaidze 1991: 582; Phimister 1993). The successors of SRANC, notably the National Democratic Party (NDP), ZAPU and ZANU also became crucial bodies through which nationalist opposition to the Act was articulated (Machingaidze 1991). For example, NPD activists entered into Native reserves and urged villagers not to attend cattle markets organised under the destocking programme and also urged peasants to disobey the instructions of agricultural demonstrators and to engage in *madiro* (freedom farming). The NDP and later ZAPU spent considerable time building local party structures in the reserves/TTLs, which were of significance during the war of liberation.

The nationalist movement became closely aligned to the revival of a cultural nationalism. Propagators of nationalism drew upon precolonial languages and cultures and reinterpreted precolonial histories as they mobilised at times across ethnic lines (Ndlovu-Gatsheni and Willems 2009). This also included the importance of songs, poems and dances as sites of struggle and means of protest (Ravengai 2016). Bessant (1994), for example, talks about Bernard Chidzero's (1957) song *Nzvengamutsvairo* and Solomon Mutswairo's poem *Mbuya Nehanada Nyakasikana*, which went beyond villages, chiefdoms and districts in their circulation. In his poem, Mutswairo was calling attention to the fact that, across colonial Zimbabwe, African people were plagued by land shortage and colonial oppression. By incorporating a widely known spirit like Nehanda in the poem, he was not only appealing to a "national" audience, but he was also creating one at the same time.

Battle weapons of the first *chimurenga* such as war-axes, swords and knobkerries took on a cultural significance and were celebrated in legitimising the nationalist project. The war-axe was an important symbol of the living spirit of African resistance (Ndlovu-Gatsheni and Willems 2009). Through cultural nationalism, African people were not only positioned on a new national map-in-the-making but were reconnected with a historical past, thereby becoming infused with a series of obligations to the contemporary nationalist struggle (Bessant 1994). The use of historical symbols, alongside a revitalised spirituality, constructed nationalism as relevant to local indigenous projects, thereby facilitating identification with the broader nationalist project. Cultural nationalism reached its apex during the celebration of the founding of the ZAPU in 1962 at Gwanzura Stadium in Highfield where different forms of African traditional dance, regalia and song graced the occasion, creating an image of a united imagined African nation (Ndlovu-Gatsheni and Willems 2009). However, the split between ZANU and ZAPU in 1963 inhibited the flourishing of cultural nationalism as a basis for building bridges within the nationalist movement.

The cultural nationalism of the 1960s, and the revival of "traditional" beliefs it provoked, raised questions on the role of spirit mediums and other forms of precolonial spirituality as well as chiefs. For Ranger (1967), for example, spirit mediums reinforced and radicalised peasant consciousness. In the Shona religion, land does not belong to the living but only to the dead (ancestors), with spirit mediums mediating between the living and the dead. Though spirit mediums articulated the view that white settlers and colonial land dispossession offended the ancestral spirits, it would be problematic to claim that they were the precursors of nationalist agitation. Nevertheless, their local legitimacy amongst reserve communities in the 1960s, at least across Mashonaland, influenced the course of the war of liberation.

With regard to the Zezuru-dominated Chiota Tribal Trust Lands near (then) Marandellas, which in the early 1960s was a People's Caretakers Council (or ZAPU) area and not a ZANU area, Fry (1976)—like others—contrasts the consensual authority of the non-hierarchically-arranged spirit mediums (based on their pronounced localised legitimacy amongst TTL villagers) with the illegitimate coercive authority of chiefs who had become increasingly subordinated to the dictates of the Rhodesian state. However, three of the chiefs in Chiota retained close relationships with spirit mediums, if only in a relationship of dependence as "chiefs need the mediums as they need a certain measure of popularity" (1976: 120). Fry (1976: 3) notes with reference to nationalism during the 1960s and early 1970s that, while "the number of people who were succumbing to spirit-mediumship was increasing [even dating back to the 1950s], churches were being burned and stoned… It became clear that 'traditional' beliefs and practices were related to the rise of African nationalism". In this way, the pre-second *chimurenga* upsurge in cultural and political nationalism amongst the people of Chiota (and beyond) found "its outlet in the cult of the spirit-mediums" (Fry 1976: 12).

From the 1930s, there was a significant breakthrough in Christian conversions in Chiota because of demands by villagers for education from the missions (notably Methodist in Chiota), which meant that "[t]he prestige that had accrued to the mediums and the old leaders [including chiefs and headmen] passed largely to the new educated elite" (Fry 1976: 112) (including Christian evangelists, teachers and businesspeople). At the same time, despite the Methodist missions taking a strong position against traditional religion, and perhaps because the Rhodesian state did not recognise them (like it recognised chiefs), spirit mediums and beliefs in them "did not die out, nor did they cease to be used; they went underground [until the early 1960s] and the open collective rituals associated with spirit-mediumship gave way to more secretly held consultations" (Fry 1976: 114). The local recognition of spirit-mediumship involved deep criticisms of colonial culture (including Christian missions), which led for instance to the burning of missionary property before the banning of ZANU and ZAPU as well as the dropping off in mission church attendance.

After the repression of the early 1960s, political activity in Chiota declined immediately and, because of this, there was general despair which led to the search for informers and sell-outs and—more specifically—a preoccupation with public accusations and identification of witches (typically women) in which spirit mediums became centrally involved. Fry (1976: 120) concludes by arguing that:

> Spirit-mediums, then are effective as a focus for nationalist sentiment because they bring the past to the present, because their ritual encourages essentially traditional activities, but above all because they are the people whose very authority is given by public opinion and who are unequivocally opposed structurally to Christianity.

But, as Spierenburg (2004) notes in her study of spirit mediums in Dande, the consensual authority of mediums did not come automatically at all times, as the pronouncements by mediums were subject to questioning, discussion and negotiation.

The position and practices of chiefs in reserves/TTLs, in relation to rising nationalism, was more controversial. Co-optation by the colonial state was a critical strategy for ensuring chiefly obedience. For instance, to avoid resistance to NLHA and ensure the support and loyalty of chiefs and headmen, the colonial government introduced favourable scales of subsidies (Machingaidze 1991). In 1961, chiefs got back their power to allocate land, their salaries were increased, and a National Council of Chiefs was set up to facilitate their manipulation. In Inyanga, one Native Commissioner reported that, when the NLHA was introduced, there was cooperation from chiefs and headmen in supplying African labour for conservation projects (Moore 2005). Some chiefs even urged the colonial government to introduce legislation to ensure local discipline (for example by banning political meetings) since they had a real interest in law and order, as their own authority was threatened especially with growing political agitation of the late 1950s. In the 1950s and 1960s, chiefs were mostly castigated by nationalists as sell-outs, who had no place in a future Zimbabwe.

But this was not a universal stance. Nyambara (2001) argues that, contrary to conventional wisdom that stresses the compliance of chiefs with government conservation measures, there is clear evidence that some chiefs and headmen refused to carry out conservation duties in their areas as required by the colonial political dispensation. Chief Mkoka's case is a clear example of defiance—he became notorious for not carrying out conservation measures in his area. In areas such as Insiza and

Mangwende, like Chief Tangwena, chiefs allied with nationalist struggles to defy the government. Therefore, some chiefs were active nationalists, even before occupying office. Others turned against the colonial government, and often to nationalism, as a result of the ongoing disregard for their demands, notably for land; and still others reluctantly obeyed nationalist dictums out of fear of retribution (Nyambara 2001). The legitimacy of chiefs, or the absence of legitimacy, was to condition the character of the war of liberation during the 1970s.

Despite the tendency to erect a dichotomy between rural and urban political consciousness, as if they were conditioned in completely different ways in colonial Zimbabwe, Raftopoulos (1999) talks about important interconnections between urban and rural spaces. His claim is based on a number of developments: the impacts of land alienation policies (particularly the NLHA of 1951) which pushed many Africans into cities, hence laying the ground for broader nationalist consciousness; the growth of African middle-class intellectuals who proposed a nationalist ideology that linked the rural and urban; and the subordination of urban-based (including working class) movements to the nationalist movement (Raftopoulos 1995). In addition, Moyana (2004) contends that the issue of land was a rallying point not only for rural people in reserves, but also for African mine workers and the urban workforce. The initial loss of land, and ongoing evictions during the colonial period, forced rural Africans into wage labour in farms, mines and urban areas where the conditions for Africans were untenable. For instance, unemployment was high and wages were low, with the labour regime buttressed by repressive labour laws such as the Industrial Conciliation Act.

These points relate to the process of proletarianisation which, in the case of colonial Zimbabwe, was characterised by a significant level of semi-proletarianisation. As noted by Duggan (1980: 229), "although the colonial state intended to divide the population once and for all between 'peasants' and 'proletariat' under the NLHA, Africans themselves clung to the urban–rural ties". Africans in urban centres were temporary residents, whether by choice or due to circumstances beyond their control (such as being denied any right to purchase urban land, housing squalor, low wages, and no pension and social security).

Evidently, by the early 1950s and indeed throughout the period in which the NHLA was implemented, urban conditions discouraged many urban workers from cutting ties to the land (Machingaidze 1991). Indeed, the migrant wage labour system continued to link the rural and the urban, with African men straddling farming and urban wage labour. Speaking about men in the Kaerezi Ranch, Douglas Moore (1998: 357) noted that "they wanted both land rights in rural areas and the opportunity for better wages in distant mines and urban industries - insisting on inhabiting 'multi-local social space'". Therefore, "grievances against the NLHA, emerged from the articulation of urban and rural routes, the relational spaces produced through migrant labor", which meant that "[t]ranslocal circuits of migrant wage labor became channels irrigating nationalist politics" (Moore 2005: 85). Even in the case of the third *chimurenga* land occupations, there was an urban dynamic to it, as urban Africans living in high-density areas of cities and towns occupied nearby land.

Overall, then, urban and rural struggles intertwined and coalesced into national political action and formations. "Discontent among urban workers was not confined to what is generally referred to as trade union questions of working conditions but involved the land question as well" (Mothibe 1993: 236). For example, the Industrial and Commercial Union (formed in 1929) was not exclusively an urban labour movement speaking out against poor housing, low wages and inadequate working conditions. It also delved into many non-labour-related and rural matters such as land shortages, racial discrimination, the violence of Native Commissioners and other issues (Msindo 2007). As Douglas Moore (2005) thus argues, nationalist-type struggles around land were articulated by Africans in all places and spaces, including farms, mines, reserves and urban industries.

3.3.2 African Women's Resistance and Nationalism

From the preceding discussions, it is clear that African resistance to Rhodesian colonialism was not homogenous and that nationalism was multifaceted, complex and convoluted. Feminist historians such as Federici (2012) have built up a formidable challenge to androcentric studies of Africa in showing the myriad ways in which women resisted and reshaped colonialism. Likewise, African women's resistance in colonial Zimbabwe needs to be brought to the fore. As a starting point, their struggles were multidimensional because they were fighting both African patriarchy and colonial authority (Chadya 2003), with the former struggles not reducible to the latter. Even before the rise of nationalism, African women contested local systems of patriarchy.

In specific relation to marriage and family, women across the reserves would sometimes refuse to marry their appointed male partners, entering into adulterous liaisons or even fleeing to missions, mines, farms and urban areas to escape marital arrangements (Ranger 1981). Women defied existing norms by not only choosing their own marriage partners, but by prioritising the formal education of their daughters and finding ways to generate income to secure a degree of autonomy from male domination (Wells 2003). Cases of runaway wives, for instance in Goromonzi, have been recorded as early as 1899 (Schmidt 1992). Single African women and girls ran away to European mission stations, submitting to the strict disciplinary codes of puritanical Christians rather than endure less tolerable circumstances at home (Schmidt 1991). In mission, mine, farm and urban locations beyond their guardians' sphere of influence, these women developed sexual and economic identities independent of lineage structures (Kesby 1996). As the authority of lineages to negotiate marital disputes was disrupted under colonialism, women abandoned their husbands for the towns and farms beyond the control of these weakened lineages (Jeater 1993). Some women also joined separatist churches and African independent churches and formed women's organisations within these churches. On this basis, in the church-based Ruwadzano Movement in the 1920s, women challenged polygamy and, in the 1930s, the Vapostori church preached against bride wealth. Barnes (1992) shows that

African women who were able to set foot in urban areas often challenged the forces of patriarchy—they travelled on motorised vehicles unaccompanied, used colonial courts and talked back to their elders. They operated illegal shebeens and brewed traditional beer to assert their economic independence or worked as prostitutes (Stott 1990).

In African reserves, women mounted resistance to various challenges brought about by colonial agrarian restructuring (Cheater 1986). From her studies in Goromonzi, Schmidt (1992) observes that there were several ways in which women sought to overcome the growing burden of household work. For instance, they refused to carry grain on their heads to markets and to pound maize, insisting that their husbands buy machine-ground maize meal. As well, women opposed the adoption of new agricultural and conservation methods which increased their workloads. With the introduction of the centralisation policy in the 1930s and later the more coercive NLHA, it became apparent that the labour burden was to fall on women as the primary agricultural producers. However, evidence (from Goromonzi for example) indicates that women openly refused to meet the new labour demands such as manuring the fields. This was based on three objections—the burden of collecting, transporting and spreading manure, the fact that manure attracted crop pests, and the promotion of weeds by manure that women were expected to then eradicate (Schmidt 1992).

In other contexts, women resisted colonial land boundaries and allocations. For instance, nearly one hundred women in Mutasa North Reserve were fined for working land outside of their allocations in 1961 (Thompson 2007). Further, women were often central to resisting colonial evictions. Tangwena women at the Kaerezi Ranch were at the frontlines of protest during several raids to enforce evictions (Moore 1998). More broadly, women's gendered tactics such as stripping naked, urinating and jeering in front of Native Commissioners, the police and soldiers, produced a colonial politics of embarrassment (Moore 1998).

Problematising the presence of African women in nationalist politics is also critical. It is clear that nationalist politics had a distinctly patriarchal agenda and that this was deeply entrenched (Raftopoulos and Mlambo 2009). Several studies expose the deeply gendered character of nationalism in colonial Zimbabwe, as was the case elsewhere (McClintock 1991). The top leadership of the nationalist movements in the 1950s and 1960s was almost exclusively male (Chadya 2003; Ranger 1981). Additionally, nationalism was insensitive to the differential character of varying sets of struggles, with gender struggles merely subsumed under racial struggles (Raftopoulos 1999) or not recognised at all. For instance, rural patriarchs envisaged nationalism, amongst other things, as a means to fortify their masculinity and the social-spatial structures that underlay it (Kesby 1996). As we will see with respect to the war of liberation in the 1970s, the nationalist-inspired historical rendering of the nationalist struggles of the preceding decades has simply asserted, incorrectly, that men and women were involved together in an inclusive manner (Scarnecchia 1996).

Questioning this rendering, of course, does not imply denying the agency of women during the days of "mass" nationalism. But it does acknowledge and recognise that, in seeking to evade both African male control and colonial subjugation, the participation of women in nationalist politics was a painful process (Barnes 1999).

For example, the male-dominated nationalist leadership seemingly deceived women, constructing political ideologies that only tacitly recognised women's demands in return for their support in the nationalist struggle (Scarnecchia 1996). In other words, the support given by male African nationalists to women's grievances was superficial (Chadya 2003). After all, addressing these grievances entailed an implicit challenge to patriarchal orders including the ones which animated nationalist politics. In this light, Scarnecchia (1996: 284) alludes to the fact that "some women even became 'enemies of the struggle' as voiced by nationalist [male] leaders when their separate social and economic demands ran counter to the immediate goals of nationalist politics". This is echoed by Ranchod-Nilsson (2003) who points out that patriarchal relations were contained in the nationalist movement and were not allowed to disrupt the movement's quest to end racial domination. Women were thus encouraged to focus on the nationalist project first before their own gendered struggles.

This effective closure of space for addressing women's grievances would at times have major implications (Raftopoulos 1995). For instance, a number of single working women were raped by male members of the City Youth League during the Salisbury bus boycott in 1956, ostensibly because these women broke the boycott by travelling on the buses. Nationalist leaders Nathan Shamuyarira and Maurice Nyagumbo argued that the rapes were justifiable as revenge for the women's defi-ance, as they were a well-deserved punishment (Lyons 1999). By this, they meant defiance of nationalism but, ultimately, the very status of these women (as single working women) defied male notions of female domesticity. By emphasising the racial dimension of the nationalist struggle, male nationalists were blind to the "frag-ments" of the nation in the form of gendered relations (Chadya 2003), and they failed to appreciate that women's experiences of the racial order were mediated by their experiences as women.

In their own often unacknowledged ways, women no doubt were significant actors in the national political struggles before the second *chimurenga*. However, at the same time, it is important to note that the category of "woman", even "African woman", is a differentiated and heterogeneous category masking a diverse range of experiences, grievances and aspirations. For urban African women in general, Dauda (2001) observes that women's political exclusion in urban settings was under the double constraint of customary law and colonially driven patriarchal attitudes. The literature on nationalist politics in colonial Zimbabwe though reveals varied nuances in the case of these women. Thus, urban women existed within various social categories—from celebrated shebeen queens (such as Elizabeth Musodzi), madams, political activists, intellectual elites, *mapoto* (cohabiting) wives and pros-titutes. Urban women were even segregated into "respectable" (married) and "non-respectable" (unmarried women) by colonial law and society (Scarnecchia 1993). Where women were positioned in the urban space and hierarchy likely conditioned their involvement in urban struggles.

Women actively participated in labour protest activities such as the Shamva mining strike of 1927 and in defiance marches such as the Salisbury bus boycott in 1956. The success of the 1948 general strike by workers depended upon the support of women (Lyons 1999). Women workers and those working for themselves were central to urban political life in relation to Charles Mzingeli's Industrial and Commercial Union. In the late 1950s, women marched for the release of incarcerated nationalist activists. Women were also arrested in 1959 during the banning of the SRANC. After the formation of NDP in 1961, women in the party (including Sally Mugabe) in Salisbury and Bulawayo organised marches. Typically, women with some acknowledged role in the nationalist parties were wives of male nationalists (Gaidzanwa 1982; Ranchod-Nilsson 2003). Coincidentally, after the protests by women, the NDP was banned, leading Bhebe (1989b) to suggest that it was women's actions which catalysed the ban and contributed to the militant turn in the nationalist movement. However, despite this and other evidence about women's urban activism, their roles have been distorted, undermined or simply regarded as supportive. In fact, this is often how women were framed by male nationalist leaders. For instance, Robert Mugabe highlighted how, during the days when the nationalist struggle was in its early stage of political formation (in the 1950s), women attended rallies where they chanted slogans, sang, danced and ululated (Lyons 1999).

It is perhaps undisputed that peasant women in reserves bore the brunt of the ravages of the colonial order; hence, as indicated, they were involved in nationalist struggles like their urban counterparts (Stott 1990; Ranger 1985; Ranchod-Nilsson 1992). Marginalised by systems of patriarchy and colonialism, nationalism provided spaces for rural women to vent their anger and frustration (Stott 1990). In recognising that women became involved in local rural struggles at times in and through nationalism, this does not imply that they were seeking to fulfil well-defined nationalist aspirations. Rather, they often used the spaces pried open by the nationalist project to address their own personal concerns around patriarchal practices.

3.4 Conclusion

In the context of land alienation and the reconfiguration of the reserves, this chapter has articulated the multiple forms of grievances and struggles in the reserves/Tribal Trust Lands as displayed by diverse social groupings (including women) in the decades leading up to the war of liberation in the 1970s. In the case of the period of emerging mass nationalism, from the mid-1950s, we also brought to the fore the existence of multiple—though connected—sites of nationalism, occurring in both urban and rural spaces. A focus on the memories, experiences, motivations, practices and aspirations of "ordinary" African people in reserves as historical subjects is an important counter-narrative to the valorisation and reification of elites in Zimbabwe's nationalist history. These historical subjects, as Munochiveyi (2011a) notes, are all too often simply parenthesised as the "masses", understood as an undifferentiated and homogenous group of supporters of the nationalist movement without any capacity for autonomous agency (Munochiveyi 2011b).

The argument, then, is that many ordinary men and women in rural Rhodesia were capable of formulating their own personalised critiques of colonialism based on multiple and diverse experiences, and this facilitated their entry into acts of resistance and struggles, either within or outside a well-defined nationalist project (Munochiveyi 2011a). These ordinary women and men expressed and practised political activism prior to the emergence of popular mass nationalism (Alexander and McGregor 1997), and they continued to do so subsequently. They were not simply led from above or outside in participating in "mass" nationalism. Certainly, under the guise of a mass nationalist project, personal interests and identities which went contrary to the struggle against the racialised colonial order were typically subordinated to the needs of the "nation". But even then, rural struggles continued to be generated in the context of the immediate conditions of existence as shaped historically.

By the early 1970s, a new sociopolitical formation arose in the Rhodesian countryside (i.e. guerrilla armies). The next two chapters consider the second *chimurenga*, primarily with regard to the practices of—and relationships between—nationalists, guerrillas and ordinary men and women in the Tribal Trust Lands.

References

Alexander J (1993) The state, agrarian change, and rural politics in Zimbabwe: case studies of Insiza and Chimanimani Districts, 1940–1990. Unpublished PhD thesis, University of Oxford, United Kingdom

Alexander J, McGregor J (1997) Modernity and ethnicity in a frontier society: understanding difference in North-western Zimbabwe. J South Afr Stud 23(2):187–201

Andersson JA (1999) The politics of land scarcity: land disputes in save communal area. Zimbabwe. J South Afr Stud 25(4):553–578

Arrighi G (1970) Labour supplies in historical perspective: a study of proletarianisation of the African Peasantry in Rhodesia. J Dev Stud 6(3):197–234

Arrighi G, Saul JS (1973) Essays on the political economy of Africa. Monthly Review Press, New York

Barnes TA (1992) The fight for control of African women's mobility in Colonial Zimbabwe, 1900–1939. Signs 17(3):586–608

Barnes TA (1999) We women worked so hard: gender, urbanisation and social reproduction in Colonial Harare, 1930–1956. Heinemann, Portsmouth, New Haven

Bessant LL (1992) Coercive development: land shortage, forced labor, and colonial development in the Chiweshe Reserve, Colonial Zimbabwe, 1938–1946. Inter J Afr Hist Stud 25(1):39–65

Bessant LL (1994) Songs of Chiweshe and songs of Zimbabwe. Afr Aff 93(370):43–73

Bhatasara S (2010) Land reform and diminishing spaces for Women in Zimbabwe: a gender analysis of the socio-economic and political consequences of the FTLRP. Unpublished Master's Thesis, Maastricht University, The Netherlands

Bhatasara S, Chiweshe MK (2017) Beyond gender: interrogating women's experiences in FTLRP in Zimbabwe. Afr Rev 9(2):154–172

Bhebe N (1989a) Benjamin Burombo: African politics in Zimbabwe, 1947–1958. College Press Publishers, Harare

Bhebe N (1989b) The Nationalist Struggle 1957–1962. In: Banana C (ed) Turmoil and tenacity. College Press, Harare, pp 50–115

Chadya JM (2003) Mother politics: anti-colonial nationalism and the woman question in Africa. J Women's Hist 15(3):153–157

Cheater AP (1984) Idioms of accumulation: rural development and class formation among freeholders in Zimbabwe. Mambo Press, Gweru

Cheater A (1986) The role and position of women in pre-Colonial and Colonial Zimbabwe. Zambezia 12:65–79

Chitiyo NK (2007) Colonial legacy of Zimbabwe's land disputes: reconceptualising Zimbabwe's land and war veterans' debate. Africa Resource Centre

Cobbing J (1977) The absent priesthood: another look at the Rhodesian risings of 1896–1897. J Afr Hist 18(1):61–84

Dauda CL (2001) Preparing the grounds for a new local politics: the case of women in two African municipalities. Can J Afr Stud 35(2):246–281

Drinkwater M (1991) The state and agrarian change in Zimbabwe Communal Areas. Macmillan Press, London

Duggan WR (1980) The Native Land Husbandry Act of 1951 and the rural African middle class of Southern Rhodesia. Afr Aff 79(315):227–239

Federici S (2012) Revolution at point zero. PM Press, Oakland

Fry P (1976) Spirits of protest: spirit mediums and the articulation of consensus among the zezuru of Southern Rhodesia (Zimbabwe). Cambridge University Press, Cambridge

Gaidzanwa RB (1981) Promised land: towards land policy for Zimbabwe. Unpublished MA Thesis, Institute of Social Studies, The Netherlands

Gaidzanwa RB (1982) Bourgeoisie theory of gender and feminism and their shortcomings with Reference to Southern African Countries. In: Meena R (ed) Gender in Southern Africa: conceptual and theoretical issues. SAPES, Harare, pp 92–125

Gaidzanwa RB (1988) Women's land rights in Zimbabwe: an overview. Occasional Paper No. 13, University of Zimbabwe, Harare.

Gaidzanwa RB (1994) Women's land rights in Zimbabwe. Issue J Opin 22(2):12–16

Hammar A (2007) The day of burning': land, authority and belonging in Zimbabwe's Agrarian margins in the 1990s. Unpublished PhD Thesis Roskilde University, Denmark

Holleman JF (1968) Chief. Council and Commissioner, Koninklijke van Gorcum, Assen

Jacobs S (1992) Gender and land reform: Zimbabwe and some Comparisons. Inter Sociol 7(1):5–34

Jeater D (1993) Marriage, perversion and power: the construction of moral discourse in Southern Rhodesia 1894–1930. Clarendon Press, Oxford

Kesby M (1996) Arenas for control, terrains of gender contestation: guerrilla struggle and counter-insurgency warfare in Zimbabwe 1972–1980. J South Afr Stud 22(4):561–584

Kramer E (1998) A clash of economies: early centralization efforts in colonial Zimbabwe, 1929–1935. Zambezia XXV(i):83–98

Kriger N (1992) Zimbabwe's Guerrilla War: peasant voices. Cambridge University Press, Cambridge

Lyons T (1999) Guns and Guerrilla Girls. Unpublished PhD Thesis, University of Adelaide, Australia

Machingaidze VE (1991) Agrarian change from above: the Southern Rhodesia Land Husbandry Act and African Response. Inter J Afr Hist Stud 24(3):557–588

Maravanyika S (2012) Local responses to colonial evictions, conservation and commodity policies among Shangwe Communities in Gokwe, Northwestern Zimbabwe, 1963–1980. Afr Nebula 5:1–21

McClintock A (1991) 'No longer in a future Heaven?': women and nationalism in South Africa. Transition 15:104–123

McGregor J (1991) Woodland resources: ecology, policy and ideology: an historical case study of woodland use in Shurugwi communal area, Zimbabwe. Unpublished PhD Thesis, Loughborough University, United Kingdom

Mlambo AS (2009) From the Second World War to UDI, 1940–1965. In: Raftopoulos B, Mlambo AS (eds) Becoming Zimbabwe: a history of Zimbabwe from pre-colonial period to 2008. Weaver Press, Harare, pp 75–114

Moore D (1998) Subaltern struggles and the politics of place: remapping resistance in Zimbabwe's Eastern Highlands. Cult Anthropol 13(3):344–381

Moore D (2005) Suffering for territory: race, place and power in Zimbabwe. Weaver Press, Harare

Mothibe TH (1993) African labour in Colonial Zimbabwe in the 1950s: decline in the militancy or a turn to mass struggle? Labour Capital Soc 26(2):226–251

Moyana VH (2004) The political economy of land in Zimbabwe. Mambo Press, Gweru

Msindo E (2007) Ethnicity and nationalism in urban Colonial Zimbabwe: Bulawayo, 1950 to 1963. J Afr Hist 48:267–290

Munochiveyi MB (2011) Becoming Zimbabwe from below: multiple narratives of Zimbabwean Nationalism. Crit Afr Stud 4(6):84–108

Munochiveyi MB (2011b) We do not want to be Ruled by Foreigners. Oral histories of nationalism in Colonial Zimbabwe. The Historian 73(1):65–87

Ndlela DB (1981) Dualism in the Rhodesian National Economy. Lund Economic Series 22.

Ndlovu-Gatsheni S (2009) Mapping cultural and colonial encounters in Zimbabwe, 1880s–1930s. In: Raftopoulos B, Mlambo AS (eds) Becoming Zimbabwe: a history of Zimbabwe from pre-colonial period to 2008. Weaver Press, Harare, pp 39–74

Ndlovu-Gatsheni S (2012) Rethinking Chimurenga and Gukurahundi in Zimbabwe: a critique of partisan national history. Afr Stud Rev 55(3):1–26

Ndlovu-Gatsheni S, Willems W (2009) Making sense of cultural nationalism and the politics of commemoration under the Third Chimurenga in Zimbabwe. J South Afr Stud 35(4):945–965

Nmoma V (2008) Son of the soil: reclaiming the land in Zimbabwe. J Asian Afr Stud 43(4):371–397

Nyambara P (2001) Immigrants, 'traditional' leaders and the Rhodesian state: the power of 'communal' land tenure and the politics of land acquisition in Gokwe, Zimbabwe, 1963–1979. J South Afr Stud 27(4):771–791

Nyandoro M (2012) Zimbabwe's land struggles and land rights in historical perspective: the case of Gowe-Sanyati irrigation (1950–2000). Historia 57(2):298–349

Palmer R (1977) Land and racial discrimination in Rhodesia. Heinemann, London

Palmer RH, Parsons N (1977) The roots of rural poverty in Central and Southern Africa. Heinemann, London

Parpart JL (1986) Women and the state in Africa. Working Paper, 117, Department of History Dalhousie University Halifax, Canada

Peters BL, Peters JE (1998) Women and land tenure dynamics in pre-colonial, colonial and post-colonial Zimbabwe. J Publ Inter Aff 9:183–208

Phimister I (1988) An economic and social history of Zimbabwe 1890–1948: capital accumulation and class struggle. Longman, Harlow

Phimister I (1993) Rethinking the reserves: Southern Rhodesia's Land Husbandry Act. J South Afr Stud 19(2):225–239

Raftopoulos B (1995) Nationalism and labour in Salisbury 1953–1965. J South Afr Stud 21(1):79–93

Raftopoulos B (1999) Problematising nationalism in Zimbabwe: a historiographical review. Zambezia XXVI(ii):115–134

Raftopolous B, Mlambo A (eds) (2009) Becoming Zimbabwe: a history from the pre-colonial period to 2008. Weaver Press, Harare

Ranchod-Nilsson SC (1992) Gender politics and national liberation: women's participation in the liberation of Zimbabwe. Unpublished PhD Thesis, North-western University, South Africa

Ranchod-Nilsson S (2003) Gender struggles for the nation: power, agency and representation in Zimbabwe. In: Ranchod-Nilsson S, Tetreault MA (eds) Women, states and nationalism: at home in the nation. Routledge, London, pp 164–180

Ranger T (1967) Revolt in Southern Rhodesia 1896–7: a study in African resistance. Heinemann Educational Books, London

Ranger T (1981) Women in the politics of Makoni District 1960–1980. Unpublished Manuscript, National Archives of Zimbabwe.

Ranger T (1985) Peasant consciousness and Guerrilla War in Zimbabwe: a comparative study. Zimbabwe Publishing House, Harare

Ravengai S (2016) Chimurenga liberation songs and dances as sites of struggle to counter Rhodesian discourse: a postcolonial perspective. In: Mangena F, Chitando E, Muwati I (eds), Sounds of life:

music, identity and politics in Zimbabwe Cambridge Scholars Publishing, Newcastle upon Tyne, pp 165–181.

Scarnecchia T (1993) The politics of gender and class in the creation of African Communities, Salisbury, Rhodesia, 1937–1957. Unpublished PhD Thesis University of Michigan, United States of America

Scarnecchia T (1996) Poor women and nationalist politics: alliances and fissures in the formation of a nationalist political movement in Salisbury, Rhodesia, 1950–6. J Afr Hist 37(2):283–310

Scarnecchia T (2008) The urban roots of democracy and political violence in Zimbabwe, Harare and Highfield, 1940–1964. Rochester University Press, Rochester

Schmidt E (1991) Patriarchy, capitalism, and the Colonial State in Zimbabwe. Signs 16(4):732–756

Schmidt E (1992) Peasants, traders and wives: Shona women in the history of Zimbabwe, 1870–1939. Baobab Books, Harare

Scoones I, Mavedzenge B, Murimbarimba F (2018) Medium-scale commercial farms in Africa: the experience of the 'native purchase areas' in Zimbabwe. Africa 88(03):597–619

Scott J (1998) Seeing like a state. Yale University Press, New Haven

Scott J (2013) Decoding subaltern politics. Routledge Press, New York

Shutt AK (1997) Purchase area farmers and the middle class of Southern Rhodesia, 1931–1952. Inter J Afr Hist Stud 30(3):555–581

Sithole M (1980) Ethnicity and factionalism in Zimbabwe nationalist politics 1957–79. J Ethnic Racial Stud 3(1):17–39

Spierenburg M (2004) Strangers, spirits, and land reforms: conflicts about land in dande. Brill Publishing, Leiden, Northern Zimbabwe

Steen K (2011) Time to farm: a qualitative inquiry into the dynamics of the gender regime of land and labour rights in subsistence farming: an example of the Chiweshe Communal Area, Zimbabwe. Unpublished Doctoral Thesis, Lund University, Sweden

Stott L (1990) Women and the armed struggle for independence in Zimbabwe (1964–1979). Edinburgh University Centre for African Studies Occasional Papers No 25.

Thompson G (2004) Cultivating conflict: agricultural 'betterment', the Native Land Husbandry Act (NLHA) and ungovernability in Colonial Zimbabwe, 1951–1962. Afr Dev 29(3):1–39

Thompson G (2007) Is it lawful for people to have their things taken away by force? High modernism and ungovernability in Colonial Zimbabwe. Afr Stud 66(1):39–77

Van Onselen C (1976) Chibaro: African mine labour in Southern Rhodesia 1900–1933. Urizen Books, New York

Van Velsen J (1964) Trends in African nationalism in Southern Rhodesia. Kroniek Van Afrika 2:139–157

Wells JC (2003) The sabotage of patriarchy in Colonial Rhodesia: rural African women's living legacy to their daughters. Feminist Rev 75:101–117

Worby E (2001) A redivided land? New agrarian conflicts and questions in Zimbabwe. J Agrarian Change 1(4):475–509

Yoshikuni T (1989) Strike action and self-help associations: Zimbabwean worker protest and culture after World War I. J South Afr Stud 15(3):440–468

Chapter 4
The Second *Chimurenga*: Early Literature and Nationalists-Guerrillas

Abstract This is the first of two chapters on the second *chimurenga*. The purpose of these chapters is to provide an overview of the historical development of the nationalist and guerrilla movements, and to identify key themes in the second *chimurenga* literature. This chapter examines the early scholarly literature on the second *chimurenga* and highlights themes embodied in this literature, but also silences and gaps which are addressed in the later literature. It then turns to a consideration of one particular theme, namely the overall character of the nationalist movements and guerrilla armies and, more importantly, the relationship between them. The chapter considers some of the temporal dynamics of the war of liberation during the 1970s, and it captures many of the complexities of the spatial diversity of the second *chimurenga* across the breadth of the Rhodesian countryside. Ultimately, in considering the second *chimurenga*, the chapter is concerned primarily with pinpointing and discussing themes which are particularly crucial in developing a comparative analysis of the three *zvimurenga* and in understanding the specificities of the third *chimurenga*.

Keywords Second *chimurenga* · Nationalist movements · Guerrilla armies · Rhodesia · Spirit mediums

4.1 Introduction

This is the first of two chapters on the second *chimurenga*. The purpose of these chapters is to provide a broad overview of the war of liberation and to identify some of the key themes in the second *chimurenga* literature. In this first chapter, we provide an overview of the early scholarly literature on the second *chimurenga* and, in doing so, highlight themes embodied in this literature but also silences and gaps which are addressed in the later literature. We then turn to a consideration of one particular theme, that is, the overall character of the nationalist movements and guerrilla armies and, more importantly, the relationship between them. This, and the following, chapter do not provide a full and comprehensive examination of the second *chimurenga*; however, they do try to cite all of the pertinent literature about this *chimurenga*. We also do not offer a complete account of the temporal dynamics

© The Author(s), under exclusive license to Springer Nature Switzerland AG 2021   69
K. Helliker et al., *Fast Track Land Occupations in Zimbabwe*,
https://doi.org/10.1007/978-3-030-66348-3_4

of the war of liberation during the 1970s, including its ebbs and flows; and nor do we seek to capture all the complexities of the diversity of the second *chimurenga* across the breadth of the then-Rhodesian countryside. However, we bring to the fore the significance of temporal and spatial variations. Ultimately, in considering the second *chimurenga*, we are primarily concerned with pinpointing and discussing themes which are particularly crucial in developing a comparative analysis of the three *zvimurenga* and in understanding the specificities of the third *chimurenga*.

4.2   Early Second *Chimurenga* Literature

In *Peasant Consciousness and Guerrilla War in Zimbabwe*, Ranger (1985) provides the first major scholarly account of the second *chimurenga* broadly and the guerrilla war more specifically, and one which was extremely influential both intellectually and politically. This included contributing, if only unintentionally, to the development of nationalist historiography. Thus, in focusing on eastern Rhodesia, Ranger's analysis is in large part uncritical of the war of liberation particularly in terms of the extent of violence enacted by the ZANU-affiliated Zimbabwe African National Liberation Army (ZANLA) guerrillas against African villagers in the Tribal Trust Lands. In reflecting upon his work years later, Ranger recognised that he reproduced, and dressed up intellectually, the rhetoric of the nationalist movement and specifically of ZANU as propagated during the 1970s and beyond. His later work on the war of liberation is therefore more firmly rooted in a historiography of nationalism, including a two-volume collection of papers about the war co-edited with Ngwabi Bhebe in 1995 (Bhebe and Ranger 1995a, b) as well as a detailed history of the Shangani area of Matabeleland co-written with Alexander and McGregor in 2000 (Alexander et al. 2000) in which there are important discussions about the activities of the ZAPU-affiliated Zimbabwe People's Revolutionary Army (ZIPRA) guerrillas. In the early literature on the second *chimurenga* more generally, there is an almost complete concentration on ZANLA at the expense of ZIPRA guerrillas, a gap which Sibanda (2005) sought to fill in his history of ZIPRA.

Ranger's (1985) book examines peasant consciousness, peasant activism and guerrilla struggles in Makoni District in now Manicaland Province, a province that borders Mozambique. He emphasises that his analysis of the war of liberation cannot be generalised in any simplistic manner to the Rhodesian countryside as a whole (Ranger 1985: ix), though he at times seems to claim otherwise (Ranger 1985: 182). His examination of the guerrilla struggle is rooted in the agrarian history of Makoni, including the record of sustained land evictions and forced resettlement by the Rhodesian state and the consequent discontent and resentment amongst reserve peasants around their agrarian livelihoods, which led to peasant protests and radicalism or a "peasant nationalism" over an extended period (Ranger 1985: 145). In this context, ZANLA guerrillas—who infiltrated the area in the 1970s—were readily welcomed and "peasant support for the guerrillas did not flag" (Ranger 1985: 181) throughout the war. The central message of the guerrillas was the principled "claim to the lost lands" ("lost" initially through colonial dispossession in the late 1800s) and this resonated with the experiential consciousness of Makoni peasants. This meant that land was "both the ideological and practical focus of resistance" (Ranger 1985: 170).

Ranger hence brings to the fore the active agency of peasants in supporting, and collaborating with, the guerrillas in Makoni as "the people were able to contribute much of their own experience and ideology" in advancing the aims of the war of liberation (Ranger 1985: 214). He also highlights the significance of spirit mediums in articulating and asserting the rights of peasants to lost land and in forging crucial linkages between guerrillas and peasants and sometimes peasant elders (chiefs and headmen)—as "one community of resistance" (Ranger 1985: 206). Further, the spirit mediums were able, at least potentially and partially, to restrain the guerrillas from any arbitrary retribution against TTL villagers who collaborated, for whatever seen, with the Rhodesian security forces. Ranger does not see any significant evidence of coercion and violence enacted against the local population by guerrillas in Makoni.

Finally, Ranger acknowledges the existence of social stratification along class lines amongst villagers living in or near the TTLs of Makoni (including the presence of elites such as Native Purchase Area farmers, master farmers in TTLs, entrepreneurs and teachers). But he argues that the Rhodesian state's agrarian and land policies undercut the possibility of significant rural accumulation and differentiation. In downplaying rural heterogeneity, he speaks of a "united countryside' (Ranger 1985: 269) in support of the guerrillas. Class differentials were thus not "determinant" as they were only "one of many forces operating to shape the course of the guerrilla war" (Ranger 1985: 230). As "the war hotted up" (Ranger 1985: 269) during the second half of the 1970s, agrarian class differences became more important with un-cooperative elites becoming subject to guerrilla punishment. The presence and relevance of age differences and of Christian missionaries for peasant-guerrilla relationships is noted at times, but there is a deadly silence about the patriarchal structuring of Makoni communities and of the guerrilla army itself. Overall, then, *Peasant Consciousness and Guerrilla War in Zimbabwe* fits neatly into a nationalist historiography of triumphant guerrillas and nationalists (and specifically of ZANLA and ZANU respectively), with the war of liberation in the 1970s understood, as noted earlier, as the interrupted continuation of the first *chimurenga* and deriving in large part from the "consistency of peasant consciousness" around land dispossession (Ranger 1985: 139).

By the time Ranger's seminal work (based originally on a series of seminars from 1983) was published in 1985, other scholarly, political and personal accounts of the nationalist movements (and less so of the guerrilla movements) of the 1970s had been published, including before independence in 1980. Ranger had also published other articles, including on the well-established nationalist elite within ZANU (Ranger 1980) and on spirit mediums and guerrillas (Ranger 1982). In 1979, Masipule Sithole (younger brother of Reverend Ndabaningi Sithole, founder and one-time president of ZANU) published *Zimbabwe: Struggles within the Struggle* (Sithole 1999) in which he details the ongoing conflict within the nationalist movements (ZANU and ZAPU), dating back to the early 1960s, with a strong emphasis on ethnic tensions and battles in explaining this conflict (see also Sithole 1980, 1984). In 1983, as well, Sithole (in the journal *Zambezia*) offered a review of six recent publications on the liberation struggle, notably the insider autobiographies by Abel Muzowera and Maurice Nyagumbo, as well as Raeburn's *Black Fire* (originally published in 1978) (Raeburn 1978) and the 1981 book titled *The Struggle for Zimbabwe* by Martin and Johnson (Sithole 1983). This latter book traces the rise and development of the guerrilla and nationalist movements (including splits within both movements, and conflicts between nationalists and guerrillas) (Martin and Johnson 1981). But, like Ranger's (1985) book, it ends up falling neatly into a nationalist historiography. In fact, the book has an approving foreword written by (then Prime Minister) Mugabe and it was extensively used in the 1980s by the post-independent Zimbabwean government as a legitimising device for recounting the bravery, heroism and sacrifices of ZANU and ZANLA, including by being used in secondary schools (Robins 1996).

For his 1985 book, Ranger used all six of these sources. Oddly, though, he never refers to Sithole's book on "struggles within the struggle", a phrase which was to be used later by other scholars in depicting far-reaching struggles between the different nationalist and guerrilla movements within the liberation struggle (Sadomba 2008b) and the heterogeneity, tension and conflict within TTL communities (Kriger 1988) during the war. Quite likely, Ranger's exclusive focus on ZANU and ZANLA and his bold claims about "one community of resistance" inhibited his identification and examination of the many contradictions within the second *chimurenga*. At the same time, Ranger refers to the broad arguments of the important study by Lan about spirit mediums and guerrillas, based on Lan's by-then available PhD thesis (completed in 1983) which was later published as *Guns and Rain* in the same year as Ranger's book (Lan 1985). The significance of spirit mediums, and their relationship to guerrillas, was central to the early writings on the second *chimurenga* (in a manner consistent with the ways in which nationalists privileged the relevance of spirit mediums). A more sustained focus on the relationship between guerrillas and Christian missions only came later. In this context, we outline the arguments by Lan before detailing criticisms of Ranger, Lan, and Martin and Johnson, as raised by Norma Kriger.

The study by Lan (1984, 1985) of spiritual mediums, peasants and guerrillas in Dande in the Zambezi Valley of northeastern Rhodesia discusses in great detail the importance of spirit mediums during the war and the ways in which guerrillas embedded themselves in the spiritual worldview of local village populations. However, it ultimately offers a rather romanticised view of the second *chimurenga* in terms of guerrilla-civilian relations. Lan, like Fry (as set out in the preceding chapter), argues that the centre of political authority in rural communities had slowly but surely shifted from the chieftainship system to spirit mediums in the years immediately preceding the second *chimurenga*, mainly because the local legitimacy of chiefs and headmen had been compromised by the Rhodesian state's imposition of indirect rule or despotic decentralism—to use Mamdani's phrase (Mamdani 1996)—in Dande and other TTLs.

Traditionally, according to Lan (1985), both chiefs and mediums owed their local status to the fact that they were legitimate representatives of *mhondoro* (i.e. dead chiefs), the mediums because of possession by the *mhondoro* and chiefs as descendants of the *mhondoro*. The senior *mhondoro* also selected chiefs through their mediums, a process which was disrupted by the colonial reconstruction of the chieftainship system as well as by the politically charged appointing and disposing of chiefs. Of particular significance to the undercutting of chiefly authority was the colonial administration's denial that witches existed and the subsequent forbidding of chiefs to handle witchcraft accusations. In this context, villagers in Dande began taking their witchcraft accusations to mediums. Hence, by the 1970s, mediums remained as the only legitimate local representatives of *mhondoro*, certainly from the perspective of TTL villagers.

As a result, when the ZANLA guerrillas first arrived in Dande, the villagers referred them to mediums. The guerrillas were "pragmatic" and "[t]heir intention was to lay down a durable foundation for long-term cooperation with the peasants. They were prepared to work with whoever they could trust" (Lan 1985: 136). Guerrillas were aware of the role of mediums in the first *chimurenga* (as this was embedded in nationalist discourse in the 1960s and formed part of their training). But they had no pre-arranged plan to work with and through them—they chose to do so because the mediums clearly held authority amongst the TTL peasantry, so that the mediums could "deliver the goods, i.e. the people" (Lan 1985: 136). From the viewpoint of the mediums, they cooperated with local ZANLA units because of "the undertaking given by the guerrillas that if their efforts should succeed, they would reverse all the legislation that limited the development and freedom of the peasantry" (Lan 1985: 136). This was significant to mediums because, as protectors of the land on behalf of the *mhondoro*, guerrillas were offering to return the land to its rightful owners.

Lan (1985: 150–151) does refer in passing to the role of Christian missions in occasionally supporting the guerrillas, but he examines almost exclusively—and presumably quite rightly—the extensive cooperation between guerrillas, mediums and villagers. Mediums thus travelled at times with guerrillas (supposedly like they travelled during the first *chimurenga*), and also advised guerrillas of the safest routes to carry weapons through the Zambezi Valley. As well, mediums worked with guerrillas in the camps in Mozambique, guiding guerrilla insurgents into Rhodesia and leading new guerrilla recruits out. Dande villagers formed secret support committees (comprising mainly of adult men) which organised the collection of food and clothing for the guerrillas. In all this, chiefs and headmen were marginalised, thereby confirming and deepening the disruption of long-standing linkages between chieftains and mediums.

Before any of this could happen, the mediums in Dande had to first explain to the villagers why armed men had entered the area, as guerrillas were not deployed in their home areas and arrived in Dande as strangers. As strangers, they were not descendants of the royal ancestors (*mhondoro*) who owned the land and, thus, they had no justifiable lineage-based presence let alone political authority locally. This was despite the fact that they claimed to speak with the authority of ZANU which, without any established organisational presence in Dande, nevertheless sought to recover all the land in Dande and beyond. Though guerrillas and mediums, ultimately, had different claims to authority, their common goals led to the wartime alliance. The effect of this alliance was that "political authority was conferred on the guerrillas" (Lan 1985: 164) through the mediums on behalf of the *mhondoro*. This therefore involved "the union of the mystical power of the ancestors with the military strength of the guerrillas" (Lan 1985: 148–149).

The mediums provided guerrillas with medicine (*bute* or snuff) to protect them in battle, based on the pre-war practice of mediums of ensuring good health and fortune for villagers. The snuff would often be rubbed on the foreheads of guerrillas so that, for example, Rhodesian security forces could not see them. But, for the *bute* to be effective, the guerrillas needed to observe a complex set of restrictions which mediums imposed on them, such as refraining from: sexual relations during active duty, contact with women during menstruation (or eating food cooked by them while in this condition); and killing wild animals in the bush. By observing these ancestral prohibitions, the guerillas would be more readily transformed from "strangers" into "royals", or from members of lineages resident in other parts of Rhodesia into descendants of the local *mhondoro* with rights to land. Lan summarises his argument as follows:

> The contribution of the *mhondoro* mediums to the guerrilla war was that they made the acceptance of the guerrillas easier, quicker, more binding and more profound by allowing this new feature [armed men] in the experience of the peasantry to be assimilated to established symbolic categories. (Lan 1985: 165)

In this way, as indicated, guerrillas displaced those other descendants of the *mhondoro* (the chiefs) as a legitimate authority in Dande.

Therefore, guerrillas increasingly assumed certain roles of chiefs, such as taking over the administration of customary law, arbitrating disputes between villagers and killing witches (as now-often identified by mediums)—with witches often seen simultaneously as sell-outs to the war of liberation. But Lan argues that the identification of witches-cum-sell-outs represented normally the manifestation of localised conflicts unrelated to the war. Hence, "some of the people accused of being witches were participants in the tensions and hostilities that follow from unexplained death and misfortune" (Lan 1985: 168), so that innocent people were "identified [as witches/sell-outs] for reasons of malice" (Lan 1985: 169). This raises a more general point highlighted by Kriger.

One of, if not, the most controversial dimension of the early nationalist historiography of the guerrilla struggle (as represented by Ranger and Lan) is undoubtedly the character of the relationship between TTL peasants and guerrillas. In particular, this relates to questions around the motivations of peasants for engaging constructively with guerrillas, the basis for peasant mobilisation and, more specifically, the levels of coercion enacted by guerrillas against villagers in TTLs.

The work of Kriger (1992) was crucial in stimulating important re-interpretations of the guerrilla struggle by offering a significantly more critical analysis animated explicitly by a historiography of nationalism. Kriger completed her PhD in 1985 (the year of Ranger's book) and published related articles (for example, Kriger 1988) prior to her book coming out in 1992. In her book, she seeks to "deromanticise or demythologise the existing portrait of ZANU's successful politicalisation of the Zimbabwean countryside during the liberation war" (Kriger 1992: 45). In her study of Mutoko District (in now Mashonaland East Province), she concludes that (more than any other actions) coercion and violence against TTL villagers by guerrillas were the key defining markers of the relationship between guerrillas and villagers. Thus, "I reject the concept of sustained popular support or voluntary cooperation between guerrillas and peasants" (Kriger 1988: 306). This led her to adopt a version of Sithole's "struggles within the struggle" argument but, in this case, with reference to the internal dynamics of TTL communities themselves.

More specifically, insofar as peasants cooperated with ZANLA guerrillas, they did so not on the basis of a deep-seated nationalist ideology focusing on the recovery of lost lands but, rather, due to the existence of local inequalities and grievances which they sought to have addressed by the guerrillas, as the new pivot of power locally. In this respect, resentment was often directed against wealthier peasants or at other local forms of hierarchy (including gerontocracy and patriarchy) in the TTLs of Mutoko. In other words, peasants "used the guerrillas for their own non-nationalist, locally centred interests" (Kriger 1988: 317). This clearly involves a profound criticism of Lan's claim about the significance of spirit mediums as well as Ranger's view of the long-standing and deepening radical peasant consciousness focused on land alienation and recovery. She also adds that "in a situation [like Rhodesia] where the guerrillas are unable to establish [properly constituted] liberated zones, guerrilla violence rather than guerrilla ideology will be the dominant mode of mobilisation" (Kriger 1988: 313).

The youth, for example, saw the guerrillas as a possible ally and established close relationships with them. In this way, with the guerrillas as a new centre of authority in Mutoko (as in Dande and Makoni presumably), the male youth in particular (known as *mujibas* during the war years) used the guerrillas tactically as a means of asserting themselves over local elders, including by berating them when they (the elders) refused to immediately obey requests by the guerrillas for material support. The youth also enacted their own claims to power by being heavily involved in the identification of alleged informers and sell-outs (mainly on flimsy grounds), who were then tortured or killed by the guerrillas. Likewise, many married women interacted with the guerrillas to bolster their influence vis-à-vis their husbands or to end domestic abuse, with guerrillas sometimes publicly berating and beating errant husbands. Hence, and consistent with the argument made in chapter three, the involvement of civilian women in the guerrilla struggle is not reducible to the nationalist project. Overall, then, Kriger (1988: 320) argues that "[t]o accuse people of not supporting the war became a nationalist disguise for a host of social and political struggles, and many simply petty personal rivalries".

Kriger's emphasis on the heterogeneity of TTL communities is clearly important, as is the recognition that peasant motivation for engaging with guerrillas is irreducible to a sweeping nationalist agenda. We make a similar argument in the case of the third *chimurenga*. The points raised by Kriger were soon taken on board by other scholars. For instance, Ranchod-Nilsson (1994) argues that the war unleashed hidden class, generational, lineage and gender tensions within the TTLs which all contributed in specific and diverse ways to shaping the trajectory of the guerrilla struggle at local levels.

Kriger's arguments though are not entirely without precedent. For instance, Martin and Johnson (1981) speak about guerrilla violence, at least with reference to press-ganging by both ZANU and ZAPU in the early years of guerrilla recruitment. This is not to deny the large-scale voluntary movement of youth out of the country, and the often complicated, lengthy and dangerous journey this entailed, as one former ZANLA guerrilla commander recounts (Mutambara 2014). As well, Ranger (1982) speaks of guerrillas and mediums existing in tension at times, including the killing of Muchetera who had claimed—falsely according to the guerrillas—to be the super-tribal medium of Chaminuka but in practice was simply a sell-out. Further, Ranger (1985) refers to ZANLA coercion in Matabeleland villages and, as noted, to increasing class-based conflict internal to rural society as the war of liberation intensified. As well, as noted earlier, Kriger (1992) qualifies her argument by implying that guerrilla coercion was more likely to take place in non-liberated zones in the Rhodesian countryside, as compared to liberated zones in which guerrillas had near uncontested authority vis-à-vis the Rhodesian state and its security forces. In reviewing Kriger's book, Ranger (1994) accepts—but not fully—her stress on guerrilla intimidation and coercion at least in the case of ZANLA. He claims though that Kriger fails to provide sufficient evidence for the existence of internal tensions and horizontal cleavages within Mutoko communities with regard to their effects upon relationships between peasants and guerrillas, while admitting that these local conflicts were likely of some importance.

In an early study, as well, focusing on guerrilla-peasant interactions during the war in the territory of chief Katerere in northern Nyanga, Maxwell (1993: 386) identifies a "diverse range of 'mobilisation' measures pursued and undertaken by guerrillas in gaining peasant support". In doing so, he looks at varying strategies of rural mobilisation during ZANLA's earliest entry and presence in TTLs as well as by the later Zimbabwe People's Army (ZIPA), the combined ZANLA-ZIPRA guerrilla army (but later mainly ZANLA) which operated in many areas of rural Rhodesia during the mid-1970s. Katerere was in large part under guerrilla control from 1976, specifically ZANLA. Importantly, Maxwell argues that guerrilla mobilisation strategies changed "over time and place" (Maxwell 1993: 363), a point which already begins to be apparent in the preceding discussion; and an issue that is central to our examination of the third *chimurenga*.

In the end, though, Maxwell (1993) argues that guerrilla ideology was weak and underdeveloped, so that mass mobilisation was "more pragmatic than ideologically determined" (1993: 363). He thus notes:

> At the commencement of the war [and in many ways throughout the war], peasant ideologies and interests asserted themselves in response to the comrades' [guerrillas'] attempts to secure the widest possible legitimacy. These often conflicting peasant agendas, based on ethnicity, social stratification, gender and generation, acted as the motor of change for guerrilla strategy. (Maxwell 1993: 363)

Lan (1985) also referred to the pragmatism of guerrillas which, in the case of Dande, led to their alliance with spirit mediums. With reference to northern Nyanga, on their initial entry, ZANLA guerrillas—according to Maxwell (1993)—soon came to realise that Shona cultural nationalism alone was an insufficient basis for gaining political legitimacy, as it held out no immediate benefits for Katerere villagers. The more secular (Marxist-slanted) nationalist ideology propagated by ZIPA guerrillas similarly lacked deep resonance in this regard. Martin and Johnson (1981: 76) as well speak of some ZANLA guerrilla commanders who were circumspect about the use and powers of mediums, such as Josiah Tungamirai, who became ZANLA's chief political commissar. However, despite his mission church background and political education, he came to realise the power and legitimacy of mediums in TTL villages such that, in opening up a new operational zone, it was necessary to approach the mediums first. Tungamirai (1995: 41) confirms this, claiming that because "peasants were shifting their loyalty from the chiefs, who were seen as puppets of the colonial rulers, to the mediums", it was tactically advisable to use spirit mediums. Again, the need to think-on-the-ground, and to develop tactics accordingly, is a feature of the third *chimurenga* land occupations.

Maxwell examines the area historically. Progressive, and relatively wealthy, Manyika immigrants from Makoni had been evicted to Katerere years before, and they occupied a higher rank than the local Hwesa and Barwe in the ethnic hierarchy. The Manyika was Christian-affiliated and looked down upon the *mhondoro*-related practices amongst the locals. When ZANLA entered the area (before ZIPA), they first contacted the African Christian elites at the missions and those in the Manyika-dominated villages. The Manyika immigrants had a strong tradition of nationalist politics but the guerrillas approached them for mainly tactical (or pragmatic) reasons, namely because the prosperous Manyika were able to provide guerrillas with food, blankets and so forth. ZANLA, in turn, made use of the Manyika to help politicise the Hwesa and Barwe; though, ultimately, they were mobilised through spirit mediums, who then played a significant role in facilitating support for guerrillas, at least amongst the older local villagers. The Hwesa youth, including younger women amongst them, did not wholly subscribe to spirit mediums and they tended to engage in the war for non-spiritual reasons. As well, not all mediums supported the guerrillas, or they did so reluctantly.

Similar to Kriger, Maxwell argues that the youth (especially male youth) "exploited the comrades' [guerrillas'] need for zealous unswervingly devoted helpers" (1993: 375), co-operating closely with guerrillas as this gave them a new-found freedom and power. These male youth sometimes beat (and stole from) older people as a challenge to local male gerontocratic power. At times, guerrillas had to assert their control over them as the new patriarchs in Katerere. Ethnicity also came into play, with the marginalised Hwesa using the guerrillas to settle old scores with the Manyika.

Maxwell adds that "[l]ike the peasantry, the guerrilla armies were not an undifferentiated whole" (1993: 377). In this regard, the first recruits of ZANLA (in the early 1970s) were ethnic Korekore with only limited missionary activity taking place in their places of origin. Because of this, they were more acceptable of *mhondoro* spirits. The later Manyika guerrilla recruits, including amongst the ZIPA guerrillas, were less amenable to spirit mediums, with local guerrilla commanders—with a mission education—sympathetic to rural Christianity. However, as noted above, spirituality and even secular ideologies were not necessarily the basis on which the different guerrilla armies engaged with peasant villagers, and vice versa.

But, for Maxwell, Kriger overstresses the non-nationalist basis for peasant mobilisation (i.e. deriving from material gain or individual calculus), as it did have an ideological dimension. Further, Maxwell argues that the oppositional relationship between chiefs and mediums (as stressed notably by Lan) should not be overplayed, as there are examples of chiefs as guerrilla sympathisers and of mediums as collaborators with the Rhodesian regime. He in fact also argues, contrary to Kriger, that "guerrilla coercion in Katerere occurred not because the comrades felt insecure but when they became too secure" (Maxwell 1993: 385), as in the late 1970s with increasing guerrilla advances.

In an extended review of Kriger's book, Robins (1996) argues that Kriger is simply unable to take full account of the multiplicity of motives, intentions and aspirations of villagers involved in the guerrilla struggle because of the overriding emphasis on coercion alone. She thus fails "to recognise the complex and subtle motivations, perceptions and actions of peasants" (Robins 1996: 86–87). Based on his own research in Matabeleland, Robins (1996) indicates that, similar to Maxwell:

[L]ocal responses were so variable over time and place that no generalisation about peasant support or coercion is possible. Instead, local villagers entered into complex, contradictory and ambiguous relationships with guerillas that often included both voluntary support and acquiescence to guerrilla coercion. Villagers deployed a rich repertoire of survival tactics. (1996: 75–76)

They did so in the context of complex encounters with security forces and guerrillas. This dilemma for peasant villagers—namely ways of survival in the context of war— was captured in the mid-1970s by the publication titled *The Man in the Middle* by the Catholic Commission for Justice and Peace in Rhodesia (CCJPR 1999), a document which is mainly about violence perpetrated by Rhodesian security forces against villagers. In a subsequent second dossier published years later, an elderly man is quoted as saying (in the light of four children in his village blown to pieces by a landmine): "We are like maize pips being ground between two stones" (CCJPR 1999: 40).

By the mid-1990s, it became clear that the scholarly literature was slowly but surely moving away from a romanticised understanding of the second *chimurenga*. In reviewing Kriger's book, Ranger (1994) notes that—in the preceding few years—a noticeable shift in the literature was taking place, away from a nationalist historiography to a historiography of nationalism, arguing that "the age of heroic interpretation is over" (Ranger 1994: 142).

The publication of the two volumes on the war of liberation edited by Bhebe and Ranger (Bhebe and Ranger 1995a, b) provided the editors with an important opportunity to reflect upon the state of this literature. They highlight for example the dominance of a nationalist historiography which focuses almost exclusively on ZANU and ZANLA: "We have been stuck with an orthodoxy—one version of the war which gives all the credit to ZANLA and none to ZIPRA, and which highlights some elements within ZANLA while denigrating others [such as ZIPA]" (Bhebe and

Ranger 1995e: 4). Further, consistent with the discussion so far in this chapter (and as noted in chapter one), Bhebe and Ranger bring to the fore that most studies until then had focused on the relationship between guerrillas and peasants, and on ideology and spirituality in this context, and not on the internal dynamics of the guerrilla armies. In other words, the guerrilla armies themselves had not been subject to significant investigation, and particularly ZIPRA (and ZIPA). In fact, more work it seems had been undertaken on the nationalist movements compared to the guerrilla armies.

As well, Bhebe and Ranger (1995e) speak about an emphasis, by the mid-1990s, on "the war experiences of women", but this is an overstatement as women's experiences, practices and perspectives received sustained attention only later. There certainly was some literature, most notably the important manuscript by Staunton (1990) which narrates the wartime lives of scores of ordinary rural villagers in TTLs. These personal accounts by women illustrate many of the themes discussed so far, and we will have reason to cite this manuscript later because it also implicitly considers many of the themes which emerged in the second *chimurenga* literature after the mid-1990s. The title of Staunton's manuscript, *Mothers of the Revolution*, tends however to imply a heroic interpretation of gallant and valiant women (as women) subservient to the needs of the liberation struggle ("mothers" of children who went off to war). If this was meant to romanticise the war or at least women in war, and this seems unlikely, it fails to do so. Rather, *Mothers of the Revolution* remains one of the most powerful and sobering depictions of despair and hope during the war of liberation, for both women and men.

Certainly, by the mid-1990s, a growing appreciation existed about the variegated dimensions of the second *chimurenga*, a point which is crucial to any consideration of the third *chimurenga* as well. This appreciation came to be embodied in examinations of the heterogeneous character of both rural communities (and their localised histories) and guerrilla armies, as well as of the many tensions and contradictions internal to both and between them. The complex combination of factors underpinning peasant motivation, mobilisation and action during the second *chimurenga* would also come out even more fully, involving factors from the national level down to the local (and even household) level.

From the mid-1990s until today, a more critical historiography of nationalism has become entrenched in the second *chimurenga* literature (Bhebe and Ranger 1995a, b, c, d, e). Together, this literature provides a wider, more in-depth and finely balanced understanding of the ambiguities, contradictions and complexities marking the war of liberation. There are two main interconnected cross-cutting themes which are particularly relevant to this manuscript on the third *chimurenga*. One of these themes, as intimated, relates to "struggles within the struggle" in shaping the form and trajectory of the war of liberation, and as embedded in the convoluted relations between guerrillas and nationalists, within the guerrilla movements, between guerrillas and peasants, and within TTL communities. A second cross-cutting theme, also intimated, refers to the significant temporal (and particularly spatial) dynamics across the countryside which are constitutive of the second *chimurenga* during the 1970s. While keeping these two themes in mind, we now discuss more specific themes, including nationalists and guerrilla armies and their relationships. Other themes, covered in the following chapter, include: women, war and patriarchy; mediums, Christianity and other forms of spiritually; memory, motivation and mobilisation amongst peasants; and the involvement of nationalist detainees and prisoners, and of African police and soldiers, in the war.

4.3 Nationalists and Guerrillas

In this section, we do not provide a comprehensive overview of the convoluted history and development of the different nationalist and guerrilla movements. Rather, we seek to offer an examination of these movements around themes which are particularly pertinent to our analysis of the third *chimurenga*: specifically, questions around divisions and outright factionalism within both the guerrilla and nationalist movements and the overall relationship between guerrilla and nationalist forces. One of the key points which we wish to raise, for purposes of our analysis of the third *chimurenga*, is articulated quite ably by Bhebe (1999). With reference to both ZAPU and ZANU, he claims that "quite contrary to popular opinion, the situation was far from being simply a matter of the politicians imposing a strategy on the [respective guerrilla] army" (Bhebe 1999: 17). He adds that, with regard to the guerrilla struggle, "no doubt ... some of the early errors were forced on the military [ZANLA and ZIPRA] by the politicians who wanted action and immediate results" (Bhebe 1999: 18). There were also clashes (or non-cooperation at the very least) between ZANLA and ZIPRA and struggles within both armies. These involved different phases of recruitment and layers of recruits leading to tensions in terms of generation, ethnicity (for example, earlier Karanga guerrillas and then Manyika recruits in ZANLA) and education (peasant youth and educated youth). Rebellion within the guerrilla armies regularly entailed a challenge to the dominance of the "veterans" (mainly but not exclusively nationalists) of the liberation struggle (Bhebe and Ranger 1995c). We illustrate these issues with particular reference to ZANU/ZANLA, making mention of ZAPU/ZIPRA only for comparative purposes.

Prior to the second *chimurenga*, and both before and after the banning of ZANU and ZAPU in the early 1960s, there were acts of sabotage undertaken by party activists without any significant military training or even the carrying of military hardware or arms. Within ZAPU, for instance, a number of activists engaged in underground activities such as arson attacks on white-owned farms and factories and the bombing of telegraph and electricity transmission poles (Dabengwa 2017). As well, ZANU activists engaged in similar acts, most famously the Crocodile Gang in which ZANU leader Ndabaningi Sithole was instrumental in sending out and assisting (Ranger 1997). Though causing the death by stabbing of a white wattle factory farmer in Melsetter in July 1964 and said by some as—in effect—launching or inaugurating the armed struggle (Raeburn 1978; Tavuyanago 2013), the Crocodile Gang was involved primarily in acts of sabotage along with the other ZANU groups and individuals operating in the 1960s. Thus, like ZAPU, ZANU's first deployments in 1964 had no arms and were led by people with little or no military training let alone expertise

(Mazarire 2017). Ranger (1997) thus describes the activities of the Crocodile Gang and others as proto guerrilla in character. In fact, ZANU historically marked the 28th of April 1966 as *Chimurenga* Day (the start of the second *chimurenga*) based on the battle of Sinoia when seven of its guerrillas were killed. Meanwhile, in the late 1960s, ZIPRA was involved in a number of battles, including the so-called Wankie and Sipolilo campaigns (Ralinala et al. 1969; Maxey 1975).

In focusing on struggles-within-the-struggle (and between mainly the nationalist movements dating back to the late 1950s), Sithole (1999) argues that "the only fundamental split" (1999: 39) was the ZAPU-ZANU split in the early 1960s. However, like many other writers, he speaks to the existence of a multiplicity of other tensions including with reference to the guerrilla movements, and between guerrillas and nationalists. But even before the guerrilla war took off the in the 1970s, there were attempts at military collaboration, such as the Mbeya Accord in 1967 which led to a loose unity agreement under the Joint Military Command; this, though, soon collapsed (Dabengwa 1995). There was also another attempt to bring together ZANU and ZAPU, in Lusaka, under the Zimbabwe Liberation Council, but this likewise failed.

In the case of ZANU, when the *Dare re Chimurenga* or War Council was formally established in 1969 and elections were used in constituting this ZANU structure, the position of Chief of Defence—a position in *Dare*—was subject to election biannually, as were all other *Dare* positions. Appointment to High Command positions in ZANLA was done though by the Chief of Defence. By 1971, the *Dare* had not a single member with any military training or experience, including the Secretary of Defence who chaired the High Command. This led to "tension between the politicians and the guerrillas" (Martin and Johnson 1981: 32), with guerrilla leader Josiah Tongogara observing that the ZANU nationalists would demand dramatic and often suicidal action by guerrillas for short-term gain in attaining legitimacy at for example the Organisation of African Unity. As Tongogara (Tongogara 2015: 60) argues: "Our first groups went into an area carrying guns on their back. They frightened away the villagers who rushed to the enemy to report our presence". The military wing (ZANLA) was itself pre-occupied with laying the groundwork for continuous and protracted struggle. As Alexander and McGregor (2004: 88) argue, guerrillas were often pawns in the internal tensions within the nationalist movement, such that their "political loyalties were put to the test".

In 1973, Tongogara became Chief of Defence. In 1974–1975, ZANU experienced a crisis in its guerrilla movement, given the arrest of the ZANLA High Command as well as of *Dare* members by the Zambian government in the context of the assassination of ZANU chairperson Herbert Chitepo in Lusaka. The assassination also entailed a crisis within ZANU itself, as it brought to the fore internal struggles within ZANU for leadership positions, particularly between Ndabaningi Sithole and Mugabe. Together, according to Sithole (1999), these problems amounted to a military-nationalist crisis.

The crisis arose soon after serious problems in ZANLA, including the first internal rebellion led by guerrilla commanders Nhari (a provincial field commander) and Badza. Nhari and his co-conspirator Badza raised critical questions about the High

Command in relation to the shortages of arms and the general neglect of guerrillas at the front, but they also criticised the luxurious living of members of the *Dare* and High Command in Lusaka. As well, the mutineers were at odds with ZANLA High Command over methods of recruitment, notably because of the 1973 Nhari-led abduction of children at St Albert's Mission as guerrilla recruits (Tendi 2017). In fomenting rebellion within ZANLA, Nhari and Badza may not have followed procedure by failing to raise their concerns first with the High Command before addressing the *Dare* directly (though the High Command was seen as the main cause of the problems outlined). Further, they killed some seventy guerrillas for refusing to join them in rebellion and detained members of the High Command and later of the *Dare* in trying to take over ZANLA headquarters and guerrilla camps in Zambia. Tongogara ordered (later ZIPA guerrilla leader) Mhanda to "crush the rebellion" (Mhanda 2011: 48) though, in the end, this task fell to Nhongo and Urimbo. Once Nhari and Badza were captured, ZANU established a disciplinary committee headed by Chitepo, who argued for demotion of the rebellion leaders. Tongogara (as Chief of Defence in *Dare*) and the ZANLA High Command, however, had the rebellious guerrillas executed without *Dare*'s knowledge, or at least ZANU was marginalised in the process.

To digress, a similar rebellion took place in ZIPRA. Tendi (2017) thus refers to ZIPRA recruits revolting in Zambia in March 1971, who complained about being inactive for a year in the camps because the ZAPU nationalist leadership was involved in petty and selfish infighting. In this context, there was a mutiny by the Mthimkhulu group within the guerrilla army (known as the March 11 Movement) which led to a split within ZAPU and the formation of Frolizi, with ZAPU commanders like Nhongo defecting to the ZANU guerrilla movement (Dabengwa 1995). This munity within ZIPRA by junior commanders led to the arrest of some senior commanders who were detained at a logistics base west of Lusaka (Dabengwa 2017). This inner-party struggle led to the official launch of ZIPRA in 1971, which "finally enabled military strategy to be developed by the military, though ZIPRA remained subordinate to the political control of the [ZAPU] party" (Brickhill 1995: 54).

In returning to the Nhari rebellion and its implications, Mazarire (2011: 578) notes that this episode "caused a good deal of friction between the political and military leaders". Sithole (1999) in fact claims that, after Tongogara had subdued the Nhari rebellion, "the nervous [ZANU] party began to evolve around him [Tongogara] and the High Command": The *Dare* had "lost control of the Party to the Chief of Defence and his men.… Defence was now in effective control of the Party" (Sithole 1999: 167). For ZIPA commander Mhanda (2011), who became deeply critical of Tongogara though simultaneously supporting the crushing of the Nhari rebellion, this concentration of excessive power in the hands of Tongogara was due to his election in 1973 as Chief of Defence on *Dare*, while also head of ZANLA's High Command. Martin and Johnson (1981) add that this entailed a greater separation between military and political affairs, "which was resented by some politicians who sought to advance their own political ends" (1981: 33). Likewise, Chung 2006—in discussing the aftermath of the Nhari rebellion—refers to "the long divided political and military leaderships" (Chung 2006: 93). Further, Tongogara "regarded the old-style nationalists as untrustworthy, corrupt, and liable to betray the military struggle for ephemeral political gains" (Chung 2006: 128). Because the military executed the rebels, the "military leadership had now gained the upper hand, but the battle between the two had not ended" (Chung 2006: 95).

We quote Mazarire (2011) at length about this. ZANU's ideology was rooted in democratic centralism. However:

> The military was subordinated to the political goals of the party, the structure of the relationship between ZANU's Supreme Council, the *Dare* (an elected body of political functionaries), and the Military High Command (an appointed body that participated in the elections of the *Dare*) tipped the balance in the military's favour. The military gained influence through the electoral system, to the point of leading and determining the affairs of the party. As a result, the ZANLA High Command became an unaccountable and undisciplined unit. (Mazarire 2011: 572)

The power of the ZANLA High Command vis-à-vis the ZANU party was expressed by the fact that its members were appointed by the Chief of Defence, yet they could vote in elections for the *Dare*.

From the ZANU/ZANLA side, the war of liberation had almost ground to a halt after the Nhari rebellion, the assassination of Chitepo, the struggle for ZANU leadership and the detention of the High Command. In this context, a new guerrilla force arose, seeking to combine both ZANLA and ZIPRA movements, namely ZIPA (Moore 1995, 2012). ZIPA arose in the light of tensions within both ZANLA and ZIPRA, and tensions between ZANLA and ZIPRA (Sadomba 2008a). Initially, ZIPA had a high command of nine ZANLA and nine ZIPRA commanders, but clashes soon followed in the camps in Tanzania, such that ZIPA increasingly became in large part a reconstructed ZANLA army. There were violent disputes between ZANLA and ZIPRA guerrillas in April and May 1976 in Morogoro and Mgagao camps in Tanzania (Tekere 2007, Mhanda 2011). After the killings at Morogoro and Mgagao in August 1976, of ZIPRA guerrilla by ZANLA guerillas subsequent to the formation of ZIPA, the united ZIPA army in effect came to end; subsequently, ZANLA was left in ZIPA under the leadership of *vashandi* commanders (Chung 2006). For Sadomba (2008a), though, the ZANLA-ZIPRA infighting and the breakup of ZIPA led to ZIPA in effect surrendering power to the nationalist elders in the restoring of order.

Former ZIPA commanders often refer to the Mgagao Declaration crafted by ZANLA officers as their foundational text, as it rejected the political efforts of détente (led at the time within ZANU by Sithole). ZIPA was born out of initiatives of the frontline states and the Organisation of African Union's African liberation committee, thus by-passing the Zimbabwean nationalist movements because of continued disunity (Sithole 1999). As Sadomba, a former ZIPA guerrilla put it, the declaration entailed "pushing nationalist politicians asunder" (2008a: 40). The Mgagao declaration formed the basis of an appeal to the Organisation of African Union's liberation committee "to support an initiative by ZANLA and ZIPRA guerrillas to resume the war without the bickering nationalists" (Mazarire 2011: 580): "The Mgagoa document set out the principle that those closest to the operations (the soldiers) should be directly involved in decisions concerning the war, without the interference of politicians" (Mazarire 2011: 580).

The declaration, released by ZANLA and ZIPRA guerrillas in Mgagao camp in Tanzania in 1976, also denounced nationalist leaders Sithole, Muzorewa and Chikerema. The declaration thus entailed, in the context of factional battles within ZANU between Sithole and Mugabe, support for Mugabe by ZIPA, just as the ZANLA High Command (including Tongogara) supported Mugabe subsequent to its release from Zambian prison after the Chitepo episode. Given the conflict taking place within ZANU, ZIPA expressed sympathy for Mugabe in hoping that the latter could be a legitimate and effective power broker for ZIPA's dealings with the ZANU nationalists (Sadomba 2008a). Mugabe was elected officially to the ZANU presidency at the Chimoio Congress in August 1977. Previously, in March-April 1977, there had been a joint meeting of ZANU's central committee and the ZANLA High Command which led to the combining of the two bodies into an expanded central committee, and this formalised the co-optation of Mugabe into the military or at least his subservience to it (Mazarire 2011). For Sadomba, with reference to the 1977 election of the ZANU central committee, the most important new element was strong representation of the military or guerrillas. But, contrary to the claim by Mazarire (2011), this was "hardly for purposes of power sharing between the guerrillas and nationalists but a strategy of the nationalists to provide a mechanism for controlling guerrillas through their commanders" (Sadomba 2008a: 54).

According to Mazarire (2011), the more ideologically inspired ZIPA movement opened up potential space for new disciplinary values as well as an initiative to establish political structures amongst rural people in the TTL operational zones. By early 1976, *vashandi* was in control 0of all the refugee and military camps in Mozambique. Under ZIPA, "combatants [guerrillas] had an opportunity to lead on both the political and military fronts. Nationalists—remote from operations—had limited control over the guerrilla forces" (Sadomba 2008a: 41). ZIPA was marked by a considerable number of educated recruits. In fact, the drafters of the Mgagao declaration were a group of intellectually inclined guerrillas, some of whom then became members of the ZIPA High Command (including Nhongo and Mhanda).

When *Dare* and High Command leaders were released eventually from prison in Zambia, "disagreements crept in as to the relationship between ZIPA and the ZANU Dare leadership" (Sithole 1999: 173). These disagreements became quite vicious after the Geneva Conference in 1976, with many ZIPA commanders rounded up and jailed in Mozambique in an effort to impose Mugabe's leadership of ZANU and reinstate Tongogara as Chief of Defence of ZANLA. ZIPA was thus purged and crushed, with Tongogara, Nhongo (though a ZIPA commander) and Mugabe central to this. Mugabe in fact was in alliance with Tongogara from 1976. Bhebe argues, with reference to the formation of the ZIPA guerrilla force, that: "By excluding the political leadership ZIPA actually raised a Third Force amongst the liberation movements" (1999: 64). Thus, the crushing of it seemed almost certain.

In this respect, Mhanda (2011) is deeply bitter about what happened to ZIPA. He notes that, at the time of Mugabe's release from prison and ZIPA's support for his leadership of ZANU, the guerrillas themselves coined the term *vashandi*. The

Mugabe "clique" (after ZIPA's demise) then used the term to label ZIPA in a negative manner. Mugabe saw ZIPA "as a blatant challenge to his authority and put an end to ZIPA", with Tongogara and Nhongo (as indicated) helping him to do so (Mhanda 2011: 153). Subsequently, Mugabe sought to "eradicate" (Mhanda 2011: 179) the contribution of ZIPA from the minds of guerrillas in camps and at the front. The end of ZIPA "paved the way for Mugabe to assert his authority over the army" (Mhanda 2011: 153). Thus, despite supporting Mugabe previously, many of the ZIPA commanders and rank-and-file later regretted this, as they also regretted collaborating with Tongogara. In this sense, post-ZIPA developments involved Mugabe "taking over command of the guerrilla movement" (Sadomba 2008a: 51) through Tongogara and Nhongo (Solomon Mujuru), and "[i]nternal and external discipline was emphasised to enforce unequivocal loyalty to nationalist authority" (Sadomba 2008a: 52).

The Nhari rebellion was not the only rebellion. In this light, another rebellion—or even coup—led by Henry Hamadziripi broke out. The revolt of the Gumbo-Hamadziripi group in December 1977 also involved some guerrillas. It threatened to cripple the war effort for most of 1978, including at the war front where serious disciplinary challenges such as desertions and "anarchism" amongst ZANLA rank-and-file emerged (Mazarire 2011). Hamadziripi had criticised Tongogara and Nhongo for supposedly blocking and undermining ZANU's control of the ZANLA army. Because of this revolt, measures were taken to re-establish order, as led by the Departments of Defence and of the Commissariat, thereby protecting and rescuing Mugabe from the threat. Hamadziripi, Gumbo and others were arrested, and a trial took place with Mugabe as presiding officer.

The character of the relationship between ZANU and the ZIPA/ZANLA is clearly open to differing interpretations, and it seems that this relationship was marked by significant fluctuation and contestation. Chung (2006: 133) appears to lean towards the conclusion that ZANU dominated the guerrilla armies, or at least that is how the relationship played itself out by the late 1970s:

> In the final analysis, it was the execution of the Nhari rebels that made it impossible for the militarists to again win the upper hand in future power struggles. When Tongogara attempted to quell opposition first by the Vashandi group led by Wilfred Mhanda and Sam Geza and later by a second group of old-style politicians led by Henry Hamadziripi and Rugare Gumbo, he was unable to destroy these two latter groups as effectively as he had done with the Nhari group. The power that he had given himself as executioner was effectively removed by the new political = leaders of ZANU, Robert Mugabe and Simon Muzenda ... As a result, as soon as the conflicts with Tongogara came to a head, Mugabe and Muzenda hastily arranged for intervention by the Mozambican authorities, who held these two groups in detention ... Had Tongogara's faction not executed their critics, they could have emerged from the war as a more powerful and coherent group.

At the same time, though, Chung (2006) indicates that Mugabe had no control over the army, claiming that Mugabe:

> Would always be dependent on the military leaders for his power base. ... The military was suspicious of being utilised as instruments and weapons to achieve political goals, and then sacrificed and discarded once these political goals were achieved. ... This inherent suspicion made them a dangerous and powerful source of opposition to any political leadership which did not take their interests into consideration (Chung 2006: 182).

Further, ZANU under Mugabe "had never been in complete control of the guerrillas" with Tongogara ensuring that "the military would remain largely outside the control of the traditional politicians" (Chung 2006: 300). Mazarire (2011) tends to argue in a similar fashion.

Beyond disagreements by different scholars, it seems that there are inconsistencies in the arguments even by the same scholar about the relationship between the nationalist movements and the guerrilla armies or, in a more sympathetic interpretation, at least attempts to trace changes in the relationship. For us, the most important point is that it is simply not possible to argue that the national movements, both ZANU and ZAPU, had unbridled and unilateral control over the guerrilla armies, or control without challenges and insubordination. It thus becomes necessary to recognise that the guerrilla armies had their own logics, imperatives and aspirations which were not reducible to the dictates of the nationalist movements. As we will show, this has significance for understanding the dynamics of the third *chimurenga*, and in particular the relationship between the ex-guerrillas (war veterans) and ex-nationalist movements (ZANU-PF) during the course of the land occupations.

4.4 Guerrilla Armies

We noted previously that, initially, the main focus in the war of liberation literature was on ZANLA at the expense of ZIPRA. This has now been subjected to correction, including the study of the Shangani area of Matabeleland by Alexander et al. (2000). This correction brings to the fore narratives about the second *chimurenga* which act as important alternatives to the official narrative propagated consistently by the ruling ZANU-PF party. In this case, the guerrilla struggles in Matabeleland (involving primarily ZIPRA) varied in important ways from the ZANLA struggles in the eastern part of the country, though the form and extent of any differences between ZANLA and ZIPRA have been the subject of considerable debate. Even the areas of infiltration and geographical spread of the two guerrilla armies across the countryside is open to debate (Sibanda 2005).

A number of scholars have sought to offer an overall periodisation of the nationalist guerilla movements and, in doing so, they raise a number of key issues. Bhebe (1999) for example speaks about three periods. Initially, up until 1970, ZAPU and ZIPRA were the hegemonic forces; subsequently—until 1974—ZANU and ZANLA became increasingly dominant, after which there were converging and diverging efforts by ZAPU/ZIPRA and ZANU/ZANLA as well as ZIPA for a number of years. In this context, Bhebe (1999) details the nationalist movements' decision to engage in guerrilla activities along with recruitment methods, establishment of bases outside the country, military training, arms logistics and infiltration into the country. In making a distinction between the guerrilla forces, Bhebe argues that ZANLA emphasised political mobilisation while ZIPRA concentrated on mainly military operations, while also having both guerrilla units and a regular army contingent.

From its early years, ZAPU had sought to form a military alliance with the MK guerrillas of South Africa's African National Congress, including joint operations in Rhodesia and cooperation in military training (MacMillan 2017). It forged significant relations with the Soviet Union from the 1960s, including receiving armoury and training (Dabengwa 2017). There was, from the beginning, a significant movement of seasoned ZAPU nationalist activists into Botswana and the onward movement of would-be recruits in Botswana to Zambia for military training, with the Special Branch in Francistown, police in Gaborone and ZAPU in Francistown and Lusaka processing them over time (White 2014). Alexander and McGregor (2004) consider the diverse ways of joining ZIPRA and crossing the border, life in transit camps in Zambia, being sent to camps in Tanzania and Mozambique for training, and then travelling back to Rhodesia and facing the Rhodesian military.

A War Council was established, consisting of five members, including the ZIPRA commander and Secretary of Defence, which represented ZIPRA on the council. But the ZIPRA High Command was subordinate to the War Council. Dabengwa (2017) argues that ZAPU's strategy was to move from small guerrilla units to larger, platoon-size units and later to a regular army. Initially, though, operations between 1972 and 1973 primarily focused on sabotage and planting land mines along roads, without any full engagement with Rhodesian security forces. The turning-point strategy in 1978 was meant to involve the deployment of regular army battalions on the northern front, across the Zambezi Valley, in order to establish semi-liberated zones in Gokwe, Lupane and surrounding areas and then occupy and defend these areas; and, as a result, allow guerrilla units to move in and liberate and occupy major cities. But Rhodesian security forces got wind of this, and it never happened.

In seeking to gain military superiority over Rhodesian forces, Bhebe (1999) argues that ZAPU operated under the assumption that it had a considerable rural following based on the underground structures carefully constructed as early as the days of the National Democratic Party (from which ZAPU arose after the banning of the party). Hence, local ZAPU political structures already existed as a basis for providing logistical support for its guerrilla forces, or resurfaced once villagers gained confidence in the guerrillas. As Sibanda (2005: 110) puts it, ZIPRA insurgents were "making contact with ZAPU underground structures and constructing some others where they did not exist". As well, at least initially, ZIPRA sought to build up its own caches for self-reliance purposes (including food, clothing and medicine supplied from Zambia) so as not to rely on TTL villagers, with all the delays, shortages and complications which might however arise from this arrangement. Later, ZIPRA increasingly relied upon local villagers for logistical support, but this was pursued through ZIPRA political commissars who worked alongside local ZAPU officials. Where possible, local party structures were used for civil administration in zones being liberated.

Likewise, for Bhebe (1999), ZANLA became engaged directly in building parallel political structures corresponding with its military sectors (involving village/cell, branch and district structures), though this was not always achieved. At times, it involved the establishment of local courts (previously administered by chiefs), which tried villagers accused of being sell-outs. Previous to this, guerrillas were killing so-called sell-outs on the flimsiest of evidence. Bhebe claims that, overall:

> ZANLA imposed more demands on ... the peasants through their Maoist political mobil-
> isation, restructuring of semi-liberated zones and through heavy dependence on civilians
> for material support than the ZIPRA which relied on the purely military approach and left
> the political mobilisation and collection of material support from the civilians to the ZAPU
> political structures and officials. But ZIPRA also weighed heavily on ... the civilians through
> its aggressive recruitment drives from 1975 to the end of the war. (Bhebe 1999: 7)

However, the extent of demands placed upon villagers also varied in terms of the character of the guerrilla presence: "The population tended to be subjected to less physical suffering wherever one of the contending forces [ZIPRA or ZANLA] was almost in full control of the area and to worst in highly contested ... zones" (Bhebe 1999: 110). Across the northern front, a considerable number of liberated zones emerged, and these were less common in the southern front where the Rhodesian forces were putting up significant resistance. For instance, south-western Rhodesia (where Bhebe conducted his research) was a fresh zone of operation for ZIPRA and therefore at first it was mostly involved in massive recruitment rather than in guerrilla activities.

Mazarire (2000) focuses on Chivi from 1976 to 1980 in also trying to periodise the war of liberation. In the case of Chivi, he reconsiders Bhebe's periodisation and highlights therefore—at least implicitly—the need for localised case study research to show both temporal and spatial variations in the activities of the guerrilla armies. What Bhebe refers to as the first phase is synonymous with the phase of rural nation-alism in Chivi, dominated by ZAPU and ZIPRA but stretching into the mid-1970s. ZAPU structures were in place in Chivi, including its youth league but, by 1975, most rural activists had been detained. This was at the time when ZANLA opened up the Gaza sector, with Chivi falling under this sector. The rural nationalists in Chivi though remained staunchly ZAPU, and ZANU/ZANLA sought to permanently exclude ZAPU from the area, leading to many ZAPU activists fleeing the area. What Bhebe considers as the end of ZANLA domination (the mid-1970s) was actually its beginning in Chivi, with ZANLA first entering the area in February 1976.

Despite ZANU'S strong anti-ZAPU stance, the ZANLA guerrillas became reason-ably popular amongst Chivi villagers because they targeted colonial instruments of oppression. The ZAPU threat, however, became a useful basis for ZANLA in identi-fying sell-outs. A culture of fear was cultivated by ZANLA guerrillas which allowed them to requisition resources of any kind and at any time, with villagers even raiding white farms for cattle to supply guerrillas. The third period in Chivi started in 1978 in the light of the internal settlement. This was marked by for example the aggres-sive and violent activities of the auxiliary forces of moderate nationalists as well as by increasing ill-discipline amongst guerrillas as the command structure became overburdened and overstretched (leading to unreasonable demands being placed on villagers, including robbing passengers of cash on buses). It was also characterised by guerrillas providing basic military training for *mujibas* in the area (such as at Mukana range near Berejena) and deploying some of them to guerrilla sections: "By this time, the majority of *mujiba* had been transformed into lower echelon guerrillas, with many of their duties having been transferred to the more juvenile sections of the community" (Mazarire 2000: 58). These local initiatives by guerrillas to train and field guerillas in the front led to *mujibas* arming themselves with AK7s and other weapons picked up from battlefields. This is not unlike what happened in parts of Honde valley, notably in the Holdenby TTL, where local young men from the start were recruited and trained as guerrillas and became active in their home area (Schmidt 2013).

Sadomba (2008a) as well provides a periodisation of the nationalist and guerrilla movements, drawing upon his experiences as a guerrilla involved in ZIPA. He likewise speaks of three periods or phases, though with reference to ZANU specifically: the Chitepo phase, 1963–1975; the ZIPA phase, 1975–1977; and the Mugabe phase, 1977 to independence and beyond. He seeks to highlight the uniqueness of ZIPA and argues that the considerable differences between the different guerrilla armies result from "the different environments through which they passed at various stages in the armed struggle" (Sadomba 2008a: 34): "Each phase is distinguished by the nature and characteristics of leadership, methods of recruitment, quality of recruits, politico-ideological thrust and level of military offensive" (Sadomba 2008a: 34). The Chitepo phase entailed: poor organisation, defective strategies, limited military training and political education, uneducated peasant recruits, and enforced conscription of recruits. In the early 1970s in the camps of ZANLA (including in Tanzania), there were not enough guns. Guerrilla trainees were often instructed to make wooden replicas of AKs and they were trained with these until they were sent back to Rhodesia. They often travelled unarmed and were only issued guns when they entered Rhodesia (White 2009). In the latter two phases, guerrilla forces became organised and militarised, though Sadomba is critical of the post-ZIPA Mugabe period.

Given the diversity of periodisation of the war of liberation, it becomes clear that the temporal development of the war was not uniform across the countryside. Any nation-wide depiction of the scale, tempo and character of the second *chimurenga* is deeply problematic, because spatial locality (including local agrarian histories) was central to the manner in which the war developed and progressed.

Finally, guerrillas gave themselves war (or *chimurenga*) names (Mutambara 2014). In identifying nearly 5,000 *chimurenga* names for specifically ZANLA guerrillas, Pfukwa (2007) considers the function and significance of this renaming process. He notes for instance that "concealing an identity [through war names] was also a process of creating a new identity" (2007: 114) with new values, attributes and possibilities which sought to reverse discursively the colonial past and to repossess the present and future. Some names though were unrelated to nationalist visions and were deeply personal by speaking to particular lived experiences. Thus, war names spoke to both national and personal deprivations and aspirations. Many of the names incorporated in some way the notion of *chimurenga*, thereby facilitating the embodiment of war in the personal identities of guerrillas. As with incorporating *chimurenga* into their very identities, labelling the struggle as *chimurenga* was significant: "Giving the conflict their own name was in itself an act of reclaiming a past that had been erased by some ninety years of colonial rule. Naming it *Chimurenga* was an act of reasserting control over ideological space" (Pfukwa 2007: 13). This also entailed reclaiming through their war practices the material space then known as Rhodesia.

4.5 Conclusion

This chapter provided an overview of the early literature on the second *chimurenga* and considered the theme of nationalist movements, guerrilla armies and their relationships. In doing so, we have sought to discuss matters which are in some way pertinent to an examination of the third *chimurenga*, which is the main focus of this volume. We conclude by raising four issues in this regard. First of all, questions arise in the second *chimurenga* literature about peasant mobilisation and the relationship between peasant villagers and guerrillas. This resonates with debates about the third *chimurenga* land occupations with reference to the mobilisation of occupiers and their relationship to war veterans. Secondly, and related to this, is the question of peasant motivations for becoming involved in the war of liberation, and specifically whether these are simply reducible to nationalist aspirations. Likewise, in the case of the land occupations, the issue of the motivational basis for people occupying white-owned farms and other landholdings is of significance. Thirdly, the relationship between guerrilla armies and nationalist movements is clearly of importance in the second *chimurenga*, as is the relationship between war veterans and the ruling party (ZANU-PF) during the third *chimurenga* occupations. Finally, and as a broader theme, the early literature on the war of liberation became marked by a recognition of diversities and contingencies, including across space. In a similar vein, in trying to come to terms with the sheer complexities of the fast track land occupations, searching for spatial variations (and other differences) is fundamental. In the following (second) chapter on the second *chimurenga*, further comparative points arise.

References

Alexander J, McGregor J (2004) War stories: Guerrilla narratives of Zimbabwe's Liberation War. Hist Workshop J 57(1):79–100

Alexander J, McGregor J, Ranger T (2000) Violence and memory: one hundred years in the 'Dark Forests' of Matebeleland. James Currey, Oxford

Bhebe N (1999) The ZAPU and ZANU Guerilla warfare and the Evangelical Lutheran Church in Zimbabwe. Mambo Press, Gweru

Bhebe N, Ranger T (eds) (1995a) Soldiers in Zimbabwe's liberation war (volume one). James Currey, London

Bhebe N, Ranger T (eds) (1995b) Society in Zimbabwe's liberation war (volume two). University of Zimbabwe Publications, Harare

Bhebe N, Ranger T (1995c) Volume introduction. In: Bhebe N, Ranger T (eds) Soldiers in Zimbabwe's liberation war (volume one). James Currey, London, pp 6–23

Bhebe N, Ranger T (1995d) Volume introduction. In: Bhebe N, Ranger T (eds) Society in Zimbabwe's liberation war (volume two). University of Zimbabwe Publications, Harare, pp 6–34

Bhebe N, Ranger T (1995e) General Introduction. In: Bhebe N, Ranger T (eds) Society in Zimbabwe's liberation war (volume two). University of Zimbabwe Publications, Harare, pp 1–5

Brickhill J (1995) Daring to storm the heavens: the military strategy of ZAPU 1976 to 1976. In: Bhebe N, Ranger T (eds) Soldiers in Zimbabwe's liberation war (volume one). James Currey, London, pp 48–72

Catholic Commission for Justice and Peace in Rhodesia (CCJPR) (1999) The man in the middle: torture, Resetlement and Eviction. Catholic Commission for Justice and Peace in Rhodesia, Harare

Chung F (2006) Re-living the second Chimurenga: memories from the liberation struggle in Zimbabwe. Nordic Africa Institute, Uppsala

Dabengwa D (1995) ZIPRA in the Zimbabwean War of national liberation. In: Bhebe N, Ranger T (eds) Soldiers in Zimbabwe's liberation war (volume one). James Currey, London, pp 24–35

Dabengwa D (2017) Relations between ZAPU and the USSR, 1960s–1970s: a personal view. J South Afr Stud 43(1):215–223

Kriger N (1988) The Zimbabwean War of liberation: struggles within the struggle. J South Afr Stud 14(2):304–322

Kriger N (1992) Zimbabwe's Guerrilla War: peasant voices. Cambridge University Press, Cambridge

Lan D (1984) Spirit mediums and the authority to resist in the struggle for Zimbabwe. In Collected seminar papers. Institute of Commonwealth Studies, 33, Institute of Commonwealth Studies, London

Lan D (1985) Guns and rain: Guerillas and spirit mediums in Zimbabwe. James Currey, London

Macmillan H (2017) 'Past history has not been forgotten': the ANC/ZAPU alliance—the second phase, 1978–1980. J South Afr Stud 43(1):179–193

Mamdani M (1996) Citizen and subject: contemporary Africa and the legacy of late colonialism. James Currey, London

Martin D, Johnson P (1981) The struggle for Zimbabwe: The Chimurenga War. Ravan Press, Johannesburg

Maxey K (1975) The fight for Zimbabwe: the armed conflict in Southern Rhodesia since UDI. Rex Collings, London

Maxwell D (1993) Local politics and the War of liberation in North-East Zimbabwe. J South Afr Stud 19(3):361–386

Mazarire G (2000) 'Where civil blood made soldiers hands unclean' rethinking war time coercion in rural Rhodesia. Reflections on the Chivi experience 1976–80. J Afr Conflict Dev 1:44–59

Mazarire G (2011) Discipline and punishment in ZANLA: 1964–1979. J South Afr Stud 37(3):571–591

Mazarire G (2017) ZANU's external networks 1963–1979: an appraisal. J South Afr Stud 43(1):83–106

Mhanda W (2011) Dzino: memories of a freedom fighter. Weaver Press, Harare

Moore David (1995) The Zimbabwean people's army: strategic innovation of more of the same. In: Bhebe N, Ranger T (eds) Soldiers in Zimbabwe's war of liberation (volume one). James Currey, London, pp 73–85

Moore David (2012) Two perspectives on Zimbabwe's national democratic revolution: Thabo Mbeki and Wilfred Mhanda. J Contemp Afr Stud 30(1):119–138

Mutambara A (2014) The rebel in me: a ZANLA Guerrilla commander in the Rhodesian Bush War, 1975–1980. Helion & Company, Solihull

Pfukwa C (2007) The function and significance of war names in the Zimbabwean armed conflict (1966–1979). Unpublished PhD thesis. University of South Africa, South Africa

Raeburn M (1978) Black fire: accounts of the Guerrilla War in Rhodesia. Julian Friedmann Publishers, London

Ralinala RM, Sithole J, Houston G, Mugabane B. (1969) The Wankie and Sipolilo campaigns. In: South African Democracy Education Trust (ed) The road to democracy in South Africa, vol 1, 1960–1970, Zebra Press, Cape Town, pp 479–540

Ranchod-Nilsson S (1994) "This, too, is a Way of Fighting": Rural Women's Participation in Zimbabwe's Liberation War. In: Tetreault MA (ed) Women in revolutions in Africa, Asia and the New World. University of South Carolina Press, Columbia, SC, pp 62–88

Ranger T (1980) The changing of the Old Guard: Robert Mugabe and the revival of ZANU. J South
    Afr Stud 7(1):71–90
Ranger T (1982) The death of Chaminuka: spirit mediums, nationalism and the Guerilla War in
    Zimbabwe. Afr Aff 81(324):349–369
Ranger T (1985) Peasant consciousness and Guerrilla War in Zimbabwe: a comparative study.
    Zimbabwe Publishing House, Harare
Ranger T (1994) Zimbabwe's Guerrilla War: peasant voices (book review). Afr Aff 93(370):142–144
Ranger T (1997) Violence variously remembered: the killing of Pieter Oberholzer in July 1964.
    Hist Afr 24:273–286
Robins S (1996) Heroes, heretics and historians of the Zimbabwe revolution: a review article of
    Norma Kriger's 'Peasant Voices' (1992). Zambezia XXIII(i):73–91
Sadomba WZ (2008a) War Veterans in Zimbabwe's land occupations: complexities of a liberation
    movement in an African Post-Colonial Settler Society. Unpublished PhD thesis, Wageningen
    University, The Netherlands
Sadomba WZ (2008b) Movements within a movement: complexities of Zimbabwe's land occupa-
    tions. In Moyo S, Helliker K, Murisa T (eds) Contested terrain: land reform and civil society in
    contemporary Zimbabwe. S&S Publishers, Pietermaritzburg, pp 144–180
Schmidt HI (2013) Colonialism and violence in Zimbabwe: a history of suffering. James Currey,
    Woodbridge
Sibanda EM (2005) The Zimbabwe African People's Union 1961–1987: a political history of
    insurgency in Southern Rhodesia. Africa World Press, Trenton, NJ
Sithole M (1980) Ethnicity and factionalism in Zimbabwe Nationalist Politics 1957–79. J Ethn
    Racial Stud 3(1):17–39
Sithole M (1983) Recent works on the Zimbabwe liberation movement. *Zambezia* XI(ii):149–159
Sithole M (1984) Class and factionalism in the Zimbabwe nationalist movement. Afr Stud Rev
    27(1):117–125
Sithole M (1999) Zimbabwe: struggles-within-the-struggle (1957–1980), 2nd edn. Rujeko
    Publishers, Harare
Staunton I (ed) (1990) Mothers of the revolution. Baobab Books, Harare
Tavuyanago B (2013) The Crocodile Gang Operation: a critical reflection on the genesis of the
    second Chimurenga in Zimbabwe. Global J Human-Soc Sci 13(4):27–36
Tekere EZ (2007) A lifetime of struggle. SAPES Books, Harare
Tendi BM (2017) Transnationalism, contingency and loyalty in African liberation armies: the case
    of ZANU's 1974–1975 Nhari Mutiny. J South Afr Stud 43(1):143–159
Tongogara J (2015) Tongogara in his own words. African Publishing Group, Harare
Tungamirai J (1995) Recruitment to ZANLA: building up a war machine. In: Bhebe N, Ranger T
    (eds) Soldiers in Zimbabwe's liberartion war (volume one). James Currey, London, pp 36–47
White L (2009) "Heading for the Gun": Skills and sophistication in an African Guerilla war. Comp
    Stud Soc Hist 51(2):236–259
White L (2014) Students, ZAPU, and special branch in Francistown, 1964–1972. J South Afr Stud
    40(6):1289–1303

Chapter 5
The Second *Chimurenga*: Guerrillas-Peasants, Spirituality and Patriarchy

Abstract In the light of the previous chapter, this chapter provides a further analysis of the second *chimurenga* by identifying and discussing a number of themes. It does so primarily in order to set a comparative basis for analysing the third *chimurenga*. Crucial in this regard is the relationship between guerrillas and villagers during the second *chimurenga*, as this speaks to the significance of the relationship between war veterans and occupiers during the fast track occupations (third *chimurenga*). One the key themes covered in the recent literature is spirituality, with an increasing move away from examining spiritual mediums and other traditional forms of spirituality, to an examination of the relationship between Christian missions, guerrillas and villagers. As well, in the Rhodesian countryside, a number of local patriarchal systems existed, which included chiefs in Native reserves and white farmers on commercial farmers. Whether intentionally or not, the very presence of male guerrillas challenged the authority of these patriarchs. Finally, in the context of patriarchy, the chapter considers the multi-faceted experiences of different groupings of women during the war of liberation.

Keywords Second *chimurenga* · Chiefs · Spirit mediums · Guerrillas · Patriarchy · White farmers

5.1 Introduction

This chapter provides a further analysis of the second *chimurenga* by identifying and discussing a number of themes, some of which were already apparent in the early scholarly literature on *chimurenga*. While the chapter seeks to cover the main themes in the literature, it does so primarily in order to set a comparative basis for analysing the third *chimurenga*. Crucial in this regard is the relationship between guerrillas and TTL villagers, as this theme speaks to the significance of the relationship between war veterans and occupiers during the fast track occupations, with the majority of occupiers coming most likely from communal areas (or former TTL areas). As well, one of the key themes in the war of liberation literature over the past couple of decades is that of spirituality, with an increasing move away from considering spiritual mediums and other "traditional" forms of spirituality, to an examination of the

© The Author(s), under exclusive license to Springer Nature Switzerland AG 2021
K. Helliker et al., *Fast Track Land Occupations in Zimbabwe*,
https://doi.org/10.1007/978-3-030-66348-3_5

relationship between Christian missions, guerrillas and TTL communities. This is one area (i.e. Christian missions) which has not received attention with regard to the third *chimurenga*, possibly because of its limited significance. As well, in the Rhodesian countryside, a number of local patriarchal systems existed, which included chiefs in TTLs and white farmers on commercial farms. Whether intentionally or not, the very presence of male guerrillas challenged the authority of these patriarchs. Finally, in the context of patriarchy, we consider the multi-faceted experiences of different groupings of women during the war of liberation.

5.2 Guerrillas and Peasants

In the last chapter, we considered the relationship between the nationalist and guerrilla movements, and the guerrilla armies themselves. In this first section, we examine what took place at the war fronts, particularly the relationship between guerrillas and rural communities.[1] This is pursued later as well in relation to questions around women and spirituality in the war of liberation.

It is important first to reiterate that there were significant spatial and temporal variations in terms of the interactions between guerrillas and TTL villagers, as raised in the preceding chapter. Thus, in discussing studies undertaken by Ranger, Lan and Kriger (and by extension, others) about differing mobilisation practices and tactics pursued by guerrillas, Alexander (1995) notes that this must be understood in terms of localised dynamics and contingencies. All areas of the Rhodesian countryside with reference to guerrilla-peasant interactions require a focus on villagers' "experience of the war (both its intensity and its longevity), their geography, their political and religious institutions, and the nature and extent of their incorporation into the colonial economy and polity" (Alexander 1995: 177). Any divergences in guerrilla mobilisation (as seen in the work by Ranger and others), including forms and levels of coercion, may reflect spatial and historical diversity rather than simply the character of the guerrilla army or army unit present in a particular rural area.

Nyachega (2017: 77, 78) demonstrates this with reference to the frontier area of Honde Valley, arguing that the valley "remained on the fringes of the colonial state" up until the 1950s (with minimal state intrusions), such that a "strong anti-colonial consciousness" had not arisen. This led to the guerrillas having to rely, at least initially, on coercion as a means of mobilisation amongst local peasants. In a deep ethnography of the Honde Valley as well, not unlike those provided by earlier scholars such as Lan and Kriger, Schmidt (2013) is able to identify and examine the local history and dynamics which conditioned rural-based struggles in the valley. In doing so, she confirms one of the key points of this manuscript about the *zvimurenga*, namely that local struggles are not reducible to nationalist-inspired narratives. For instance,

---

[1] A decade ago, Ranger (2010: 244) argued, at least in relation to Bulawayo, that "[h]istorians have not yet explored the interactions between the cities and the guerrilla war and there will doubtless be a hidden story to tell". We are not aware of any subsequent literature to date.

even in the 1960s in the case of agitation around land, "grievances were expressed in vernacular mode, not in nationalist discourse" (Schmidt 2013: 119). Likewise, in the case of the war of liberation, support for the guerrillas most probably entailed "a radically local agenda of defending a frontier mode of life instead of nationalist mobilisation" (Schmidt 2013: 2).

It may also be the case that neither diversity across areas nor differences between armies were the crucial determining factor in mobilisation tactics. Temporal issues may be the key factor, as indicated in the previous chapter. Thus, in order to kick-start the war so to speak, there was significant forced recruitment by both ZANLA and ZIPRA in the early years. As Mhanda (2011: 18) notes, in order to obtain recruits initially, guerrilla armies "had to resort to various unorthodox means, i.e. abductions, kidnapping and luring recruits with promises of jobs and education". It is hence crucial to be sensitive to the possible periodisation of the war (as discussed in the previous chapter) in terms of its altering intensity as well as changes in the strategies and tactics of guerrilla armies with, for instance, many areas of the countryside experiencing a deterioration of relations between guerillas and TTL communities in the last couple of years of the war (2013). However, to emphasise, temporal dynamics were often locally specific.

Tungamirai (1995), a guerrilla commander, in fact argues that recruitment went through three phases—voluntary recruitment (of exiles in neighbouring countries), then press-ganging (by both ZIPRA and ZANLA) and then voluntary recruitment again (and now from within the country). But, as Douglas Moore (2005: 192) argues in his study of situated practices and struggles in Kaerezi in Nyanga along the Mozambican border: "Whig histories often suggest progressive teleologies that unfold, inexorably, towards triumphalist futures. Metaphors like the "rising tide of African nationalism" lend nature's inevitability to history's progress. Kaerezi's situated struggles, in contrast, emerged from contingency rather than certainty". The importance of local contingencies and situated practices, across both time and space, imply that any temporal changes in guerrilla mobilisation strategies (or any broader changes in the war of liberation) did not occur in any fixed linear sense, at least certainly not at a national level.

It is also the case that, while ZANLA and ZIPRA may have had (generally speaking) different methods of mobilisation, these forms of mobilisation were subject at times to change arising from direct encounters between different guerrilla armies in bounded territories. Hence, as Tungamirai (1995) indicates, both ZANLA and ZIPRA operated in parts of Midlands, Matabeleland South and Mashonaland West, and contested strongly for recruitment in these areas. The presence of both armies in the same area, sometimes simultaneously and as separate forces, was in fact not uncommon, including in Matopos (Bhebe and Ranger 1995) and Mberengwa (Bhebe 1999). The stories of women from different TTLs, as Staunton's interviews show (Staunton 1990), also highlight the presence of ZANLA and ZIPRA guerrillas operating in the same areas at the same time. On occasion, units from the different guerrilla armies established boundaries to mark out their respective zones of operation but, at other times, they entered into direct conflict. There was fighting between ZANLA and ZIPRA units, including in the ZANLA stronghold of Maranda Tribal Trust Lands which led to the deaths of between five and ten guerrillas in September 1977 (Fontein 2015).

One key issue raised in the war of liberation literature is alleged differences between the different guerrilla armies, including in relation to logistical support within TTL communities. For instance, it is claimed that ZIPRA was able to draw upon existing ZAPU political structures, while ZANLA had to ensure that political committees supportive of ZANU were established. Ndlovu-Gatsheni (2017) notes that the latter was particularly the case throughout Mashonaland. This meant that ZANLA first had to dismantle well-established ZAPU structures in operational zones before it could form its own and effectively start fighting. In this regard, ZIPA guerrillas in the mid-1970s observed that ZANU lacked ZAPU's organisational continuity (based on its political structures from the early 1960s) as well as well-structured coordination between its guerrillas and TTL villagers (Ranger 1985; Mazarire 2011).

Below we first discuss the ways in which ZIPRA guerrillas were able to draw upon existing local ZAPU structures in mobilising peasants. For example, Alexander and McGregor (2017: 58) argue that ZAPU's long-standing committee structures in Rhodesia even meant that potential guerrillas were often recruited in Rhodesia through the party and were politicised before they joined the armed struggle. Brickhill (1995) confirms this, noting that ZAPU had significant support in most parts of the countryside, certainly in areas where ZIPRA came to operate, but also in areas where ZANLA and ZIPA operated. With the first ZIPRA recruits coming from refugee camps in neighbouring countries, about 60% of recruits had been members of functioning branches of the ZAPU Youth League before they left the country. Most of these were from towns and cities as ZAPU clandestine structures survived more fully in urban than in rural areas. Many recruits had parents who were involved in illegal party structures.

Alexander and McGregor (2004), in providing retrospective war stories of ex-guerrillas from ZIPRA, refer back to their earlier work with Ranger on memory and violence in the deep forests of Matabeleland (Alexander et al. 2000). In this earlier work, accounts by villagers of the war stressed ZAPU's lengthy and uninterrupted political history. This, Alexander et al. (2000) argue, gave authority to unarmed villagers in their relationships with ZIPRA's young, AK-wielding soldiers. Thus, a deep history of local and rural nationalisms existed, including the enduring presence of ZAPU branches or cells in the countryside. This ZAPU-inspired nationalism often facilitated the establishment of close relationships between ZIPRA guerrillas and rural communities. Alexander et al. (2000: 160) thus emphasise:

> The importance and depth of explicitly nationalist commitment, and the crucial role of the rural party [ZAPU] in mediating between guerrillas and civilians; nationalism was the legitimising ideology of the war in the Shangani, and nationalist identifications and institutions were strengthened and broadened over time.

But, in line with the later literature on the second *chimurenga*, they note that "there were important divisions and differences amongst rural social groups, between them and [ZIPRA] guerrillas, and within guerrilla ranks" (Alexander et al. 2000: 160). The basis for these differences, and the ways in which they were handled by both villagers and guerrillas, had important and varied implications for the kinds of relationships formed between specific rural groupings amongst the peasants and ZIPRA guerrillas.

The accounts of the ex-guerrillas (Alexander and McGregor 2004) highlight, in their collaboration with villagers, the constant and not always successful efforts to spread the material demands of guerrillas equitably amongst members of the local community, to broaden and unify nationalist mobilisation, and to configure and enforce acceptable uses of violence. In this regard: "Civilian ZAPU leaders elaborated norms of law and order, they sought to enforce a moral economy of supply, and they struggled to establish a socially inclusive political practice" (Alexander and McGregor 2004: 83). According to the ZIPRA guerrillas, there was no need to make use of *pungwes* (all-night vigils) as they could simply rely on the long-established ZAPU committees. Guerrillas were typically equipped with the name of a local party leader before arriving in an area, and were readily accepted once their identity was confirmed.

This was unlike ZANLA guerrillas, who might have resorted to violence to assert themselves. Because ZAPU party structures survived, thereby attracting recruits for the guerrilla army and organising logistical support for guerrillas, ZIPRA did not have to make new claims for legitimacy as ZANLA had to do (Brickhill 1995). Brickhill argues that this meant that ZIPRA did not need to select and use *mujibas* and *chibwidos* (young females) for support, or to appeal to non-political forms of legitimacy such as spirit mediums. However, Sibanda (2005) claims that *mujibas* (or *umjibha*) were central to ZIPRA's intelligence-gathering system, conveying messages between underground ZAPU cells and guerrillas, and also politicising local populations. Further, Alexander et al. (2000) note that ex-guerrillas, on reflection, tend to emphasise that party structures had to be formed or revived, while party leaders speak of an uninterrupted heritage of the structures well into the 1970s.

Certainly, where ZAPU never existed, local structures had to be established. Thus, in areas where ZIPRA guerrillas first operated such as Hwange and the north-west more broadly, where no political organising took place during the open nationalist period, political structures had to be formed. Further, ZIPRA likely had to rebuild structures even in well-organised areas. Plus, there were instances where ZIPRA guerrilla units did seek legitimacy through Matopos shrines and mainstream churches. Overall, though:

> ZIPRA guerrillas relied a great deal on party branches and party contacts for support during the war. Food, medical supplies, transport and intelligence was provided to the guerrillas by their local party contacts. [And, as noted], [w]hen ZIPRA guerrillas were about to enter a new operational area, it was customary for the ZIPRA commissar to seek out the local party contacts before the guerrillas were deployed in the area. (Brickhill 1995: 68)

If no party structure existed, the party commissar sought out individual party members and, together, they resurrected defunct party branches. Additionally, "[h]y operating through the party structures the guerrillas were able to avoid being drawn into … local conflicts and divisions" (Brickhill 1995: 70).

Alexander et al. (2000) examined Lupane and Nkayi in the Shangani areas of Matabeleland where, from the protection of the forests, ZIPRA guerrillas would launch attacks. They agree with Brickhill that ZIPRA guerrillas could rely on local ZAPU structures in Shangani. Nkayi and Lupane both had long-established and close links with urban areas, so there were urban–rural networks facilitated by labour migrants and bus operators. In Nkayi and Lupane, party structures had survived the repression of the 1960s and 1970s but to varying degrees. At the same time, during the mid-1970s, "local ZAPU leaders' efforts to expand branch and district committees, and to improve communication with the central party through provincial levels, was greatly stimulated by the arrival of small groups of guerrillas" (Alexander et al. 2000: 160). But, they go on to argue that guerrillas did in fact "appeal to" spiritual authorities like spirit mediums "in order to deal with injury and illness, fear and danger, and ZAPU's rural nationalists were themselves not so secular" (Alexander et al. 2000: 160).

McGregor (2009) examined specifically the Tonga in Binga in the Zambezi, including in light of their forced displacement due to the building of the Kariba dam. The prevailing Tonga cultural nationalism articulated with ZAPU's overall political message, with ZAPU organisers (in opening up branches for ZAPU) stressing: the marginalisation of Tonga language and culture, and the horrific effects of the Kariba displacement (as these were the issues central to the Tonga people), as well as the broader anti-colonial message based on national-based grievances. McGregor (2009: 130) notes that "[t]he process of nationalist mobilisation in the borderlands [along the Zambezi] is notable for its similarities with other ZAPU/ZIPRA areas, where civilian party structures were set up in advance of guerrilla incursions". So, at least in Binga (though less so in neighbouring Hwange), ethnic and nationalist mobilisation proceeded together and were closely linked. ZAPU committees and activists in Binga close to the Zambian border facilitated and supported guerrilla crossings of the Zambezi River, as crossings needed the local Tonga people's intimate knowledge of the river. Also, there was a network of branches of ZAPU in Binga which would host ZIPRA guerrillas.

Overall, then, much of the literature on ZIPRA stresses its existing political structures and its non-coercive modes of mobilisation. Where pre-existing support structures did not exist, local villagers—whether based on coercion or voluntary cooperation—soon established support committees for the guerrillas. To the contrary, in the case of ZANLA, pre-existing support systems rarely existed, at least not those aligned to ZANU. At times, in entering an area where ZAPU structures existed, ZANLA guerrillas sought to literally eliminate ZAPU activists (Staunton 1990). ZANLA, as discussed in the previous chapter, made use of spirit mediums and other spiritual authorities as an entry point into TTL communities and as a means of ensuring local support amongst villagers.

Guerrilla discipline amongst ZANLA forces, as it did no doubt amongst ZIPRA forces as well, varied over time and across space, and between guerrilla units. Mhanda (2011) claims, for example, that ZIPA had to discipline certain guerrilla units for intimidating and beating villagers, as such action went against the expected conduct of guerrillas. He likewise claims that, after ZIPA, there was a loss of discipline amongst ZANLA fighters, particularly by harassing peasants in those war zones with intense fighting. In the case of ZIPRA, at least in the Shangani areas of Matabeleland,

ZAPU branch chairmen at times had to remonstrate with guerrillas (and local youth) about the burning of schools and dip tanks, as this was considered as vandalism and such infrastructure was important to the lives of villagers (Alexander et al. 2000) including in a post-colonial future. In the case of the Honde Valley, by 1978 and 1979, insufficiently trained "rogue guerrillas" (Schmidt 2013: 189) were rampant.

In a manner which resonates with the second *chimurenga*, war veterans (as ex-guerrillas) during the third *chimurenga* engaged with villagers in the communal areas (ex-TTLs) in seeking to mobilise occupiers. They had to make appeals in mobilising occupiers and, once on a farm, they had to establish forms of collaboration and sustenance to maintain the occupying force. Likewise, appeals were made by the guerrillas in trying to stir TTL villagers into supportive action. On occasion, urban workers linked up with their TTL relatives in supporting guerrillas, at least in the case of TTLs near cities and towns (such as Chiweshe TTL). These workers would travel to their TTL homesteads on a regular basis to obtain a list of requirements for the guerrillas, and return later with the necessary supplies. Though the appeals during the war of liberation were in large part founded on a nationalist narrative (as it was during the fast track land occupations), villagers were not moved by nationalism alone.

Certainly, guerrillas sought to draw upon the historical memories and social-spatial experiences of peasant villagers throughout the countryside. This did not necessarily mean that a fully formulated undifferentiated nationalism existed within TTL villages, and that a broad reference to the nationalist project was sufficient for gaining traction amongst villagers. In his study of Vhimba (now Vumba), Hughes (2006: 72) argues that, during the war of liberation, "Vhimba residents embraced a narrow, hectare-focused form of nationalism". Those who inhabited the national parks in the area remember the war as mainly a squatters' movement against state-backed conservation. Rhodesia had simply ceded this indefensible strip of low-lying territory along the Mozambican border to the guerrilla army, with ZANLA guerrillas occupying Vhimba in 1978 and using it as a base for attacks on white farms. This was in effect a semi-liberated zone, in which relations between guerrillas and villagers were highly cooperative, so that there was thus no "struggles within the struggle" as depicted by Kriger in Mutoko. Overall, "Vhimba residents concentrated on overrunning and overturning the boundary of Chimanimani National Park" (Hughes 2006: 72), with guerrillas promising them more land.

Likewise, Fontein (2015) considers the localised basis for involvement of villagers in the war of liberation and how guerrillas engaged with villagers on this basis. Local displacements had taken place because of the construction of the Mutirikwi (then Kyle) Dam which was completed in 1961. As Fontein argues (2015: 231–232): "The war engaged with African pasts not obliterated by the dam through the materialities of landscape, as sacred mountains and caves became guerrilla bases, and local clans agitated to return to lands, graves and ruins from which they were evicted sometimes only two decades before". Guerrillas established bases on the scared mountains (Fontein 2009), in part because they were difficult for Rhodesian security forces to access, and they launched attacks on white farms and tourist hotels from these mountains. In this way, similar to the argument by Douglas Moore (2005), "the war took place *in*, *with* and *against* the active and affective manifestations of past and enduring African and European remakings of Mutirikwi's landscape" (Fontein 2015: 240, emphasis in original). Coming as *"vatorwa* (strangers) into active autochthonous landscapes" (Fontein 2015: 242), guerrillas needed the knowledge and support of local villagers. This support was not reducible to sweeping nationalist claims about land reclamation, though these claims resonated in the main with local aspirations about spiritual landscapes.

Though considering the lives of political detainees and inmates, the work by Munochiveyi (2013, 2014) highlights the key point raised by many of the scholars cited in this section, that is, the importance of "people's *individual* passages within the meta-narrative of Zimbabwean nationalism" (Munochiveyi 2014: 10, emphasis in original). In this sense, a broad nationalist rhetoric as such did not appeal in unmediated form to TTL villagers, as if a homogenised category of people—"peasants"—acted upon a homogenous set of grievances in engaging with guerrillas during the war of liberation. Hence, it becomes necessary to bring to the fore the "personalised nature of their [villagers'] political commitments to African nationalism" and therefore to "understand the growth and proliferation of nationalism as it was understood, debated, and embraced by ordinary men and women" (Munochiveyi 2014: 31). These entailed deeply localised experiences and grievances of marginalisation and frustration, leading to the rise of what Munochiveyi (2014: 34) refers to as "personalised nationalisms", in which each nationalism "sprang from his [or her] own immediate situation—his own grievances, his own hopes, his own past, his own present" (Munochiveyi 2014: 42).

These points, we will show, are particularly pertinent to the local reasoning behind the land occupations during the third *chimurenga*, thereby contributing to understanding the occupations without necessarily referring to the ruling ZANU-PF's party machination and mobilisation. For the war of liberation, insofar as these embodied local nationalisms were prevalent in villages across the Rhodesian countryside, villagers had their own strong motivations for supporting guerrillas. But this also entailed the existence of motivations rooted in local discomfort, tensions and conflicts.

As Kriger (1988) highlighted in her work, intra-community tensions or "struggles within the struggle" at village level were quite common, and these were important with reference to the ways in which locals negotiated relationships with guerrillas and vice versa. Fontein (2015) notes this as well, arguing that "[l]ocal jealousies meant businessmen and shopkeepers were unusually vulnerable to accusations of witchcraft and/or selling-out: the two were often conflated" (2015: 245). The sell-out identity was constructed situationally, so that any villager could potentially be labelled a sell-out, even if seen simply in the vicinity of Rhodesian security forces, with the assumption that the villager is an informer. The dilemmas faced by villagers, in trying to negotiate their way through the war under very trying circumstances, is captured (as noted previously) by the Catholic Commission for Justice and Peace in Rhodesia, with its notion of "the man [or woman] in the middle" (CCJPR 1999). As the president of the commission, Bishop Lamont, notes in his introduction to the CCJPZ report by way of quoting from one villager: "If we report to the police, the terrorists kill us. If we do not report, the police torture us ... We just do not know what to do" (CCJPR 1999: 3). Daneel also speaks about the "divided loyalties of the rural population", and argues that the actions of guerrilla fighters under the stress of war were "sometimes coercive and discriminatory" (1995b: 9) in relation to peasant villagers.

In the case of Hurungwe district, Marowa (2009) considers the diverse constructions of the sell-out identity and how so-called sell-outs were often tortured and killed by ZIPRA guerrillas. Alleged sell-outs included those employed in Rhodesian state departments, and women who searched for tobacco at a nearby white farm (hence, considered as fraternising with the enemy). They even included villagers who crossed the Musukwe River, which ZIPRA had declared as a moral and military border, with those undertaking an eastward crossing labelled as likely flirting with the enemy (as Rhodesian security had a base at Kenyungo Mountain to the east of the river) (Marowa 2015). Importantly, as well, ZIPRA considered villagers who—perhaps out of jealousy—supplied false information about others (so that the latter would be declared as sell-outs) as themselves sell-outs for misleading the guerrillas.

In their study of Lupane and Nkayi, Alexander et al. (2000: 168) argue as follows about sell-outs worthy of vengeance:

Whether, and which, [Rhodesian] civil servants should be considered legitimate wartime targets, and how they should be killed, were ... a source of debate amongst guerrillas and local nationalist leaders. ZAPU leaders, often in alliance with guerrilla commanders, generally acted to contain killings; local nationalism tended to be incorporative and humane, to prefer personal rather than categorical evaluations of loyalty.

Accusations of selling-out were therefore deeply controversial, with a distinction often drawn between sell-outs and witches. In this respect, both ZIPRA and ZAPU leaders strove to "control violence against witches precisely because it undermined support for guerrillas" (Alexander et al. 2000: 173). So, unlike what Lan (1985) tends to claim, killings of witches and sell-outs by guerrillas was not necessarily popular with villagers in Lupane and Nkayi. There were genuine sell-outs (liaising directly with the enemy) and killing these sell-outs was justified by ZIPRA and ZAPU leaders. But, as noted, the proof underlying such claims were often flimsy. Generally, the killing of witches was not allowed by ZIPRA but this often happened (again on flimsy rationales) and sometimes with the approval of the local ZAPU chair.

Tribal Trust villagers often recognised that accusations of selling-out and witchcraft were based "on individual animosities motivated by greed and jealousy. Such animosities were very local and personal, sometimes intra-familial" (Alexander et al. 2000: 173). Accusations were quite arbitrary such that, unlike Kriger's position, these could not be neatly reduced to class, gender or generational conflict (though they often did have such elements). For example, better-off peasants were thought to be receiving money from the Rhodesian security forces for acting as informers. Women in particular were vulnerable, if only because large numbers of men had left the rural areas, or stayed in urban centres, for a variety of reasons. In this respect, women note how jealousies and tensions played themselves out during the war, and how stories about others were often fabricated and then reported to the guerrillas.

Staunton (1990: 77) records the thoughts of Thema Khumalo, who said: "The sell-out business was a way that some people used during the war to get rid of others they didn't like". In a similar way, Meggi Zingani highlighted: "People who were jealous also caused problems. Some people did not like others who were better off than themselves" (Staunton 1990: 127).

The fact though that the Rhodesian government set up protected and consolidated villages is testimony to the support received by guerrillas, irrespective of the basis for this support. The first protective villages arose in the second half of 1973 in the Zambezi Valley. In what is now Mashonaland Central Province, between 43,000 and 47,000 villagers were moved from their homesteads on Chiweshe Tribal Trust land into protected villages, and another 13,500 in Madziwa Tribal Trust lands (CCJPR 1999). At the same time, peasant villagers resisted their forced removal to these villages. One of the main reasons given by government for these villages was "to isolate civilians from the insurgents and so to starve the latter into surrender" (Weinrich 1977: 220). The tactic of starving the guerrillas, as noted, was also used by the BSAC to thwart the rebellion during the first *chimurenga*.

As well, the Rhodesian security forces destroyed any crops outside the villages (called "keeps") to cut the guerrillas off from any food source (Weinrich 1977). In his examination of southeastern Zimbabwe, where the Gonakudzingwa Restriction Camp was located, Hove (2012) also speaks about the scorched earth policy of Rhodesian security forces, by which they destroyed crops of villagers even living outside protected villages, thereby seeking to force guerrillas into submission or cause villagers to withdraw support from the guerrillas. In response, people fled into inaccessible areas such as the Musimbiti Forest of the Gonarezhou National Park to escape security force harassment.

For those in "keeps", as Weinrich (1977) notes, the conditions were quite horrific with women subject to sexual abuse by the District Assistants overseeing the villages. Sometimes guerrillas climbed the fences of villages or crawled under them to receive food and other assistance, and at times they attacked and burnt villages. Furthermore, it was not unusual for suspected guerrilla sympathisers to be interrogated and tortured in protected villages (Chenaux-Repond 2017). But those confined to protected villages, as throughout the Honde Valley, showed significant initiative and ingenuity in maintaining their support for guerrillas (by for example, women in particular smuggling food to them). They also engaged in everyday livelihood activities including illicit beer-brewing and informal trading (Msindo and Nyachega 2019). The Rhodesian government sought to "soften" the effects of living in protected villages through the work of female Development Workers and Women Advisors employed by the Ministry of Internal Affairs (formerly the Native Affairs Department) (Chenaux-Repond 2017).

5.3 Spirituality, Guerrillas and Peasants

As noted, from the start there was a significant focus in the second *chimurenga* literature on spirit mediums and guerrillas, and only later did a clearer emphasis on spirituality other than spirit mediums arise. Arguments about the significance of spirit mediums continue to be significant in the literature, but there has been a discernable shift away from an exclusive focus on them, including a stronger focus on Christian missions and guerrillas.

Daneel (1995a, b) for instance not only considers spirit mediums but also examines the *Mwari* cult with reference to the second *chimurenga*. Daneel (1995a) argues that the *Mwari* cult was kept intact after the first *chimurenga* but it worked underground with a range of networks remaining in existence. With the rise of African nationalism and the visits of Joshua Nkomo and other nationalist leaders to the Matonjeni shrines, "the cult's revival [in the 1960s] signalled rising expectations of *Mwari*'s involvement in the pending liberation struggle" (Daneel 1995a: 226). As with the first *chimurenga*, Daneel in fact posits direct involvement of the *Mwari* cult during the war of liberation: "The gist of Mwari's messages throughout the struggle was: full condonation of militancy and support for the ZANLA and ZIPRA fighters who were trying to regain the lost lands" (Daneel 1995a: 227).

Guerrillas regularly requested Mwari's power and guidance in the war with, for example, those guerrilla fighters (both ZANLA and ZIPRA) operating in or near the Matopos area consulting the oracle and receiving strategic directives. As well, guerrilla fighters from further afield (at a distance from Matopos) visited the shrines, and cult messengers collaborated with guerrillas in their home districts. Like Ranger's argument about the first *chimurenga*, spirit mediums networked with the Cult during the war of liberation, with mediums sometimes visiting the Matonjeni shrines to confer with *Mwari* on *chimurenga* issues. In this way, the Mwari cult had a far-flung influence with *Mwari*'s directives even feeding into "the wider communication system for the attention of ZANLA's high command in Chimoio" (Daneel 1995a: 220) in Mozambique.

In the case of spirit-mediumship specifically, mediums advised guerrillas who operated within the borders of their own spirit provinces, or were even drafted into guerrilla units and moved far and wide with the fighters on the battle front. Mediums also insisted that guerrillas make direct contact with the Matonjeni cult centre. In his work titled *Guerilla Snuff*, which is a semi-fictional account of the war of liberation, Daneel (1995b) highlights the relationship between ZANLA guerrillas and mediums specifically in the Gutu area. Guerrillas were offered, and took snuff, from mediums in recognition of the protective powers of the spirit world. Douglas Moore (Moore 2005: 245) notes that along the Mozambican border, medium-guerrilla interaction was significant: "Beyond warning guerrillas about the presence of Rhodesian forces, ancestral spirits guided fighters through forests, helped hunters track wildlife, and offered spiritual and material assistance during the war". At the same time, Daneel argues that chiefs were not invariably bypassed by the guerrillas.

Daneel's work shows the important of *Mwali/Mwari* cult shrines to ZANLA guerrillas in Shona-speaking Masvingo district and beyond, including the Matonjeni shrine in the Matopos reserve. Thus, even in Matabeleland, ZANLA turned to the shrines, if only because they could not draw upon existing party structures, like ZIPRA could. In this regard, as Ranger and Ncube (1995) argue, Daneel has always seen the Matopos shrines as a Shona legacy in the zone of Ndebele conquest. But Daneel does not address spirituality and ZIPRA in Matabeleland. Twenty years ago, Bhebe (1999: vii) expressed deep concerned "about the way only the ZANU side of the struggle was unfolding [or being written about] while the ZAPU side remained an almost completely unchartered territory". This is, he argues, a key reason why there had been such a strong focus on ZANLA, spirit mediums and guerrillas while, in his area of interest (Matabeleland broadly), the *Mwari* cult is very pervasive. In this light, Ranger and Ncube (1995) examine war and spirituality in southern Matabeleland.

Ranger and Ncube (1995) note that the literature until the mid-1990s had seemingly concluded that, for Matabeleland, "traditional religion" had played no part in the guerrilla struggle while Christian mission churches were invariably attacked or abandoned. They consider one possible explanation for this, namely that the branches of ZAPU had continued to exist after the organisation's banning in the 1960s, as we discussed earlier. Hence ZIPRA could draw upon support and recruits through ZAPU's elders along with the party's youth league and women's organisation, all of which was based on an entrenched ideology and practice of rural nationalism. Meanwhile, as indicated, ZANLA had to construct its own support networks and regularly used mediums in legitimising its presence amongst villagers. They argue however that, during the 1960s, there was in fact a "fusion of urban cultural nationalism and the rural revival of the Mwali cults" (Ranger and Ncube 1995: 43). ZAPU nationalist ideology incorporated the power and discourse of African religion without this necessarily involving "merely a co-option of the reputation of the cult by myth-making nationalists" (Ranger and Ncube 1995: 44). So, by the time of the war of liberation, it might be expected that ZIPRA guerrillas would show respect to *Mwari* shrines given the populist (religion-infused) nationalism in rural areas of Matabeleland.

Ranger and Ncube (1995) show that ZIPRA, like ZANLA, guerrillas drew upon the shrines as well, and that they supported the religious–cultural nationalist programme of local villagers. Thus, there were local denunciations to the guerrillas about "sell-outs" who refused to obey the injunctions of the *Mwari* cult. Along with the guerrillas, pressure by cult adherents was placed upon the congregations of the Christian mission centres which had been deliberately sited close to the shrines. In southern Matabeleland, by 1979, the missionary presence had almost completely vanished, with mission stations and schools closed and Catholic and other missionaries killed. Sibanda (2005) though cites evidence of rural churches in southern Matabeleland, such as the Brethren in Christ Church and the African Episcopal Church, providing sanctuary and support to ZIPRA forces. Insofar as there was guerrilla hostility to certain missions, this was likely because local villagers were hostile to them, just as guerrillas had to take the *Mwari* cult seriously because people did.

Again, this shows the pragmatic and tactical perspective adopted by guerrillas when it came to peasant mobilisation, as discussed in the previous chapter. As Bhebe

(1999) notes more broadly in his study of Christianity and guerrillas, "[t]he guerrillas, who were always pragmatic and used whatever institutions were powerful enough to advance their war efforts, quickly cooperated with the *mhondoro* and used their influence to mobilise the peasants effectively" (Bhebe 1999: 37). Further, the attitude of local villagers to a particular mission was critical to what happened to that mission during the war (Ranger and Ncube 1995).

In southern Matabeleland, as in other parts of the country, there were many reasons why tensions existed between villagers and missions, with land being the major grievance. But a strong reliance by certain missions on police for protection of property also alienated locals. Police protection in southern Matabeleland arose in part because of the closeness to the Botswana border, with guerrillas launching attacks in Rhodesia and then crossing the border. However, churches in this region were marked by a theological conservatism and simply did not support guerrillas as they did for example in eastern Zimbabwe. In examining Lupane and Nkayi in the Shangani areas of Matabeleland, Alexander et al. (2000) confirm guerrilla apathy to mission Christianity. ZIPRA guerrillas closed churches, prevented worship and cast Christian missionaries as sell-outs, despite their often-Christian backgrounds. Political education undoubtedly had an effect on the worldviews of guerrillas, but this caused problems with many local ZAPU leaders who were avid church attendees. While they remained open, Catholic missions did provide medicine, clothing and other supplies to guerrillas (Alexander et al. 2000).

Much of the literature on spirituality and the war of liberation, as indicated, has focused increasingly on Christian missions. The role of Christianity in the third *chimurenga* occupations remains in large part unknown. Certainly, as Bhebe (1999: 73) argues with regard to the war of liberation: "In academic discussions so far particular attention has been devoted to the remarkable interaction between guerillas and traditional religious institutions". There were very diverse relationships during the war between missions, local villagers and guerrillas, which were contingent on local histories of mission-village relationships as well as the attitudes of guerrilla units entering a particular area. The ensuing discussion of Christian-related churches seeks to, once again, emphasise the sheer levels of contingency, fluidity and diversity within the war of liberation, which cannot be captured by any monolithic account of the second *chimurenga*—a point central to our argument about the third *chimurenga*.

In *Guerrilla Snuff*, Daneel (1995b) considers the role of the African independent churches (Zion-Ndara Apostolic church and Marange's Apostolic church) and its prophets during the second *chimurenga*, who preached and baptised with holy spirit. The Zion prophets would declare to the guerrillas: "Take to the mountains and the caves. There the God who led you out of Egypt, the one who started showing us the way to freedom already in the days of Mkwati, Kaguvi, Mashayamombe, the heroes of Zimbabwe's first *chimurenga*, will keep you safe" (Daneel 1995b: 189). The Zion Apostolics provided food to the guerrillas and even at times identified witches who would then have to face the wrath of guerrilla commanders. Further, sick guerrillas would come to the prophets for prophetic diagnosis (including the whereabouts of the Rhodesian security forces), for the laying on of hands for guidance and strength, and for the sprinkling of their AK rifles with holy water. However, very heavy demands were placed on the church by the guerrillas and, on occasion, the prophets had to protect villagers from aggressive guerrillas. Other African-initiated Christian churches, such as what is now known as ZAOGA (founded by Ezekiel Guti), rejected the cultural nationalism of the nationalist movements, the violence of the guerrillas and the ongoing prevalence of ancestral worship. But they remained in large part neutral during the war of liberation. As a mainly urban-based church, and "[w]ith few church buildings in rural areas, ordinary ZAOGA members continued to meet and pray in their homes, usually unhindered by the guerrillas" (Maxwell 2006: 99).

The history of mainstream mission churches in Zimbabwe is marked by considerable controversy but also, as indicated, by remarkable diversity. Many churches, at least before the second *chimurenga*, had a long and pronounced relationship with the colonial state. This dates back to the time at which missions received considerable tracts of land. For instance, by 1925, the British South African Company had given 325,730 acres of land to church missions and the missions had purchased an additional 71,085 acres (Murdoch 2015: 12). There is evidence of considerable support by Christian missions for the Rhodesian government, including by the Salvation Army to the Smith regime, if only because the regime contributed funds to its churches, schools and hospitals. In 1971, the Salvation Army broke with the Christian Council of Rhodesia (aligned with the World Council of Churches) over its support for African land rights, liberation and majority rule. However, in the case of the Salvation Army, "its African membership and many of its missionary officers increasingly protested this stand" (Murdoch 2015: 129).

In this regard, as Bhebe (1999) notes, the leadership of many Catholic and Protestant churches in Rhodesia eventually came out in partial or full support of the liberation war and many local Christians and their parish leaders did likewise. Catholic Bishop Lamont was the most prominent missionary to do so (Linden 1980). Others though, including some mission stations in the Bulawayo Diocese of the Catholic Church, "maintained cordial relations with [Rhodesian] Security Forces" (Linden 1980: 242), and indeed relied on protection from nearby securitised white farmers. It was often the mission stations themselves, in diverse places across the countryside, which were at the coalface of the guerrilla struggle, as was the case in the Honde Valley in eastern Zimbabwe (Nyachega 2017).

Certainly, during the war of liberation, some missionaries were killed. For instance, guerrillas (presumably from ZIPRA) are suspected to have killed two British Salvation Army women missionary teachers at Usher Secondary School near Figtree, east of Bulawayo, in June 1978 (though the Selous Scouts of the Rhodesian security apparatus may have been responsible) (Murdoch 2015). Additionally, many African Salvationists were killed, sometimes brutally, because of their supposed sell-out involvement in the Salvation Army (and sometimes specifically because they would not deny their faith) but also because they were members of the Rhodesian army. The challenge faced by the Salvation Army was exacerbated by the fact that white commercial farms surrounded Usher's primary and secondary schools, so that the mission was in part isolated from, and without crucial connections to, villagers in the area with, as well, no Salvation Army church for local villagers at Usher.

In the end, though, history and local contingencies influenced relationships between guerrillas and mission churches (Murdoch 2015), including a local mission's "standing within the rural community" or the "quality of relationships with local people" (Griffiths 2017: 145). Again, as we show more fully below, the variegated character of the war of liberation comes to the fore.

Bhebe (1999) examines the history of the Evangelical Lutheran Church in southwestern Rhodesia (notably Mberengwa and parts of Mwenezi, Beitbridge and Gwanda), which was the dominant Christian denomination in the area, including in relation to the provision of schools and health facilities. Bhebe's emphasis is on mission pastors, teachers and headmasters and their relationship with both ZANLA and ZIPRA guerrillas, which was complicated by the fact that the two guerillas forces "became involved in severe confrontations" (Bhebe 1999: 121). By the 1970s, the Lutheran church had experienced a high degree of indigenisation especially in the localising of its mission officials and workers. The Bulawayo-based church leadership was alienated and detached from its rural parish base, with even the bishop "effectively distanced ... from the majority of his parishes throughout the war" (Bhebe 1999: 139).

Overall, the central church council remained neutral vis-à-vis the main contending forces in the war, such that pastors, evangelists, teachers and health workers attached to local missions (most of whom had permanent homes in the area) determining the conduct of the church during the war. In this sense, the church's working arrangements with both security and guerrilla forces were forged at parish level. Hence, "it was ... the courageous and wise lay-persons who influenced and directed local relationships between congregational and parish structures and the liberation movements" (Bhebe 1999: 141). The organisational form of the church, involving a degree of de-centralised authority, ensured that local missions could act in this way without seeking approval from the council or bishop. In this respect, Bhebe (1999) goes on to show significant diversity between missions in the Western and Eastern deaneries in terms of mission-guerrilla-villager relationships. The pragmatism of guerrilla units also becomes clear, including the presence of anti-Christian guerrilla units liaising with spirit mediums which nevertheless did not hinder ongoing parish worship.

Two other studies are worth discussing to highlight contingency and diversity further.[2] Maxwell (1995) examines the Pentecostal Elim Mission in Katerere, northern Nyanga in eastern Rhodesia which was well-established and locally rooted. In 1977, though, there was a move of missionaries from Katerere to Vumba where the mission had no long-term opportunity to establish relations with the local community; hence, soon after the move, relations with guerrillas became deeply problematic. Elim mission in Katerere, over the years, had developed a solid standing with local villagers, and this standing along with the role of Christian African staff at the mission was crucial in protecting the missionaries from harm by guerrillas. More

[2]This comes out most vividly by the account of Catholic missions (in particular, St. Albert's Mission, St Paul's Mission and Avila Mission) by McLaughlin (1996), including the complex and dynamic interactions between missionaries, local church leaders and members, guerrillas and the surrounding communities.

generally, "[t]he relative strengths of popular Christianity around mission stations ... influenced their standings with guerrilla bands coming into the area" as did "[t]he standings of missionaries in the eyes of local communities" (Maxwell 1995: 70). These factors were present at Katerere but not at Vumba, at which missionaries were murdered in 1978, which is discussed in gripping detail by the son of two of the Pentecostal missionaries (Griffiths 2017). In demonstrating the pragmatism of guerrilla units, Maxwell (1995: 82) concludes that "[i]n areas ... where the brokers of popular religion were Christians, guerrillas were forced to seek legitimacy from priests and Black pastors, rather than spirit mediums".

McLaughlin (1995) analyses the Catholic Avila mission, which was able to establish cordial relationships with guerrillas. When ZANLA first entered the area in the early 1970s, its guerrilla units approached any community members as the point of entry, whether headman, mediums or Christian clergy, in order to win over the wider community. As with Maxwell's study, McLaughlin shows that guerrillas acted tactically by approaching whoever they considered as most influential locally. From 1972–1974, they mainly worked with spirit mediums and not Christian missions, with relations between guerrillas and missions strained at that time. Later, ZIPA became the main guerrilla presence in the area, and they demonstrated pragmatism as well. Its guerrillas initially chanted anti-Christian slogans and opposed all forms of spirituality. With reference to spirit mediums, ZIPA claimed that they gave counterproductive directives including not to fight until certain rituals were performed, which "would hamper the development of our war" (Mhanda 2011: 114). But, again, ZIPA was not against mediums per se, such that they adopted a "pragmatic approach" (McLaughlin 1996: 131) in consulting mediums when necessary.

It was not necessarily a question of guerrilla units, in mobilising villagers, turning to either missions or indigenous spiritual authorities like spirit mediums. In areas where Christian missions were wholeheartedly in support of the guerrillas, a spiritual and political alliance emerged between spirit mediums and the missions, though not articulated explicitly as such. As a Sister at a Catholic Mission (Avila) claimed: "We were united with the spirit mediums... The spirit mediums gave the guerrillas security. The Church gave them food, clothing and medicine" (quoted in McLaughlin 1996: 132).

5.4 Men: Chiefs, *Mujibas*, Soldiers, Detainees and White Farmers

In this section, we consider the position of a diverse range of males, notably chiefs, *mujibas*, soldiers and detainees, with chiefs and soldiers often finding themselves caught in the middle rather than unambiguously on one side of the war of liberation. We end by considering white (male) farmers and their black labourers, an issue quite crucial to the third *chimurenga*.

While much of the war of liberation literature highlights the significance of spirit mediums and other local "traditional" spirit authorities (and Christian missions) in collaborating with the guerrillas, and downplays any role for chiefs, there is evidence which suggests that the position of chiefs was more uncertain than the literature broadly asserts or assumes. Douglas Moore (2005: 191) for instance notes, in his study of Nyanga, that Chief Rekayi Tangwena (selected by kraal heads to be chief in 1966), was "a subversive populist and self-proclaimed chief who fused nationalist politics, land rights and labour militancy". Of course, Tangwena, who (we noted) supported the guerrilla movement very proactively and forcefully, is often seen as the exception.

However, Fontein (2015) highlights in his study of Masvingo: "We must avoid positing mediums and guerrillas in simple opposition to discredited chiefs and headmen 'co-opted' into state structures" as "the roles, strategies and motivations of chiefs were complex and diverse" (2015: 250). Of the four chiefs in the Chiweshe TTL, one chief visibly supported the guerrillas and one spoke out against them, while the other two appeared impartial though likely assisted the guerrillas behind-the-scenes (Manungo 1991). Likewise, chiefs and headman in Buhera sought to "negotiate" their way through the war, much like ordinary villagers did, in order to manage their survival in the face of the main contending warring armies (Rhodesian and guerrilla armies) (Ndawana and Hove 2018). Chiefs at times were the first port-of-call when guerrillas entered an area, and one chief attended *pungwes* and mobilised his villagers there. Though these writers make a similar distinction to that of Manungo, they further note that the position of many chiefs was "dynamic and often changing as dictated by prevailing circumstances" (Ndawana and Hove 2018: 120).

For Fontein (2015), it is incorrect to claim that the chieftainship in Masvingo simply lost rural legitimacy, no matter how far chiefs were caught between state pressures and nationalist mobilisation. Ex-guerrillas thus recollect going to see village elders, such as chiefs and headmen, for instance to move through their area of jurisdiction safely. Fontein does not deny the existence of many guerrilla attacks on chiefs, and these attacks have been recorded for many parts of the countryside. Thus, in Lupane and Nkayi, Alexander et al. (2000) note attacks on Chiefs Madliwa and Sivalo in late 1977, as part of a broader process of beatings, rapes and burning of property. Generally, chiefs and headman seemed to be legitimate targets of ZIPRA guerrillas and ZAPU in these areas, but lower-level kraal heads were considered by the ZAPU party as possible allies (indeed, some were local party chairs) which guerrillas came to accept as trustworthy (Alexander et al. 2000).

Alexander (2006) examines the areas of Insiza and Chimanimani and demonstrates the need to underplay conflict between guerrillas and chiefs. Chiefly authority continued into the 1960s, and "the charge that they [chiefs] became discredited and irrelevant grants far too much success to Rhodesian officials' traditionalist project" in the 1960s (Alexander 2006: 83–84). The killing of chiefs during the war had more to do with repudiating and undercutting any association with the colonial state, rather than an outright rejection of chieftaincy:

> The Rhodesian state did not 'win' the struggle for chiefs' allegiance... The individuals
> who occupied chiefly office brought with them diverse political views. ... Some chiefs were
> nationalists before occupying office; others turned against government, if not to nationalism,
> as a result of the disregard for their demands, notably for land; still others reluctantly obeyed
> nationalist dictums for fear of retribution. Nationalists preferred recruiting chiefs to attacking
> them: they were not opposed to chieftainship per se, or to customary ideology, which formed
> an increasingly important component of cultural nationalism, but to its use in the service of
> government. (Alexander 2006: 84)

This speaks to the relevance of spatial and historical differences in understanding the divergent positioning of chiefs during the war of liberation, in ways perhaps similar to the existence of rebellious and loyalist chiefs during the first *chimurenga*.

Alexander (2006) goes on to argue that the chieftaincy system was used by TTL villagers to make demands, voice criticisms and undermine state authority until the guerrilla war intensified in the late 1970s. By then, making any form of contact with the state (including chiefs) became less possible and in fact quite dangerous. In certain instances, chiefs simply turned a blind-eye at the presence of guerrillas. In Insiza, at least up to the mid-1970s, when guerrilla movements and underground meetings intensified, chiefs did not report this to the colonial authorities. Other chiefs clearly sided with the guerrillas. Chief Maduna became increasingly active in ZAPU in the late 1970s, was detained for failing to report the presence of ZIPRA guerrillas, was sent into restriction and then deposed. In Chinimamani, James Ngani was appointed acting chief in 1973, and he was deposed in 1977 also on charges of supporting guerrillas.

It is also the case that the relationship between guerrillas and mediums was problematic at times. Daneel (1995b) recounts the story of the spirit medium Lydia Chabata, who was summoned by a guerrilla company commander and accused it seems unjustly of being a witch and prostitute. She had been instrumental through the *mhondoro* Sekuru Chabata of detecting and condemning many witches who were then executed at *pungwes*, and now she faced the same ordeal: "She had rarely taken time to think about the proven guilt of those luckless people. Did all of them deserve death?" (Daneel 1995b: 146). At times, spirit mediums felt the wrath of guerrillas because of disputes with the local chief, such as in the case of the Dirikwi medium in Buhera who was beaten by guerrillas based on the false claim by Chief Nyashanu that the medium was a traitor (Ndawana and Hove 2018). This is also brought out by McLaughlin (1996: 240), who warns us that "ZANLA's written records [to which she had access] point to much more ambiguity and tension between ZANU and traditional religion than previous studies have indicated". She argues that the use of spirit mediums was a very divisive issue within ZANLA and ZANU and that, by the end of the war (at least in her study of Catholic missions), missions were more valued than spirit mediums by ZANLA guerrillas.

In this context, guerrilla units (from both ZIPRA and ZANLA) became new loci of (patriarchal) power in TTLs, particularly in the case of semi-liberated and liberated zones, and in a way supplanting the power and authority of chiefs. In their respective areas of operation, most noticeable in liberated zones, the guerrilla armies sought to develop and indeed did developed new forms of political authority and civil administration, at least embryonically (Foley 1993). In the late 1970s, as Dabengwa (1995: 35) claims, ZIPRA tried to "restore civil administration in the liberated zones" after the literal departure of the Rhodesian state's local apparatuses. In certain instances, innovative health care systems—initiated between guerrillas and TTL villagers—began to be firmly established in liberated zones (Mavhunga 2015) and indeed elsewhere.

Certainly, there is evidence, such as in Lupane and Nkayi, of guerrillas establishing some very basic localised authority in displacing government power (Alexander et al. 2000). ZIPRA's turning-point strategy of 1978, which was meant to involve the introduction of regular army forces to defend liberated zones, entailed the creation of new administrative systems in rural areas, with ZAPU party structures required to assume key administrative functions in the liberated territory (Brickhill 1995). The undercutting of state authority was witnessed by guerrilla support for freedom farming in the 1970s, including the destruction of the spatial and agrarian intrusions of the state, as happened in Lupane and Nkayi in the Shangani areas of Matabeleland (Alexander et al. 2000).

But, Alexander (1995: 178–179) argues that "new administrative, political and service structures were rarely established in the course of the war. ... [P]olitical organisation [for example, support committees for guerrillas] served primarily military goals and was not used by the guerrillas themselves to achieve social transformation within political communities". In this sense, the destruction of Rhodesian state authority in TTLs was not envisaged by guerrillas or villagers as a simultaneous process of social and political reconstruction in which alternative forms of politics and sociability would arise in a pre-figurative way. As we show later, similar forms of political power arose on the occupied farms during the third *chimurenga* without any alternative polity and society envisaged.

Just as the relationship between guerrillas and chiefs was a dynamic and complex one, so was the interaction between guerrillas and *mujibas*, with guerrillas needing at times to stamp their authority over *mujibas*. With the power that they often possessed, in and through their relationships with guerrillas (Manungo 1991), *mujibas* "actively turned traditional relationships with elders [including chiefs] on their head" (Maxwell 1995: 85). With reference to the Shangani areas of Matabeleland, *mujibas* were often guilty of "abusing their power" (Alexander et al. 2000: 165). But local ZAPU chairmen tried to keep these youths in check, liaising with guerrilla commanders in doing so.

These male youth were often called upon to do the most dangerous tasks such as scouting for Rhodesian soldiers and carrying messages. The significance of *mujibas*, in this regard, was often highlighted by the women interviewed by Staunton (1990). She cites for example the story of Agnes Ziyatsha, who speaks about how the *mujibas* and even the *chimbwidos* were the ones making false claims about villagers to the guerrillas, so that "we were afraid of them" (1990: 171). As another woman (Daisy Thabedhe) indicated, guerrillas began to realise that information provided about sellouts by the *mujibas* (and others) was often false. So, guerrillas would assert their presence more forcefully by undertaking their own investigations before labelling a villager a sell-out or witch. In certain ways, as we discuss later, ZANU-PF youth played a similar role to *mujibas* during the fast track land occupations, with these youth taking matters into their own hands at times.

We consider briefly (typically male) nationalist detainees and prisoners, along with African police and soldiers. African police and soldiers, perhaps like state-sanctioned chiefs, found themselves in an extremely awkward position, serving the Rhodesian state by enforcing colonial security while often having deep roots (and families) in TTLs. Like during the first *chimurenga*, such low-level African servants of the state contributed to undercutting the second *chimurenga*. In examining this matter extensively, Stapleton (2011) argues that "[y]oung African men who joined the army [in the Rhodesian African Rifles, RAR] in the late 1950s and early 1960s were sometimes already imbued with the nationalist spirit of the time" (2011: 172). The Rhodesian state often considered African troops as unreliable and as an unacceptable security risk, so that African troops were housed away from the urban townships, such as at Inkomo Barracks. It was only very late in the war of liberation that white fears about African troops were finally outweighed by a rapidly deteriorating security situation, though nationalist sympathy within the RAR had probably lessened by the late 1960s and 1970s. The paranoia during the 1970s about the untrustworthiness of African soldiers was hence likely unfounded as there was "limited evidence of professional black ... soldiers assisting nationalist forces or deserting to them" (Stapleton 2011: 180).

While off-duty, African soldiers could be subjected to intimation by guerrillas or their rural homes might be destroyed. Village families with members who were soldiers were divided at times, as their family members in TTLs supported the guerrilla struggle. African soldiers were considered by villagers as particularly ruthless. Added to the mix in the late 1970s were the paramilitary auxiliary units of moderate nationalist leaders who aligned themselves with the Rhodesian state, including Muzowera and Sithole. In many parts of the countryside, they were loathed by civilians for their brutality.

Also, of some significance to the guerrilla war effort were political detainees and inmates, as discussed by Munochiveyi (2014). Munochiveyi (2014: 3) argues that there is a "multiplicity of historical subjects germane to the history of Zimbabwe's liberation struggle". But he adds that "because of the heavy scholarly focus on the guerrilla war, the centrality of the 1960s activism and urban violence in the shaping of Zimbabwean nationalism ... has never been highlighted" (Munochiveyi 2014: 46). The ZANU-PF narrative of the second *chimurenga* emphasises the heroic role of (particularly ZANLA) guerrillas. As we noted in discussing the *chimurenga* monologue more broadly in chapter one, this marginalises and silences other historical subjects—including prisoners and detainees of the Rhodesian regime. However, post-1980, ZANU-PF politicians trumpet their prison or detention experience in trying to fit within the pantheon of liberation heroes.

In the context of the guerrilla war, peasant men (and admittedly women) found themselves facing the confinement laws of Rhodesia, with large numbers of villagers detained and imprisoned. Those who ended up in prisons came from very diverse social backgrounds (including nationalist leaders and ordinary villagers), with the testimonies of women by Staunton (1990) regularly referring to the husbands of these women (often ZAPU activists) arrested and placed in prison. Indeed, in the 1970s, Rhodesian security forces rounded up many villagers on alleged charges of either intending to join the guerrillas or aiding young people to slip out of the country to become guerrillas. In terms of indefinite detention, detainees were often confined to remote centres across Rhodesia, such as Wha Wha and Gonakudzingwa camps.

Munochiveyi (2013: 286, 296) argues that "[d]etainees were self-aware of their roles as symbols and vanguards of the struggle for liberation" and that they "developed their own ways of governing their spaces, which drew upon the political hierarchies of their own political formations". In this respect, the high-level detainees at Gonakudzingwa camp from 1964 to 1974 became involved in "an extended experiment in self-government" (Alexander 2011: 21), as a kind of nationalist project in self-government. Detainees at times also left their restriction areas to meet with local village communities (Nkomo 1984). It was at Sikombela camp, which consisted of ZANU's leadership, that the first nationalist agreement and resolution to wage a guerrilla war was apparently made; and it was at the same detainee camp where Sithole was in effect replaced by Mugabe as ZANU president. African guards were the Achilles heel of the camp system and sympathised with political prisoners, helping to smuggle out letters including to (ZANU chairperson) Chitepo in Zambia. Thus, though it may be difficult to understand the contribution of political prisoners to the liberation struggle, the prison (like the guerrilla zone) was a terrain of resistance, confrontation and struggle (Munochiveyi 2013).

Ultimately, the key patriarchs during the war of liberation, particularly within the TTLs, were the guerrillas and not the chiefs and *mujibas*, or even local ZAPU branch chairmen. The guerrilla armies and nationalist movements, as depicted in the previous chapter, were led and dominated by men. Overall, women were marginalised yet they bore the brunt of the intensity of the war, including sometimes caught between the warring male Rhodesian soldiers and male guerrillas. Other patriarchs existed though in the countryside, namely white commercial (male) farmers. Before considering women and patriarchy during the war of liberation, we briefly examine guerrillas in relation to white farmers and their labourers.

The Rhodesian countryside was in large part divided into two separate though interconnected spatial areas, namely the TTLs and the white commercial farming areas. While TTLs existed under communal tenure, white agriculture was based on freehold title. White farms were deeply privatised and despotic spaces ruled by the male white farmer (Hartnack 2016) under what Rutherford calls "domestic government" (Rutherford 2001). In relation to the war of liberation, white farmers (and their sons) were central to defending the countryside, including by way of becoming active soldiers in the Rhodesian army on either a part-time or full-time basis. Like women in TTLs who often found themselves alone without husbands during the war, the wives of white farmers likewise stayed alone for extended periods. This was because of male military and police service call-ups, during which time wives managed and cared for the farm, as Alexander Fuller details from personal experience (Fuller 2002). In areas designated by the Rhodesian government as "supersensitive", white farmers received call-up concessions in the later years of the war (Godwin and Hancock 1993).

White farmers were subject to regular attacks by guerrilla units, and sometimes with the assistance of *mujibas*, farm labourers and TTL villagers. These attacks entailed, for example: ambushing and killing of farmers; planting of land mines on farm roads; mortar and rocket attacks; cutting perimeter fences and stealing cattle (at times for consumption by guerrillas and TTL supporters); destroying crops; setting fire to pastures; and stealing supplies at farm-run stores. Further, in the context of the establishment of protected villages, food access for guerrillas declined, and they starting seeking food from farm worker compounds. In late 1976, the Rhodesian government initiated Operation Turkey whereby white farmers were expected to ration food to their farm labourers to minimise this risk (Cilliers 1985).

The possibility and ferocity of attacks depended in part on the local reputation of a particular farmer, with "obnoxious neighbours or bad employers" (Samasuwo 2000: 228) reported to the guerrillas by nearby TTL villagers and farm labourers respectively. At times, there was tension between white farmers and nearby missions (including St. Albert's Mission run by Jesuits), because of claims that missionaries were harbouring guerrillas (McLaughlin 1996; Linden 1980). Attacks on African farmers in Native Purchase Areas were also common, though guerrillas also had bases and held *pungwes* in these areas (such as Dewure Purchase Area) (Mujere 2012).

Many farmers, in areas infiltrated significantly by guerrilla forces (such as Melsetter and Mtoko), abandoned their farms, thereby heightening the risk of further incursions. Remaining farmers spent considerable funds on upgrading security equipment and measures on their farms and particularly around their homesteads. At local levels, they banded together to form their own private militias, anti-stock units and defence and reaction teams (normally liaising with Rhodesian police and military), in seeking to defend each other against guerrilla incursions (Godwin and Hancock 1993). On strategically placed farms, Rhodesian security forces set up bases, including "a mini-JOC, Joint Operations Centre" on Jim Barker's farm called Tango Base (Barker 2007: 176); or farms held fuel to be used for refilling planes used by the Rhodesian Air Force. For particularly vulnerable (labelled as "frontline") farms, members of the Rhodesian army's Guard Force were stationed. As well, there arose the formation of "black militia squads" consisting of "usually farmhands who have received semi-military training" (Grundy and Miller 1979: 12). These squads would often guard farm compounds, which the farmer had normally security fenced, to prevent guerrilla access to the farm labourers. Additionally, inter-farm radio systems were put in place and monitored 24 h a day, such as the one in Centenary incorporating 64 white farmers.

Though with vastly different histories and forms of existence, farm labourers and their families (like TTL villagers) might be understood as "caught in the middle". White farmers, in part because of suspicions on their part about the loyalty of their labour force, would request their labourers to report any guerrilla activity in the area. At the same time, guerrilla units sought the support of farm labourers, sometimes holding *pungwes* at night in farm compounds to garner such support. In the Honde Valley, ZANLA guerrillas insisted, not always successfully, to have workers withdraw their labour from the tea estates, such as Eastern Highlands Plantation Limited

(Nyachega 2017). The guerrillas were "ambivalent" about farm labourers: "Farm workers were closely identified with the farmers, particularly as farmers organised some workers into auxiliary defence forces" (Rutherford 2001: 39). Acts of intimidation were common at the *pungwes*, and perhaps were even more essential than in the TTLs. This is because a large proportion of farm labourers on a particular farm may be of foreign origin. Even though perhaps born in Rhodesia, they did not have any identifiable TTL which they could call home; and guerrillas typically viewed them as not entitled to land.

Guerrillas had to counter the paternalistic dependence of labourers on the white farmer and any sense of loyalty and belonging which came with this, despite the despotic rule of the farmer. For example, guerrillas had "problems persuading the Malawian immigrant workers who were the majority employed in the farms to give them food" (Bhebe 1999: 40). Though such problems may not have arisen from worker dependence and loyalty, this certainly meant that support from labourers was not necessarily forthcoming. Further, this dependence became increasingly tenuous during the war of liberation, as the coercive character of the farm labour regime increased because farmers sought to extract maximum labour capacity from their workers, at the lowest cost imaginable, in the face of productivity and liquidity crises. In this context, desertion by farm labourers took place on a regular basis (sometimes encouraged by guerrillas), and the willingness of labourers to cooperate with guerrillas (as informers for instance) increased. There is evidence of significant support by labourers for guerrillas, including the supply of food, and occasionally with the blessing of the farmer (Staunton 1990)—with the story of Garfield Todd, farm owner and former Rhodesian prime minister, regularly cited in this regard. Guerrilla units often wrote short letters to farmers requesting their assistance, "sometimes pinned to a fence or delivered by an old African employee" (Grundy and Miller 1979: 26); or just warnings requesting that they leave, as many of them did.

As we will see from later chapters, there are significant parallels between what happened on white commercial farms during the war of liberation and the third *chimurenga*. Of course, unlike the presence of war veteran-led occupiers on white farms during the latter *chimurenga*, white farmers did not experience the direct physical presence of guerrillas. Further, while the ZANU-PF government seemingly turned a blind-eye to the fast track land occupations, the defence of farm white space was supported by the Rhodesian government in the 1970s. Nevertheless, parallels are clear, including with respect to the troubled relationships between farmers, labourers and guerrillas (or war veteran and other occupiers).

5.5   Women and Patriarchy

In the 1970s, the nationalist movement (at least ZANU) regularly portrayed women as playing a crucial and at times equivalent role to men in the guerrilla struggle—as guerrillas—during the second *chimurenga*, though acknowledging that women guerrillas were far less in number than male guerrillas (Frederikse 1982). For instance, in a speech before an international audience in the late 1970s, female guerrilla commander Teurai Ropa (the *chimurenga* name for Joice Mujuru) spoke about the existence of patriarchy and feudal-minded male chauvinists amongst the oppressed African population of Rhodesia, but then went on to say: "Gone are the days when all women [as heirs of Mbuya Nehanda] did was to sew, knit, cook, commiserate with and mourn for their fallen soldiers. Now they too participate in the fighting ... Our women's brigade is involved in every sphere of the armed revolutionary struggle" (ZANU Women's League (1974: 20–21). Brickhill (1995) claims that ten percent of ZIPRA guerrillas were women, in large part incorporated into ZIPRA's women's brigade. Such claims were often highly propagandist in character. In the end, the second *chimurenga*—with the deeply militarised dimension to it—was profoundly patriarchal in both discourse and practice as depicting men as agents of heroic change.

Further, besides the literature discussed below which speaks directly about women and patriarchal arrangements during the war of liberation, a considerable portion of the scholarly literature on the war has been, and remains, gender insensitive. Even Fay Chung's autobiography of the liberation struggle (Chung 2006), titled *Re-living the Second Chimurenga*, is described in this way, as if Chung's self-identification is with a masculinised female-ness (Gwarinda 2016). Ngoshi (2013), in a comparative analysis of the autobiographies of Chung and ex-ZANU leader Edgar Tekere (Tekere 2007), tends to argue otherwise. Tekere "projects a militaristic masculinity" (Ngoshi 2013: 120), primarily pertaining to his own heroic role in the liberation struggle, but also with reference to militarised feminine identities such as guerrilla commander Teurai Ropa (Bull-Christiansen 2004).

But, for Ngoshi (2013), Chung's work tends to challenge this hegemonic masculinity by examining women collectively in terms of the pain and vulnerability they suffered (Staunton 1990). In this way, according to Ngoshi, Chung questions the masculinist articulation of the nationalist project as the diverse activities of women during the war disrupted accepted images of femininity. However, we question Ngoshi's conclusion that Chung's narrative compromises and destabilises the manliness of war, and that it goes beyond Tekere's arguments in this regard. Ultimately, what appears in some of the academic literature, at least implicitly, is a militarised masculinity of war as the quintessential form of hegemonic masculinity both honoured and desired (Parpart 2015), and into which deserving women may gain entry and thereby qualify as heroes too.

In 2000, Ranger wrote that, in terms of actual practice during the 1970s, "the problem of gender was never analysed or confronted within ZANU-PF ideology or guerrilla education" (in a foreword to Nhongo-Simbanegavi 2000: xvii). The original studies of the war of liberation (including Ranger's work) raised questions around gender at times (Kriger 1992), but these studies were in the main silent on the gendered and indeed patriarchal character of the war of liberation. Subsequent to the early work of for example Weiss (1986) on women leaders in Rhodesia, there has been a burgeoning literature on women in the second *chimurenga*, including with reference to the guerrillas themselves, to women in refugee camps (in notably Mozambique) and to the experience of women villagers in the Rhodesian countryside (including in protected villages) during the war. This literature clearly fits within the historiography of nationalism because of its critical edge.

In her important work on gender and the war of liberation, Nhongo-Simbanegavi (2000) examines the motivations of females in joining the guerrilla armies (mainly ZANLA) and their experiences in base camps outside the country and in the operational areas inside. She concludes that "[a]s young women joined the armed struggle they passed over to the patronage of a new set of patriarchs, the nationalist leaders" and, "[a]s in the camps, women [guerrillas] in the operational zone remained in spaces carved out by men" (Nhongo-Simbanegavi 2000: 17, 83). Female guerrillas in the operational zones performed stereotypically defined womanly roles, such as carrying weapons and ammunition for male guerrillas, or cooking and doing laundry for them (Parpart 2015). Patriarchal practices within the guerrilla armies were not significantly challenged, as other literature readily confirms, though there may have been differences between ZIPRA, ZANLA and the short-lived ZIPA.

Mhanda (2011), as a former commander of ZIPA, argues that it was ZIPA's policy not to send female guerrillas to the front on combat missions as it had sufficient male fighters, but he refers to women fighters transporting materials such as weapons into Rhodesia from Mozambique. At the same time, he argues that male guerrillas did everything for themselves and that they never expected women fighters to cook and do laundry for them (Mhanda 2011: 129). This though is part reality, part romanticisation, as Mhanda seeks to draw a clear dividing line between ZIPA and ZANLA. In a similar way, Parpart (2015) suggests that ZIPA sought to challenge the hegemonic militarised masculinity prevailing in ZANLA and ZIPRA armies, or at least to open it up to gender equality, but ZIPA was soon crushed by Mugabe's ZANU. Earlier, the leaders of the 1974 rebellion within ZANLA, that is the Nhari rebellion, accused ZANLA commanders of predatory sexual conduct against women, but this rebellion was likewise violently suppressed.

All this is a far cry from the way in which "women's frontline roles" were "glorified" (Lyons 2004: 161) by nationalists. In the case of refugee and guerrilla camps in Mozambique, Chung (2006) paints a disturbing picture of the lives of women, with women subjected to the "traditional feudal [gender] attitudes" (2006: 125) of ZANLA guerrilla commanders. Citing the case of Pungwe III military camp specifically, she argues that commanders used young women as "warm blankets" (calling them to their accommodation for "night duty") (Chung 2006: 127). The attitude of ZANLA commander Tongogara and his sub-commanders, with regard to their requests for sexual favours, was likely one reason why Nhari and Mhanda's *vashandi* "both attracted very large numbers of women" and why ZANLA's top commanders were regarded with "revulsion by many women guerrillas" (Chung 2006: 127, 125). At the same time, senior ZANLA women also took their pick of men from the newly arrived young male recruits in the camps.

The motivations for females becoming guerrillas were highly diverse and complex. Lyons (2004) for example argues that some women joined the liberation war as guerrillas for ideological and nationalist reasons (noting the case of Margaret Dongo), but others did so "to facilitate their own agency as women or for personal reasons" (Lyons 2004: 114). Likewise, O'Gorman (2011)—in her study of

Chiweshe—refers to the importance of women's "personal awareness and experience" and how the "political slogans of the [nationalist] revolutionaries ... tapped into the felt grievances of black women living under colonial rule" (O'Gorman 2011: 83). But, as we noted more broadly with regard to villagers' motivations for involvement in the second *chimurenga*, "their [women's] motivations for participation were mixed" and included "personal struggles that are not directly, or necessarily, anti-colonial, for example, family and household tensions" (O'Gorman 2011: 85). We highlight such personal projects, and the gendered roles of women occupiers as well, in the case of the fast track land occupations.

The work of Ranchod-Nilsson (2003) provides important perspectives in this respect. She stresses how the practices and agendas of women, multiple and inconsistent as they might be, were an integral part of the war of liberation, which itself was not unified or characterised by uniform nationalist goals. In her earlier work, she observes that rural women's agendas were rooted in the reconfigurations of gender relations during the colonial period, and interwoven with their class circumstances as villagers (Ranchod-Nilsson 1994: 64). Based on fieldwork in Wedza District, she highlights that "women had their own imaginings about the nation that suggested they had clear visions regarding how they wanted their lives to change as a result of their involvement" (Ranchod-Nilsson 2003: 170). In doing so, they were often left to manage rural homesteads on their own because husbands may be working as migrants elsewhere. As well, it was not unusual for men to go to town to avoid security force harassment. As a result, they left their wives behind to keep the homestead afloat in times of deep political and personal turmoil. Simultaneously, women regularly saw their sons (and at times daughters) just disappearing, assuming that they went to become guerrillas.

As well, guerrillas were taking teenage girls as lovers, so that some women if possible sent their daughters to town. Women in Lupane and Nkayi, voluntarily or not, entered into relationships with guerrillas, about which local ZAPU chairmen expressed concern (Alexander et al. 2000). In Bulilimamangwe District in the southwestern part of the country, and elsewhere, guerrillas raped young women, and many children were born out of rape (Dube 2016). Though sexual liaisons between guerrillas and local women (and girls) were common, Dzimbanhete (2015) suggests that the extent to which these arose from coercion on the part of the guerrillas should not be overemphasised. Further, when ZANLA commanders (even in Mozambique) heard of incidences of rape and sexual abuse in operational areas, the responsible guerrilla was disciplined normally, as such conduct went against ZANLA's code of conduct. Just as *mujibas* used their association with guerrillas for purposes of asserting their authority amongst villagers, *chibwidos* could attain a degree of independence from lineage control if involved in a plutonic relationship with guerrillas.

Importantly, Ranchod-Nilsson (2003) provides a temporal analysis of the liberation struggle and how different categories of women participated. For instance, during the early phase in the 1960s, wives of nationalists were very active in nationalist political organisations but, as the liberation war gained momentum, young uneducated women crossed borders, at first simply carrying supplies and later trained as armed guerrillas. Added to the temporal analysis, her work additionally exposes the spatial peculiarities that shaped the participation of different women. For example, Wedza had high rates of male out-migration so that guerrillas relied abnormally heavily on women not only to cook and clean but to ferry supplies and weapons and act as informants. Many women were either arrested or subjected to violence for cooking for guerrillas and other "crimes" associated with coming into contact with guerrillas. Also, particularly in ZIPRA areas of operation where long-established structures of ZAPU remained in place, women sometimes were local ZAPU activists who provided logistical support to the guerrillas (Staunton 1990).

Overall, the role and activities of female villagers in operational areas were structured in a largely patriarchal manner, with (particularly older) women becoming the "mothers of the revolution" as defined in a traditional gendered sense. With the inevitable establishment of guerrilla bases in the rural areas, where *pungwes* were held at night, women in particular were expected to cook and deliver food to the guerrillas (Staunton 1990). With specific reference to guerrilla bases established in sacred places (after liaising with spirit mediums), "[a]ncestors prohibited women" (Chitukutuku 2019: 327) from these places at times. In one instance, discussed by Chitukutuku (2019), the presence of female guerrillas camped at Mumurwi Mountain was held to be responsible for a major defeat inflicted against ZANLA by the Rhodesian army.

But women were often caught between the demands of guerrillas and Rhodesian security forces and suffered violence from both sides. Women involved in the Federation of African Women's Clubs, which sought to instill hygienic practices amongst African women in rural homesteads and was regularly seen as engaged in a colonial-inspired project, were labelled at times as sell-outs, though many of the club members actively supported the guerrilla struggle (Burke 1996). The female Development Workers and Women Advisors under the Ministry of Internal Affairs were also branded as sell-outs and subjected to intimidation and threats, and even death (Chenaux-Repond 2017).

Chadya (2007: 32) highlights the flight of civilian women from rural areas to Harare during the war, as a form of "gendered war-induced migration" arising from the militarisation of the countryside. Women became "agents of their own survival" (Chadya 2007: 45) by moving, often with children, to Harare for a complex array of reasons which were often very personal (including leaving behind a troubled or broken marriage).

But the sheer ferocity of the war between the Rhodesian army and guerrillas also led at times to personal animosity between women. One woman, Margaret Viki, is quoted as saying:

> You could be killed for nothing, just because someone wanted you dead, and told lies about you… It was common, especially amongst women, for them to sell out on each other. For instance, if one worked very hard and acquired some possessions, one could be branded a 'witch' by those who were jealous. The freedom fighters were against witchcraft, so that was a sure way of having someone killed. (Staunton 1990: 152)

This was part of the broader personalised conflicts, in the context of the sell-out label, which we have discussed already.

In the case of women in protected villages during the war, at least in Chiweshe, this might lead to challenges around local patriarchies and especially "the masculinity of adult village men" (Kesby 1996: 573), as these men lost their authority over young women given the Rhodesian state's system of controls within the villages. In many cases, women outnumbered men significantly in the protected villages, such that "women and [female] juveniles seized the increased sexual and other opportunities in the absence of their male guardians" (Schmidt 2013: 192), leading to adultery and sex work. This was not unlike the increased sexual autonomy of *chimbwidos* (outside of protected villages) in their encounters with guerrillas.

What Kesby and others do not explore about women during the second *chimurenga* is the implications for women of the, at least partial, undercutting of the authority and legitimacy of specifically the chieftainship system. Historically, this system did not simply benefit white rule, as it entailed an unspoken alliance between the colonial state and rural male patriarchs designed to, amongst other issues, control the movement and sexuality of women. Insofar as women, during the second *chimurenga*, were able to express and assert agency, the extent to which this was due to the illegitimacy of colonial-backed chiefs and headman remains unclear.

The overall conclusion of the literature on patriarchy and the second *chimurenga* seems to be that women were not mere victims of the war of liberation, but were active participants in it. Further, through their practices, they sought regularly to pursue personal, and not, nationalist aspirations. Though women in many ways repositioned themselves and used the social spaces opened up by the war to their advantage, this did not necessarily involve any meaningful or even medium-term undermining of local systems and practices of patriarchy, certainly if the position of women in independent Zimbabwe is taken into consideration. Any claim that the third *chimurenga* occupations undercut patriarchy would also be problematic, as we show later.

5.6   Conclusion

This chapter has highlighted key themes in the war of liberation scholarly literature, many of which have parallels with the third *chimurenga* despite the vast differences between these two *zvimurenga*. Throughout this chapter, we have sought to show the sheer complexity, diversity and fluidity of the second *chimurenga* and the significant variations which existed spatially, and for a variety of reasons. Certainly, in the case of the relationships established by different guerrilla armies and units with for instance TTL villagers (including chiefs and spirit mediums) and nearby Christian mission stations, there was a reasonable degree of pragmatism and tactical decision-making which were deeply situational and contextual. Guerrillas, and other local actors, maneouvred their way through the landscape of war, leading at times to cooperation, at times to accommodation, and at other times to tension. Tensions also arose within TTL villages, and for very personal and experiential reasons. None of this is understandable, or at least not fully understandable, by framing the war in terms of sweeping claims about a nationalist project cascading down to local levels.

This is similarly the case with regard to women and the ways in which different categories of women negotiated their way through the war or somehow beyond the clutches of the war.

We argue, in chapters seven and eight, that a similar conclusion can be (and should be) reached about the third *chimurenga* land occupations, thereby arguing against any claim to the effect that the occupations are reducible to ruling party machinations. In order to set the stage for this argument, we turn to important developments between 1980 and the year 2000 in the following chapter.

References

Alexander J (1995) Things fall apart, the centre can hold: processses of Post-War political change in Zimbabwe's rural areas. In: Bhebe N, Ranger T (eds) Society in Zimbabwe's liberartion war (volume two). James Currey, London, pp 175–191

Alexander J (2006) The unsettled land: state making and the politics of land in Zimbabwe 1893–2003. James Currey, Oxford

Alexander J (2011) Nationalism and self-government in Rhodesian detention: Gonakudzingwa, 1964–1974. J South Afr Stud 37(3):551–569

Alexander J, McGregor J (2004) War stories: Guerrilla narratives of Zimbabwe's liberation war. Hist Workshop J 57(1):79–100

Alexander J, McGregor J (2017) African soldiers in the USSR: oral histories of ZAPU Intelligence Cadres' Soviet Training, 1964–1979. J South Afr Stud 43(1):49–66

Alexander J, McGregor J, Ranger T (2000) Violence and memory: one hundred years in the 'Dark Forests' of Matebeleland. James Currey, Oxford

Barker J (2007) Paradise plundered: the story of a Zimbabwean farm. Jim Barker, Harare

Bhebe N (1999) The ZAPU and ZANU Guerilla Warfare and the Evangelical Lutheran Church in Zimbabwe. Mambo Press, Gweru

Bhebe N, Ranger T (1995) Volume Introduction. In: Bhebe N, Ranger T (eds) Soldiers in Zimbabwe's liberation war (volume one). James Currey, London, pp 6–23

Brickhill J (1995) Daring to storm the heavens: the military strategy of ZAPU 1976 to 1976. In: Bhebe N, Ranger T (eds) Soldiers in Zimbabwe's liberation war (volume one). James Currey, London, pp 48–72

Bull-Christiansen L (2004) Tales of the nation: feminist nationalism or patriotic history? defining national history and identity in Zimbabwe. Nordiska Afrikainstitutet, Uppsala

Burke T (1996) LifeBuoy men, lux women: commodification, consumption, and cleanliness in modern Zimbabwe. Leicester University Press, London

Catholic Commission for Justice and Peace in Rhodesia (CCJPR) (1999) The man in the middle: torture, resetlement and eviction. Catholic Commission for Justice and Peace in Rhodesia, Harare

Chadya J (2007) Voting with their feet: women's flight to Harare during Zimbabwe's liberation war. J Can Hist Assoc 18(2):24–52

Chitukutuku E (2019) Spiritual temporalities of the liberation war in Zimbabwe. J War Cult Stud 12(4):320–333

Cilliers JK (1985) Counter-insurgency in Rhodesia. Croom Helm, London

Chenaux-Repond M (2017) Leading from behind: women in community development in Rhodesia, 1973–1979. Weaver Press, Harare

Chung F (2006) Re-living the second chimurenga: memories from the liberation struggle in Zimbabwe. Nordic Africa Institute, Uppsala

Dabengwa D (1995) ZIPRA in the Zimbabwean war of national liberation. In: Bhebe N, Ranger T (eds) Soldiers in Zimbabwe's liberation war (volume one). James Currey, London, pp 24–35

Daneel M (1995a) Mwari the liberator—Oracular intervention and Zimbabwe's quest for the 'Lost Lands'. Missionalla 32(2):216–244

Daneel M (1995b) Guerrilla snuff. Baobab Books, Harare

Dube T (2016) Shifting identities and the transformation of the Kalanga people of Bulilimamangwe District, Matabeleland South, Zimbabwe c. 1946–2005. Unpublished PhD thesis, University of the Witwatersrand, Johannesburg

Dzimbanhete J (2015) Sexuality in the war zones during Zimbabwe's war of liberation: the case of the Zimbabwean African National Liberation Army. J Pan Afr Stud 8(8):52–59

Foley G (1993) Progressive but not socialist: political education in the Zimbabwe liberation war. Convergence 26(4):79–88

Fontein J (2009) The silence of Great Zimbabwe: contested landscapes and the power of heritage. Routledge, London

Fontein J (2015) Remaking Mutirikwi: landscape. Water and belonging in Southern Zimbabwe. James Currey, Oxford

Frederikse J (1982) None but ourselves: masses vs. media in the making of Zimbabwe. Heinemann, London

Fuller A (2002) Don't let's go to the dogs tonight: an African childhood. Picador, London

Godwin P, Hancock I (1993) Rhodesians never die. Macmillan, Northlands

Griffiths S (2017) The axe and the tree. Monarch Books, Oxford

Grundy T, Miller B (1979) The farmer at war. Modern Farming Publications, Salisbury

Gwarinda M (2016) Reliving the Second chimurenga: a nationalist female perspective. J Literary Stud 32(2):108–120

Hartnack A (2016) Ordered estates: welfare, power and maternalism on Zimbabwe's (once White) Highveld. Weaver Press, Harare

Henkin Y (2013) Stoning the dogs: Guerrilla mobilisation and violence in Rhodesia. Stud Conflict Terrorism 36(6):503–532

Hove M (2012) War legacy: a reflection on the effects of the Rhodesian Security Forces (RSF) in south eastern Zimbabwe during Zimbabwe's War of Liberation 1976–1980. J Afr Stud Dev 4(8):193–206

Hughes D (2006) From enslavement to environmentalism: politics on a Southern African Frontier. University of Washington Press, Seattle

Kesby M (1996) Arenas for control, terrains of gender contestation: Guerrilla struggle and counter-insurgency warfare in Zimbabwe 1972–1980. J Southern Afr Stud 22(4):561–584

Kriger N (1988) The Zimbabwean war of liberation: struggles within the struggle. J South Afr Stud 14(2):304–322

Kriger N (1992) Zimbabwe's Guerrilla War: peasant voices. Cambridge University Press, Cambridge

Lan D (1985) Guns and rain: guerillas and spirit mediums in Zimbabwe. James Currey, London

Linden I (1980) The Catholic Church and the struggle for Zimbabwe. Longman, London

Lyons T (2004) Guns and Guerrilla Girls: women in the Zimbabwean National Liberation struggle. Africa World Press, Trenton

Manungo K (1991) The Zimbabwe war of liberation, with special emphasis on the role peasants played in Chiweshe district. Unpublished PhD thesis. Ohio University, United States of America

Marowa I (2009) Construction of the 'Sellout' identity during Zimbabwe's war of liberation: a case study of Dandawa Community of Hurungwe District, c1975–1980. Identity, Cult Polit Afro-Asian Dialogue 10(1):121–131

Marowa I (2015) Forced removal and social memories in north-western Zimbabwe, c1900–2000. Unpublished PhD thesis, University of Bayreuth, Germany

Mavhungu C (2015) Guerrilla healthcare innovation: creative resilience in Zimbabwe's *Chimurenga*, 1971–1980. Hist Technol 31(3):295–323

Maxwell D (1995) Christianity and the war in Eastern Zimbabwe: the case of Elim Mission. In: Bhebe N, Ranger T (eds) Society in Zimbabwe's liberation war (volume two). University of Zimbabwe Publications, Harare, pp 58–90

Maxwell D (2006) African gifts of the spirit: pentecostalism and the rise of a Zimbabwean Transnational Religious Movement. James Currey, Oxford

Mazarire G (2011) Discipline and punishment in ZANLA: 1964–1979. J South Afr Stud 37(3):571–591

McGregor J (2009) Crossing the Zambezi: the politics of landscape on a central African Frontier. James Currey, Suffolk

McLaughlin J (1995) Avila Mission: a turning point in church relations with the state and the liberation forces. In: Bhebe N, Ranger T (eds) Society in Zimbabwe's liberation war (volume two). University of Zimbabwe Publications, Harare, pp 91–101

McLaughlin J (1996) On the frontline: Catholic Missions in Zimbabwe's liberation war. Baobab Books, Basel

Mhanda W (2011) Dzino: memories of a freedom fighter. Weaver Press, Harare

Moore D (2005) Suffering for territory: race, place and power in Zimbabwe. Weaver Press, Harare

Msindo E, Nyachega N (2019) Zimbabwe's liberation war and the everyday in Honde Valley, 1975 to 1979. South Afr Hist J 71(1):70–93

Mujere J (2012) Autochthons, strangers, modernising educationists, and progressive farmers: Basotho struggles for belonging in Zimbabwe 1930s–2008. Unpublished PhD thesis, University of Edinburgh, UK

Munochiveyi M (2013) The political lives of Rhodesian Detainees during Zimbabwe's liberation struggle. Int J Afr Hist Stud 46(2):283–304

Munochiveyi M (2014) Prisoners of Rhodesia: inmates and detainees in the struggle for Zimbabwean liberation, 1960–1980. Palgrave, New York

Murdoch N (2015) Christian warfare in Rhodesia-Zimbabwe: the Salvation Army and African liberation, 1801–1991. The Lutterworth Press, Cambridge

Ndawana E, Hove M (2018) Traditional leaders and Zimbabwe's liberation struggle in Buhera District, 1976–1980. J Afr Mil Hist 2:119–160

Ndlovu-Gatsheni S (ed) (2017) Joshua Mqabuko Nkomo of Zimbabwe: politics, power, and memory. Palgrave Macmillan, London

Nkomo J (1984) Nkomo: The story of my life. Methuen, London

Nyachega N (2017) Beyond war, violence, and suffering: everyday life in the Honde Valley borderland communities during Zimbabwe's liberation war and the RENAMO insurgency, c.1960–2016. Unpublished MA thesis, Rhodes University, South Africa

Ngoshi H (2013) Masculinities and femininities in Zimbabwean autobiographies of political struggle: the case of Edgar Tekere and Fay Chung. J Literary Stud 29(3):119–139

Nhongo-Simbanegavi J (2000) For better of worse: women and ZANLA in Zimbabwe's liberation struggle. Weaver Press, Harare

O'Gorman E (2011) The front line runs through every woman: women & local resistance in the Zimbabwean liberation war. Weaver Press, Harare

Parpart JL (2015) Militarized masculinities, heroes and gender inequality during and after the nationalist struggle in Zimbabwe. NORMA 10(3–4):312–325

Ranchod-Nilsson S (1994) "This, too, is a Way of Fighting": Rural Women's participation in Zimbabwe's Liberation War. In: Tetreault MA (ed) Women in revolutions in Africa, Asia and the New World. University of South Carolina Press, Columbia, SC, pp 62–88

Ranchod-Nilsson S (2003) Gender struggles for the nation: power, agency and representation in Zimbabwe. In: Ranchod-Nilsson S, Tetreault MA (eds) Women, states and nationalism: at home in the nation. Routledge, London, pp 164–180

Ranger T (1985) Peasant consciousness and Guerrilla War in Zimbabwe: a comparative study. Zimbabwe Publishing House, Harare

Ranger T (2010) Bulawayo Burning: the social history of a Southern African City 1893–1960. Weaver Press, Harare

Ranger T, Ncube M (1995) Religion in the Guerrilla War: the case of Southern Matabeleland. In: Bhebe N, Ranger T (eds) Society in Zimbabwe's liberation war (volume two). University of Zimbabwe Publications, Harare, pp 35–57

Rutherford B (2001) Working on the margins—Black Workers, White Farmers in Postcolonial Zimbabwe. Weaver Press, Harare

Samasuwo N (2000) 'There is Something about Cattle': towards an economic history of the beef industry in Colonial Zimbabwe, with special reference to the role of the state, 1939–1980. Unpublished Doctoral thesis, University of Cape Town, South Africa

Schmidt HI (2013) Colonialism and violence in Zimbabwe: a history of suffering. James Currey, Woodbridge

Sibanda EM (2005) The Zimbabwe African People's Union 1961–1987: a political history of insurgency in Southern Rhodesia. Africa World Press, Trenton, NJ

Stapleton T (2011) African police and soldiers in Colonial Zimbabwe, 1923–80. University of Rochester Press, New York

Staunton I (ed) (1990) Mothers of the revolution. Baobab Books, Harare

Tekere EZ (2007) A lifetime of struggle. SAPES Books, Harare

Tungamirai J (1995) Recruitment to ZANLA: building up a war machine. In: Bhebe N, Ranger T (eds) Soldiers in Zimbabwe's liberartion war (volume one). James Currey, London, pp 36–47

Weinrich A (1977) Strategic resettlement in Rhodesia. J South Afr Stud 3(2):207–229

Weiss R (1986) The Women of Zimbabwe. Kesho Publishers, London

ZANU Women's League (1974) Liberation through participation: women in the Zimbabwean revolution, writing and documents from the ZANU and the ZAPU Women's League. National Campaign in Solidarity with ZANU Women's League

Chapter 6
Post-independence Land Reform, War Veterans and Sporadic Rural Struggles

Abstract This chapter discusses the intervening period between the second and third *zvimurenga* by focusing on developments central to the rise of the fast track land occupations in the year 2000. A central consideration for this period is the Zimbabwean state's failure to shift fundamentally the colonial land and agrarian structure, with the land reform programme failing to de-racialise the countryside in terms of landholdings. Alongside this stalled land reform programme were two further developments which facilitated the emergence of the third *chimurenga*. On the one hand, large numbers of ex-guerrillas from the war of liberation were marginalised in the post-1980 period and they began to mobilise and organise in a manner which led to the eventual formation of a national war veterans' association which expressed discontent with the Zimbabwean state and ruling party. On the other hand, because of minimal land reform, as well as ongoing land pressures and livelihood challenges in the communal areas, villagers often in alliance with war veterans increasingly began to occupy land in the 1990s in a deeply localised way. By the late 1990s, the stage was set for another large-scale episode of land struggles.

Keywords Zimbabwe · Land reform · Land occupations · War veterans · White farmers

6.1 Introduction

This chapter discusses the intervening period between the second and third *zvimurenga* by focusing in particular on developments central to the rise of the fast track land occupations in the year 2000. Thus, it is not a sweeping political economy of Zimbabwe in the period from 1980 to 1999. For our purposes, a central consideration for this period is the Zimbabwean state's failure to shift fundamentally the colonial land and agrarian structure, with the land reform programme failing to de-racialise the countryside in terms of landholdings (Mbaya 2001). Alongside this stalled land reform programme were two further developments which, combined with the first, facilitated the emergence of the third *chimurenga*. One the one hand, large numbers of ex-guerrillas from the war of liberation were marginalised in the post-1980 period and they began to mobilise and organise in a manner which led

© The Author(s), under exclusive license to Springer Nature Switzerland AG 2021
K. Helliker et al., *Fast Track Land Occupations in Zimbabwe*,
https://doi.org/10.1007/978-3-030-66348-3_6

to the eventual formation of a national war veterans' association which expressed discontent with the Zimbabwean state and ruling party. On the other hand, because of minimal land reform, as well as ongoing land pressures and livelihood challenges in the communal areas (formerly, Tribal Trust Lands), communal area villagers often in alliance with war veterans increasingly began to occupy land in the 1990s in a deeply localised way. By the late 1990s, it seemed that the stage was set for another large-scale episode of land struggle (or, more broadly, a *chimurenga*).

6.2   Land Redistribution in the 1980s

At the dawn of Zimbabwe in April 1980, the post-colonial government inherited a racially divided agrarian and land structure. About 4500 white commercial farmers owned roughly 15.5 million hectares of land (39% of the total land in the country). More than a million African households in now communal areas post-1980 had access to only about 16 million hectares. Undoubtedly, one of the most—if not the most—acute and difficult question confronting the new government at independence was land. This was because of the political, social, emotional, spiritual and economic importance of land to both white and black (or African) people. Palmer (1977: 246) adds: "The problem will not be an easy one to resolve. The continuing stranglehold of the land division of the 1890s, … will … impose constraints on future land and agricultural policies".

Given the deep history of land alienation and ensuing agrarian policies prior to 1980, it is not surprising that land redistribution featured prominently in the rhetoric of the nationalist movements (ZANU and ZAPU) and the new ZANU-PF government (Cousins 2003). When ZANU-PF led by Robert Mugabe came to power, it made land redistribution a high priority. In reality, the government in the first twenty years of independence (until the year 2000) gave far less priority to land redistribution than its manifestos and policy platforms would suggest, and for a range of reasons.

For instance, in 1980 the white agricultural commercial sector produced about 80% of national agricultural production (in terms of value) and 90% of formally marketed agricultural commodities, while also employing a third of the total labour force in the country (de Villiers 2003: 7). Armed with these statistics, white farmers successfully argued against significant land redistribution during the first fifteen years after independence. Over the first decade, the government also had to abide by the principles of the independence agreement (the Lancaster House Agreement) which restricted land redistribution to market-led reform based on the willing-seller, willing-buyer model. Further, in the early 1990s, the government adopted a typical neo-liberal restructuring programme which reinforced the prioritisation of market-led land redistribution. Lastly, there was widespread skepticism amongst agricultural policymakers about undercutting large-scale farming and the prospects for small-scale agriculture on redistributed farms in ensuring agricultural productivity and food security. The irony of the whole situation is that white commercial farmers, who had vigorously fought against independence, became the most protected "species" in the new Zimbabwe.

Guided by the land clauses contained in the Lancaster Agreement, the government embarked on an overly ambitious land reform project after independence. Kinsey (2004: 1671) points out that "the new government honoured its liberation war promises by swiftly launching a land resettlement programme based initially upon land abandoned [by white farmers] during the war". The market-based constraint of the Agreement meant that, for much of the 1980s, only limited land acquisition and resettlement took place. The main objectives of this initial phase of land reform were to: reduce conflict by transferring white-held land to black people; provide agricultural opportunities for ex-guerrillas and landless people; relieve population pressure in communal lands; and maintain levels of agricultural production (de Villiers 2003). The plan initially targeted the resettlement of 18,000 households over five years. In 1981, the number increased to 54,000 and in 1982 it further escalated to 162,000, to be resettled by 1985 (Palmer 1990: 169; Masiiwa 2005: 217). Palmer (1990), however, notes that by the end of the decade (July 1989), only 52,000 families (around 416,000 people) had been resettled. In terms of the amount of land transferred, this represented only 16% of the agricultural land owned by whites at independence. The number of farms acquired dwindled during the decade: it peaked in 1982 (600); two years later, the relevant figure dropped by 93%; and, for the second half of the 1980s, an average number of 48 farms were acquired per year (Kinsey 2004: 1690). The targets tended to be highly ambitious given the low capability of the government to implement such complex programmes (Bratton 1990). Even President Mugabe noted that "[i]t just proved to be impossible, because it was beyond our management and our resources" (*The Herald*, 27 October 1989).

Nevertheless, the first ten years of land reform focused on providing land for landless and land-short people, including those living in the former TTLs. In other words, the objectives of land reform entailed addressing racial inequalities through land redistribution, as this was seen as critical for fostering social and political stability. Land reform was rehabilitative and hence beneficiaries were drawn from the following groups of people, in order of priority: refugees and other people displaced by the war (this category was comprised of refugees who had fled the country during the war of liberation, as well as urban refugees and former inhabitants of protected villages); the landless (including those who had no or little land to support themselves and their dependents); and those with inadequate land to maintain themselves and their families (Government of Zimbabwe 1981). The beneficiaries were expected to be married or widowed (with dependents), aged between 18 and 55 years, physically fit to enable them to make productive use of the land, and not employed in the formal sector of the economy. Communal farmers moving to resettlement areas were supposed to give up their land rights in those areas, but there was rarely any adherence to this. After 1984, experienced Master Farmers were added to the list provided they were willing to forgo their land rights in communal areas (Marongwe 2008).

The Lancaster Agreement's stipulation of willing-buyer/willing-seller transactions was clearly responsible in part for hampering the possibility of extensive legal land acquisition in the 1980s (Andrew and Sadomba 2006). In addition, the government seemingly lost all but rhetorical interest in the entire land issue, only mentioning it at election times. Kinsey (2004) points to several pieces of evidence which show that the government had lost interest in land reform, including: even before the matching land grants (for land purchases) from the British came to an end in the late 1980s, government's budgetary appropriations for resettlement—always inadequate—began to decline; and the government failed to submit a new proposal to the British when the initial phase of the programme finished, that is, without the British funds being exhausted.

There are various interrelated factors which might explain why the land question went quiet in the mid-1980s. For example, not all land acquired between 1980 and 1983 was suitable for resettlement and, after this period, not many white farmers were willing to sell their land or were only selling marginal land (Palmer 1990). However, there were several other legal modes of land acquisition possibly available. These included a land tax, reparations and reclaiming historic subsidies given to farmers by the colonial government, which the Zimbabwean government did not pursue and utilise (Kinsey 2004). The scrupulous compliance of the government to the Lancaster Agreement, in the view of Madhuku (2004), was shortsighted and in the end severely compromised land reform. Another reason for the failure of the government to act on land resettlement was the severe droughts of the mid-1980s, which led to some villagers who had moved into resettlement areas returning to communal areas (Palmer 1990). The post-independence agricultural boom in communal agriculture (the so-called peasant miracle), which in fact only took place in a few areas, also misled the government into believing that such agricultural production increases would enhance rural livelihoods without the need for extensive resettlement.

The role of the white-dominated Commercial Farmers Union (CFU) in slowing the pace of resettlement cannot be underplayed since its members assiduously courted the government, travelled with the prime minister (president from 1987) on foreign trips and ensured that it was influential in debates on agriculture (Palmer 1990). Indeed, the first minister of agriculture post-1980, Dennis Norman, was a past president of the Rhodesia National Farmers Union (later CFU) and thus farmers' interests were represented at the highest level within the government (Bratton 1990). While the CFU did not reject land reform out-of-hand (Weiner et al. 1985), it skillfully argued and lobbied that a rapid land reform process would lead to a massive decline in agricultural production and threaten vital export earnings resulting in significant employment losses. Not only would agriculture suffer, but investor confidence and foreign exchange earnings would be dented by such a move.

Thus, the strategy of the CFU entailed not only direct lobbying of the Zimbabwean state on agricultural and land policy, but also contributing more generally to "an atmosphere of risk aversion by stressing the importance of commercial agriculture" (Herbst 1990: 56) relative to small-scale farming (Sachikonye 2003). Von Blanckenburg (1994: 12), in his sympathetic empirical work on white commercial farmers in Zimbabwe (Helliker 2006), concludes that resettlement in Zimbabwe would be "a negative rather than a positive experience". More specifically, he argues that the export performance of large farms is of "strategic importance" (von Blanck-enburg 1994: 27) and that small-scale agricultural performance in resettlement areas is "weak" (von Blanckenburg 1994: 32). The CFU's position was very persuasive to a government that was all too aware that, even if it wanted, it had no capacity to redistribute as much land as it had promised or the amount that communal farmers expected. The guaranteeing of commercial agriculture's predominance, in practice, meant the continued existence of the distributional imbalance in land and agriculture.

Post-independence, the CFU expanded its ranks to include approximately 300 African commercial farmers (including ten government ministers), and thus it entrenched its interests further within the government. In this regard, resourced Africans were acquiring land through the private market, with about one million hectares transferred through this approach (Palmer 1990). There were, as well, a sizeable number of farms belonging to the state which were leased to black elites in terms of the Agricultural Land Settlement Act (Chapter 37), and such farms could have been repossessed without violating the letter of the new constitution. Tshuma (1997: 60) notes, however, that "[m]ost of the leases were only terminated after the expiry of the entrenched constitutional clauses [ending in 1990] in an attempt to transfer them from white farmers to mostly black politicians, civil servants and influential people under a tenant farmer scheme". Leases were allocated without any transparency, with no advertising to the public on the availability of these farms (Moyo 1995).

The British government became sceptical of land reform in Zimbabwe more broadly, partly because of fears that land beneficiaries were not drawn from the appro-priate beneficiary groups and that land allocation was subject to abuse by ZANU-PF. The government at times refused to purchase farms on offer, allowing them to enter into the private land market so that wealthy blacks with government connec-tions could purchase them (Goebel 2005). Consequently, about 7% of the large-scale commercial farms were African-owned by 1986 (Alexander 1994: 337). This represented visible and real attempts by the government to restructure the existing agrarian space to pave way for the allocation of land to members of the governing elite (Marongwe 2008). In this sense, the failure of land reform during the first ten years was related as well to the secretive manner in which black political heavyweights were acquiring land for themselves.

In terms of budgetary appropriations to resettlement, in 1983 and under pres-sure from the World Bank, Great Britain and other Western donors, the government tightened its budget (Palmer 1990). It seemingly made more political sense to the government to cut back on land reform than on the social programmes (education and health) in which it had invested heavily. The financing of land reform was one of the major stumbling blocks in the first decade of independence (de Villiers 2003). The British undertook to assist in the financing of land reform provided its contribu-tion was met on a pound for pound basis by the Zimbabwean government. However, the Lancaster Agreement did not contain a detailed and enforceable commitment from any of the foreign donors to actually contribute to land reform. The British had promised £75 million and the United States a further US$500 million but, by 2000, the government had only received £35 million in total. The commitment to

Zimbabwe by the British government was at best suspicious and at worst myopic and misguided. Even though limited, more white-held agricultural land became available on the market than the government was able to acquire in the 1980s, though this likely related primarily to administrative incapacities rather than a sheer absence of available financing.

Nevertheless, in the mid-1980s, land prices also started to increase. It became a seller's market perversely to the detriment of land reform. Given the modest international support for land acquisition from the former colonial power (Britain), it might be understandable that the government did not vigorously address the land question. In particular, post-acquisition settlement of farmers ended up being extremely resource-intensive. The infrastructural, technical and financial demands were beyond what the government could realistically achieve. Most funding was spent on land acquisition, leaving minimal post-acquisition funds for ensuring basic service provision on the redistributed farms. The support from overseas donors towards post-acquisition settlement was nonexistent. The government had to bear the burden for paying half of the funds for land acquisition and the full costs of resettling small-scale farmers.

In 1985, government enacted the Land Acquisition Act. Modifications were introduced in the assessment of compensation for land purchased, shifting from explicit reference to the willing-seller/willing-buyer market-based compensation to fair and reasonable compensation. Another important addition was the redefinition of under-utilised land, reducing the period of under-utilisation from five to three years, and placing in the hands of the courts the identification, based on multiple factors, of under-utilised land (Madhuku 2004). Another creative addition was that of the "right of first refusal" to the state, where the state would have the first option to buy any agricultural land available on the market. In large part, these innovations were based on the recognition that the land reform programme, as constituted, was becoming deeply problematic as a basis for recovering alienated land.

6.3   Land Reform (1990–1997)

In April 1990 the restrictions imposed by the Lancaster House agreement expired and the land issue rose to the fore again (Marongwe 2008). This time, however, it became an electioneering ploy as an opposition party under former ZANU-PF stalwart Edgar Tekere emerged. The land issue became a rallying point with Mugabe again promising a revolutionary land reform programme. The British wanted a continuation of market-based land reform and were unwilling to support reform without this assurance. The government though was now faced with the first opportunity to redress the land question in its own way, without Lancaster House restrictions. It issued a policy paper in January 1990, which culminated in an amendment to Section 16 of the constitution in order: to enable the government to acquire any land (including utilised land) for resettlement purposes; to require only fair compensation to be paid and within a reasonable time; and to abolish the right to remit compensation out of the country as required by Sections 5 and 6 of the constitution (Mushimbo 2005). Farms belonging to multiple farm-owners and absentee landlords, and under-utilised or derelict land contiguous to communal areas, were targeted first for redistribution. Payment in any currency of choice was abolished. Combined, this entailed a break from the Lancaster House Agreement.

This led to the Land Acquisition Act of 1992 which was consistent with the 1990 policy paper. Its tenets are summarised by Nmoma (2008) as follows: payment for land acquired was to be in local currency only; the government could now compulsorily acquire land which was being fully utilised; and the willing-seller/willing-buyer principle was in effect discontinued. The Act would enable the government to acquire some 5.5 million hectares of the eleven million hectares of land still held by white farmers. However, the underlying constitutional amendments (on which the legislation was based) culminated in a constitutional crisis in 1991, when Chief Justice Anthony Gubbay challenged parliament's power to amend the constitution, by invoking the essential features doctrine. Chief Justice Gubbay asserted the power of the courts and their prerogative to oversee the legal aspects of land reform and he warned that the courts might invalidate the constitutional amendment. The amendment, which revoked white farmers' right to appeal to the judiciary to determine the amount of compensation, made land redistribution a political rather than a legal matter (Mushimbo 2005). The legal debate on land reform became polarised and, in the end, Justice Gubbay was asked to step down as a judge (van Horn 1994).

The irony is that the drastic change in policy did not benefit the communal area farmers and landless people. Rather, a further scramble for land began amongst the black political elites as corruption became rampant (de Villiers 2003). Although introducing the new reforms as a means of empowering communal villagers, "the ruling elite have made little more than token resettlement of the landless peasant farmers on acquired land" (Makumbe 1999: 14). Makumbe (1999: 15) goes on to note that "the elites have made effective use of the Land Acquisition Act 1992 to feather their own nests". Various critics (Goebel 2005; Moyo 1995) argue that redistribution of land continued to lack transparency, and was marked by regional, ethnic and class biases that favoured elite Africans from the regions and ethnic groups dominant within ZANU-PF. A report in the *Guardian Weekly* (30 June 1996: 4) highlighted: "Two years later [after 1992] it was revealed that the first farms compulsorily purchased had been allocated to cabinet ministers, top civil servants and army generals". Similarly, *The Herald* (1 July 1996: 10) noted: "A proper commission of inquiry should be appointed to look into, and establish the veracity of allegations made at the weekend that senior government officials in Masvingo have taken over a farm earmarked for resettling peasants, and that they are helping themselves to the farm".

The 1992 Land Acquisition Act empowered the president to acquire agricultural land compulsorily and set out the procedure in accordance with which acquisition would take place. Once written notice had been given to a farm owner that the farm fell within the acquisition category, the land could not be sold or permanent improvements made on it. The notice was for one year and, as soon as it was published, ownership was automatically transferred to the state even though compensation had not yet been settled. The responsible Minister had discretion to designate any land as agricultural and therefore compulsorily acquired in the public interest. The Minister was required to specify the purpose for which the land was to be acquired and the period within which it was intended to be acquired, and this period could not extend beyond ten years.

The government also set up a Compensation Committee, which determined the amount to be given to white farmers whose land had been designated. The committee would take into consideration factors such as the character and cost of improvements made by the farmer as well as the type and size of landholdings, with any dispute regarding compensation referred to the administrative court. The Act required the owner to receive half of the compensation within a reasonable time (within a year) and the remaining balance within five years (de Villiers 2003). Many scholars (Matondi 2007; Moyo 2000c; Makumbe 1999; Marongwe 2008) highlight that, despite the drastic and controversial measures spelt out in the Land Acquisition Act, the whole process of acquisition and resettlement remained frustratingly slow during the 1990s.

Reasons for the snail-paced character of land reform in the 1990s are multiple and complex but include government's lack of will and funding, corruption, and class biases that increasingly favoured black business people rather than communal villagers. There was a clear weakening in government's commitment to broad-based resettlement, in part because of the shifting class interests of the ruling party (Goebel 2005). Overall, the land transfer programme shifted towards distributing land to capable farmers rather than peasants (Moyo 1995). Indeed, while the Land Acquisition Act of 1992 seemingly marked a break from market-based land reform programme, the structural adjustment programme (known as ESAP) implemented in Zimbabwe ensured the continuation of, and further support for, large-scale white commercial agriculture. Zimbabwe officially embarked on structural adjustment in October 1991. Since 1980, the World Bank had been Zimbabwe's largest donor and was able to exert critical pressure on government policies (Goebel 2005). The process of adjustment (backed by the World Bank) meant the withdrawal of any sustained interest in land redistribution as neo-liberal policies, which promoted commercial agriculture, took root (Gibbon 1995). Land reform therefore soon became a "hostage of measures intended to reduce budget and balance of payments deficits" (Tshuma 1997: 58).

Intriguingly, then, land reform under ESAP worked against the spirit and clauses of the Act of 1992, as it had an "increasingly market oriented conception of Zimbabwe's land question" (Moyo 2000c: 9). As a result, it led to what Sachikonye (2003: 231) terms an "interlude" in addressing land inequalities, with the government disengaged from land reform. The period from the late 1980s to 1996 witnessed declining redistribution as black commercial farming was increasingly promoted. In this way, ESAP dovetailed neatly with legitimising agricultural accumulation amongst an aspiring black landed class. The government constrained its land acquisition programme by not increasing budgetary allocations for this purpose and by not using more extensively the compulsory land acquisition and price fixing mechanisms it had enacted in

1992 (Jowah 2009). Overall, minimal land redistribution took place, at a land acquisition rate of approximately 40,000 ha per annum; with less than 20,000 households
receiving land between 1990 and 1997 (Sachikonye 2003: 231).

In the meantime, the status of women in communal areas post-1980 did not change
significantly over the first two decades of independence (Gaidzanwa 1994; Goebel
2005; Steen 2011; Makura-Paradza 2010; Chakona 2012). Though, for most of this
period, chiefs were not officially recognised, in practice they were central to local
systems of patriarchy prevailing in communal areas. Male patriarchs, including male
household heads, tended to dominate in terms of control over (and access to) land,
with women typically having secondary rights to land, limited rights of inheritance
and the burden of domestic responsibilities. In this regard, there was some land-access
variation across marital status, specifically, between widowed, divorced, married and
single women.

6.4   Immediate Pre-fast Track Period (1997–1999)

The period from 1997 to 1999 is often interpreted as a critical watershed, or time
of uncertain transition, in the land reform programme in Zimbabwe. It is the period
which directly provided impetus to the land occupations of 2000. In 1998 the government began what it called the second phase of its land reform initiative. The objectives
of phase two were spelt out as follows: acquire five million hectares of commercial
agricultural land for redistribution; resettle 91,000 families as well as youth gradu-
ating from agricultural colleges (and others with demonstrable experience in agriculture) in a gender-sensitive manner; and reduce the extent and intensity of poverty
amongst communal area families and farm workers by providing them with adequate
land for agricultural use. At the same time, in 1997, there was a controversial attempt
by the state to acquire over 1471 commercial farms (Marongwe 2002). These farms
were designated for acquisition but were later reduced to 841 farms following the
de-listing of over 400 farms. Most of the remaining farms were removed from the
designation list after owners appealed to the courts; and government failed to respond
to the appeals within the legally defined period and subsequently discontinued the
acquisition process. This, coupled with an ongoing lack of funding (Matondi 2007),
led to serious frustrations over the pace of land reform amongst the landless especially
marginalised veterans of the guerrilla war.

A 1998 Donor's Conference in Harare saw the government trying to speed up land
reform, and under a land programme based on democratic principles and inclusivity.
Forty-eight major countries including Great Britain, the United States and South
Africa, as well as donor organisations such as the World Bank, attended. But, as
Masiiwa (2005) notes, donor unwillingness to fund land reform in Zimbabwe was
underscored at the Harare conference. As part of the phase two land reform, government unveiled a programme requiring US$1.9 billion (about ZW$42 billion). To
the disappointment of the Zimbabwean government, the donors only pledged about
ZW$7339 million, just a drop in the ocean.

The conference though considered a number of issues and agreed on certain principles: transparency, respect for the rule of law, poverty reduction, affordability, and consistency with Zimbabwe's wider economic interests. It advocated for a broadened and more flexible approach to land acquisition and resettlement, and strengthened stakeholder (including civil society) consultations and partnerships. This became embodied in the Inception Phase Framework Plan of 1998–1999. There was also consensus around: addressing gender issues in relation to access to and control of land; proportionate representation on land reform decision-making structures and streamlining of land policies such as around land taxation, subdivision and tenure (Mushimbo 2005). In the end, this restructuring of the land reform programme was overtaken by events in early 2000.

Overall, it seems that the years from 1997 to 1999 were marked by significant ambiguities and contradictions in state policy and practice on land reform, which in part reflected tensions within the ruling party. As Moyo notes: "Throughout the mid-1997 to 1998 period of compulsory land acquisition, the government publicly appeared not to be providing room for negotiation, when in fact negotiations and trade-offs with stakeholders had long been underway" (Moyo 2000a: 28).

On the one hand, there was extensive land policy formation by the state, as well as intensive "policy dialogue activities" and intermittent "high profile negotiation[s]" (Moyo 2000b: 2, 3) on land reform taking place between government and various non-state bodies including the CFU. This was part of a broader "politics of broadly-based policy dialogue and negotiation" (Moyo et al. 2000: 186) that the ruling party had tentatively adopted in the late 1990s. On the other hand, and consistent with developments from the early 1990s, the government made the shock announcement that 1471 commercial farms (on prime land) would be acquired on a compulsory and urgent basis. From Moyo's perspective, this "adoption of a centralised method of compulsory land acquisition" was "instigated" by "the war veterans" (Moyo 2001: 314). Simultaneously, ZANU-PF was also under pressure from its own radical nationalist wing to adopt this route.

This ambivalence and inconsistency embedded in land reform initiatives was an outward expression of the dilemmas facing the government in responding to deepening crises in the late 1990s, including trade union and civic movement opposition and a downturn in the economy. Despite the seeming embourgeoisement of the ruling party and the land reform programme, it appeared that—within the ruling party and state—there was a "political hardening of the radical nationalist social forces and an escalation of demands to resolve land reforms as a matter of sovereign right, pride and reparation, rather than as a mere matter of poverty alleviation" (Moyo 2000b: 5) or even, for that matter, agricultural productivity.

While these developments were unfolding, other (even more) dramatic events were taking place: a series of isolated land occupations combined with war veterans asserting their power vis-à-vis the ruling party. Together, these events likely contributed to the playing out of the tensions existing within the ruling party.

6.4.1 Land Occupations

Land problems festered after 1980 as communal villagers faced land pressures and increasingly grew disillusioned by the slow pace of land reform. Soon after independence, a significant number of land occupations driven and led by communal villagers took place (200 alone in 1980), primarily on farms which had been abandoned during the war of liberation (Chavhunduka and Bromley 2013); and hundreds of occupations were to occur subsequently (Marongwe 2002). In this regard, Moyo (2004) notes that, from 1980 to 1984, government's land redistribution programme was pursued mainly on the basis of these occupations, in which communities in effect squatted on abandoned land and became self-selected beneficiaries of land reform. This in many ways supplanted the official land beneficiary waiting lists. Thus, at times, community-led occupations were accepted by government and it was forced to purchase the land and regularise and legitimise the occupations. Alexander (2003) argues that these early occupations were initiated by local ZANU-PF party leaders who effectively supported some form of ongoing grassroots nationalism. The response of the national government to these early land occupations, and to later ones as well, was varied and contextual. Thus, "[w]hereas, the GoZ [Government of Zimbabwe] used forced evictions to restrain this approach from 1984, evidence abounds on the complicity of Zanu PF and GoZ officials, and war veterans, in land occupations during the entire first 20 years" (Moyo 2004: 7) after independence.

Sufficient evidence exists showing that, from the mid-1980s, the government tended to treat any new occupations as illegitimate, with the occupiers defined officially as squatters (Marongwe 2002; Moyo 1995) and thus liable to be forcibly removed or evicted from the land by police, which did in fact happen and often violently (Herbst 1990). Nevertheless, sporadic and isolated occupations continued, but there was a general lull between 1988 and the mid-1990s. A significant number of occupations, approximately 200, then arose in 1995 (Marongwe 2002) and many others over the next few years; unlike in the preceding years, the government was not as active in evicting the occupiers (de Villiers 2003). Thus, pressure on government from occupations increased in the late 1990s (Sachikonye 2003), including the Chikwaka farm occupations in Goromonzi district spearheaded by war veterans and chiefs (Sadomba 2008). This occupation, as with many others, was based on land restitution claims: that is, reclaiming land on ancestral grounds. For example, Alexander (2003) shows how villagers in Dende communal areas occupied Munenga claiming restitution of land they had lost under colonialism.

In the late 1990s, in Matabeleland, 200 families occupied four farms in Nyamandlovu. The communal villagers occupying these farms came from Tsholotsho and Nyamandlovu, and others came from resettlement areas established in the 1980s such as Irisvale and Zimdabule, where sons of the land beneficiaries struggled to access land. In the mountainous Muzura area of the Nyakapupu small-scale commercial farming area in Guruve, around 200 families occupied land held by the state. With regard to the occupation of Mbalabala Ranch in Umzingwane District, again sons from older resettlement areas were amongst the occupiers. As well, 700 people occupied Longdale Farm near Masvingo town and, in the same province, a significant number of war veterans (36 in total) occupied the state's Mkwasine Estate in Chiredzi (Marongwe 2002). A number of occupations took place while the donors' conference was being held.

The occupations from 1980 to 1999 ebbed and flowed, with Moyo (2001, 2005) seeking to identify different phases in the occupations. We quote Moyo (2001) at length about the details of the shifting character of the occupations over time (Moyo 2001: 321–322):

> The first phase of land occupations can be termed, 'low profile, high intensity'. These occurred throughout the country, from 1980 to 1985... These early land occupations were led by landless communities led by war veterans, the ZANU (PF), 'dissidents' (especially in Matabeleland), and by other traditional leaders, such as the spirit mediums. They were tacitly supported by ZANU (PF) and PF ZAPU structures albeit without the public flaming of the political basis of the mobilisation.

> The period between 1985 and 1996 witnessed what we can call in relative terms 'normal low intensity occupations'. ...During the 1990s, landless communities increased 'illegal' occupations and poaching of natural resources on private, state and 'communally' owned land, and in urban areas. Local 'squatter' communities made themselves beneficiaries by occupying mainly abandoned and under-utilised land, most of which was in the liberation war frontier zone of the Eastern Highlands. This 'community led' occupation approach resulted in a process of informal land identification. The central government purchased the occupied land at market prices, thereby formalising the occupations in what came to be known as 'normal intensive land reform'.

> However, the government has used forced evictions to restrain land occupations, especially during the transition to the liberalised economic policy framework. The brutality with which these evictions were carried out, both by police and farmers, were reminiscent of the colonial era. This was coupled by an increase in the violent attacks against illegal occupants by property owners, particularly white farmers often operating with implicit or explicit state approval. ...In any case, the squatter policy failed to reduce land occupations, mainly because of legitimacy problems at the local level. ...

> The last phase of high intensity and high profile land occupations began in 1997, ...In September 1997, the more high profile community-led land occupation approach re-emerged and isolated land occupations started to occur, with the explicit aim of redistributing land from white farmers to landless villagers and war veterans. These occupations reinforced existing low profile land occupations throughout the country. They came in waves, starting with just about thirty cases in 1997, mostly on farms, which had been identified for compulsory acquisition. The squatters later 'agreed' to 'wait' for their orderly resettlement and in some cases were evicted by the government in 1998.

Besides this temporal variation, the occupations were marked by considerable spatial differences, with some areas devoid of any occupations. It should be added as well that, in addition to these occupations, there were constant incursions by communal villagers onto nearby white commercial farms and other landholdings to access grazing, wood and other natural resources, and to engage in illegal gold panning. Further, there continued to be contestations within the communal areas, including disputes around boundaries, grazing rights and land access. Below, we discuss certain occupations which took place.

For Chimanimani, Alexander (2003) provides an analysis of how "squatting" or unofficial occupations increased from 1981, with chiefs sometimes leading the occupations. There were also government officials who were sympathetic to the occupations and shielded "squatters" from arrest by the police. While government tended to be sympathetic to "squatting" in this region, elsewhere across the country the new official discourse about "squatters" was emerging in which occupations were viewed as destructive, criminal and a major environmental and economic threat. Thus, as indicated, the second half of the 1980s saw the emergence of a state machinery geared towards coercing "squatters" off occupied land, which took place through the creation of Squatter Control Committees (Alexander 2003). These committees showed government's commitment to ruthlessly dealing with what was viewed as a growing "squatter problem".

In Lupane, Thebe (2017: 202) provides an analysis of land occupations before 2000 steeped in localised concepts of *siphile* or land self-provisioning and "radical land restitution (of land previously annexed from people by the local authority for a pilot grazing project)". This account provides a nuanced understanding of land occupations within a cultural and historical context of what is termed *madiro* (defined as both freedom farming and unauthorised development of settlements). Chaumba et al. (2003) argue that *madiro* was, for many communal area farmers, a form of protest and a way to be seen and recognised by the government. Wels (2003) outlines how, in the Save Conservancy area from August 1998, occupations took place on a regular basis and grew increasingly aggressive as they were accompanied by threats to game guards. Earlier occupiers amongst the Matsai people on part of the conservancy (Angus farm) had been ruthlessly removed by the police in 1986, burning their huts (Wels 2003).

Probably the best-known case of occupations in the pre-2000 period is by the Svosve people in Mashonaland East who in 1998 occupied four farms. Marongwe (2008) states that land restitution claims were at the centre of the Svosve farm occupations. This was buttressed by the lack of transparency in earlier land reforms before 1997. Many villagers were angered by the murky beneficiary process at Pinewood Farm—only three people from Svosve were allocated plots on the 600-ha farm and, in the main, "outsiders" accessed redistributed land. Similar occupations at this time were driven by complaints around marginalisation of communities from land redistribution projects.

Andrew and Sadomba (2006) and Sadomba (2008) also speak about the occupations in Svosve. They claim that war veterans from Harare instigated the occupations and contacted local Svosve war veterans in the process. After three days of discussions over whether or not to occupy farms, local leaders in Svosve, including chiefs and spirit mediums, mobilised others to occupy Igava farm, with war veterans remaining in the background during the process. They refer as well to occupations in Goromonzi three months later. The first Goromonzi farm to be occupied was targeted because the farmer (named "Shiro") allegedly committed atrocities against prisoners of war—both captured guerrillas and local villagers from Chikwaka—during the second *chimurenga*. Another ten surrounding farms were occupied, including by a local chief in Goromonzi, councilors, war veterans in nearby communal areas, civil servants in Goro Rural District Council, communal area villagers, and farm workers. These occupations then spread rapidly to encompass commercial farms around Chinamora communal areas about 29 km away.

The land occupations during the first two decades of independence were very diverse in terms of their composition, motivation and leadership. Nelson Marongwe (2002: 23), in speaking about the 1998/1999 occupations, argues that they were mainly community-led:

A combination of villagers and farm workers played an important role in the occupations. The major concerns by the villagers were delays by the Government in resettling them and the fact that, in general, they were not informed of the land reform programme. The proximity of the farms to their homes, poor relations between farmers and neighbouring farms and historical land claims by the communities pushed them to occupy farms.

Occupations prior to the year 2000 arose at deeply local levels, and because of localised experiences, concerns and aspirations. This is not to deny that localised-recognised ruling party leaders were involved, because at times they were (just as they were involved, as we will show, during the third *chimurenga*). For instance, in Chegutu District, political friction between two senior ZANU-PF officials led one of them to incite villagers to occupy a state farm being leased by the other (Marongwe 2002).

The third *chimurenga* occupations, though, are seen as fundamentally different from the pre-2000 occupations. Like Goebel (2005), Nelson Marongwe (2002) and others, Chitiyo and Rupiya (2005) claim that the fast track occupations were a "state-sponsored exercise", unlike "the previously sporadic and spontaneous grassroots farm invasions by landless peasants" (2005: 359) before the year 2000. In fact, the contrast between fast track and earlier land occupations is constantly highlighted in the fast track literature. Thus, Alexander (2003: 199) argues that there were key differences between occupations in the early 1980s and the third *chimurenga* ones:

Grass-roots nationalism in the early 1980s had encompassed a desire both for the return of land, and for an accountable, responsive state. Zanu (PF) in 2000 promised the land, but at the prices of an extreme and violent political intolerance that severely undermined the long-standing popular aspirations for a 'good' state... It was not a revived pre-independence nationalism that lay behind the wave of occupations in 2000, but a far narrow one.

Sithole et al. (2003) also draw a clear line between the fast track occupations in the year 2000 and those in the late 1990s. For these scholars, earlier land occupations which occurred in Svosve and Nyamandhlohvu (and which sparked other occupations in areas such as Nyazura) reflect genuine disenchantment by communal area villagers with the ruling party and the slow pace of land reform before fast track. In the case of the fast track occupations, genuine grievances by "genuine" peasants in effect were replaced by "manufactured" peasants comprising state and party-financed militias (including war veterans, unemployed youths and local party supporters).

Despite the claims by critics of the third *chimurenga* land occupations that these occupations were centrally organised, we claim (in following chapters) that they were also situational and founded on the contingencies of local dynamics.

6.4.2   War Veterans

In understanding the third *chimurenga* occupations, it is necessary to trace the genesis of war veterans and their activities between 1980 and 1999. After independence in 1980, and based on the ceasefire brokered at the Lancaster House Agreement, the Zimbabwean government grappled with creating a new functional military bringing together the Rhodesian Army, and ZANLA and ZIPRA guerrilla fighters (Mazarire and Rupiya 2000). Chitiyo and Rupiya (2005: 339) note that around 65,000 soldiers from the three armies were available for integration into the Zimbabwean army and the remaining 30,000 former guerrillas were to be processed through demobilisation and hopefully resettled in food and production enterprises which would serve the standing army. The government created a Demobilisation Directorate in 1981 focusing on three options for demobilisation for Rhodesian security force members and ex-guerrillas, including: "an involuntary option under which lapsing Rhodesia Security Forces contracts were not renewed, a voluntary package of four months' salary plus a monthly stipend of Z$185 for two years, and a disabled rehabilitation centre for special cases" (Jackson 2011: 384).

The demobilisation process itself was complex and fractured leading to serious conflicts at the assembly points where guerrillas gathered. There were deadly incidents such as on November 9 and 10 in 1980 when ZIPRA and ZANLA soldiers fought leaving 55 people dead and 550 injured (Kriger 2003). Also, between 7 and 10 February 1981, 197 people were killed at one assembly point alone. Conflicts and fractures between ex-ZIPRA and ex-ZANLA guerrillas continue to this day. Muzondidya and Ndlovu-Gatsheni (2007) refer to the seeming ethnic conflicts and divisions between ZANU and ZAPU (including during the war of liberation), which contributed to problems pertaining to demobilisation and reintegration.

Nyathi (2004) shows how these tensions were carried into the period known as *Gukurahundi* (from 1982 to 1987) where a specially trained force of ex-ZANLA fighters killed an estimated 20,000 people in the Matabeleland and Midlands regions, including ex-ZIPRA fighters. Many villagers were maimed while some were raped or sodomised and tortured (Zimbabwe Human Rights Association 1982; CCJPR 1999). This provided the backdrop to "the historical ethnic polarisation that characterised Zimbabwe's political [space]" and it, for some time, "prevented the establishment of a national war veterans lobby group".[1] *Gukurahundi* appeared to be consistent with earlier attempts by ex-ZANLA leaders and forces to eliminate ZIPRA during the demobilisation process at assembly points. After the 1983 arrest of ZAPU leadership, there was widespread use of violence against former ZIPRA guerrillas within the Zimbabwe National Army, coupled with segregation, disarmament and disappearances, and an overall downplaying of ZIPRA's role in the liberation struggle (Jackson 2011).[2] It was only after the 1987 Unity Accord between Robert Mugabe

---

[1]https://www.theindependent.co.zw/2017/11/25/war-veterans-love-hate-affair-mugabe-endgame/.

[2]The ongoing bitterness between ZANU and ZAPU, dating back to the pre-1980 period and including of course *Gukurahundi*, was reflected in the response of Ndebele people and former

and Joshua Nkomo that a national all-inclusive war veterans' national organisation could be formed despite diverse political, regional and ethnic affiliations.

Overall, the Demobilisation Directorate failed to come to terms with the huge task before it. As Dzinesa (2000: 6) concludes:

Notwithstanding the existence of a dedicated Demobilisation Directorate, there were programmatic and institutional gaps. These included a lack of broad and consistent socio-economic profiling of combatants, the failure to implement financial management skills training for the many ex-combatants inexperienced in handling (demobilisation) money, incompetent and corrupt directorate staff, an absence of elaborate and workable business or cooperative support mechanisms and the lack of proactive monitoring mechanism. The majority of the ex-combatants enterprises collapsed due to these factors while agro-based enterprises were also hard-hit by drought.

The government initiated the Zimbabwe Reconstruction and Development Conference to facilitate war veterans' reentry into education and retraining but this was largely a failure as the type of training provided was ill-equipped to the needs of war veterans. Another failed initiative to provide a viable future for war veterans was Operation SEED initiated by the Zimbabwean state in 1981. Operation SEED is an acronym for the "Operation of Soldiers Employed in Economic Development". It was designed to encourage ex-combatants to swap their guns for picks and shovels and to work on land acquired by the government for that purpose.

There was a serious crisis of expectation as large numbers of demobilised ex-guerrillas became destitute and the majority were left out of the benefits of independence, including access to land. In fact, after 1980, a range of ZANU-PF politicians began to monopolise the political and cultural capital of their liberation war status for purposes of capital accumulation, while most ex-guerillas were sidelined socially and economically (Sadomba 2008). Musemwa (2011), in writing about a novel by Charles Samupindi, outlines the major issues facing veterans (ex-guerrillas) of the liberation struggle in Zimbabwe years after the war. In the novel Samupindi poses three critical questions facing disillusioned war veterans by the 1990s: "Was the war worth fighting for? What has happened to the combatants? Who will validate, who will acknowledge the memories that they carry with them?" (cited in Musemwa 2011: 122). This relates to ex-guerrillas' disillusionment with the events of the post-independence period, about which many veterans living in squalor wrestled (Mashike 2005). It was deeply disappointing for veterans, after participating in (and surviving) a horrific armed struggle, to have their plight forgotten after the war ended.

The process of demobilisation included as well the paying of a monthly allowance to ex-guerrillas, but the process was fraught with fraud (Kriger 2003). Those who accepted demobilisation (and thus were not integrated into the army) were given US$185 per month for the next two years but this was largely inadequate for many ex-guerrillas who had to start new lives after the war. By the end of 1982, the new government was taking steps to place ex-combatants in public and private institutions, though they found it difficult to be accepted within private companies because

ZIPRA guerrillas to the ZANU-PF's government's 'Capturing a Fading National Memory Project' initiated in 2004, which Dombo (2019) suggests was aligned with patriotic history.

they often threatened business owners (Kriger 2003). In promoting cooperatives, the government also embarked on land redistribution programmes that targeted war veterans. Nyathi (2004) notes that, in Manicaland, 250 farms were handed over to war veterans for cooperative projects; however, these failed dismally leading the government to repossess the farms by 1995. Though supported to some extent by government, the agricultural and other cooperatives amongst war veterans faced many serious operational challenges, including recurrent droughts for those that were in agro-based cooperatives under the Zimbabwe Producer and Marketing Organisation (Musemwa 1995). The cooperatives suffered from a lack of technical and managerial skills amongst war veterans, which was made worse by limited government assistance and unfavourable economic conditions especially with the adoption of the structural adjustment programme post 1990.

For female war veterans, the marginalisation and exclusion was deeper, given the double-edged sword of patriarchy and stigma associated with being women fighters. During the war, women guerrillas were seen, at last in nationalist rhetoric, as equals and they assumed that independence would bring about gender equality. As Groves (2007) argues, citing a number of authors:

> Gender politics appeared increasingly institutionalised within ZANU, allowing the party's Women's League to declare publicly that 'for the revolution to triumph in its totality there must be emancipation of women' …. [Nationalist propaganda] 'carried pictures of formidable-looking women, often in situations defying traditional notions of femininity' …. The emancipation of women was possible, … 'thanks to the social change that was set in motion by the armed revolution'; women were playing a 'dynamic role in the national liberation movement'.[3]

However, after independence, women war veterans quickly found out that independence would not bring about emancipation. Popular art such as the movie *Flame* help in understanding the social stigma which faced women guerrillas, who were largely seen as unmarriageable, had minimal skills to create livelihoods and were largely excluded from the opportunities under the demobilisation process as compared to their male counterparts. Schmidt (2001: 416–417) concludes that:

> Despite ZANU-PF's claim to be an emancipatory force for women, the movement's rhetoric was not matched by reality. In both military and civilian life, women continued to be subordinate to men… As "natural" teachers and nurturers, women rallied support amongst the local population and nursed the sick and injured. Relatively few women served as guerrilla fighters. With the exception of a limited number who were connected to powerful men, women were generally excluded from positions of power and authority.

Sadomba and Dzinesa (2004) thus argue that absence of a gender-sensitive demobilisation and reintegration policy resulted in the marginalisation and exclusion of female ex-guerrillas in the military, social, political and professional spheres.

This gendered marginalisation of female war veterans played into the growing discontent amongst veterans with the ruling ZANU-PF party. By the 1990s the majority of war veterans were angry and vengeful (Nyathi 2004). The demobilisation

---

[3]http://www.e-ir.info/2007/12/13/the-construction-of-a-%E2%80%98liberation%E2%80%99-gender-and-the-%E2%80%98national-liberation-movement%E2%80%99-in-zimbabwe/.

and reintegration processes had been a huge failure. In the case of female veterans, nothing of significance had been done to prepare their integration into a society that had deep-seated sentiments against women fighters. Even the male veterans found it difficult to reintegrate and gain the same respect they once had as guerrillas. As Nyathi (2004: 69) put it, "[e]x combatants had to struggle with an identity crisis: they knew they were 'winners' in terms of the revolution but what did they have to show for it?" This crisis of expectation, of aspirations unfulfilled, entered into combined with an ongoing colonial agrarian structure alongside significant accumulation of wealth by black elites, including former nationalists and guerrillas embedded in the higher echelons of the ruling party and the state.

In this context, war veterans began to organise themselves into a national body. McCandless (2005) provides a detailed overview of the history of the Zimbabwe National Liberation War Veterans Association (ZNLWVA). The association was formed in 1989, amidst government suspicion about it, mainly to advocate for the plight of war veterans who felt neglected and ignored by politicians (Kriger 2003). Its formation took place after ex-guerrillas were officially designated as veterans of the war. At the inaugural meeting of the association with then-president Robert Mugabe in 1992, war veterans expressed deep concern about the configuration of power within Zimbabwe, as a seeming alliance existed between ZANU-PF and captains of industry (Sadomba 2008: 157). The war veterans proclaimed to Mugabe that the state was run by "opportunists and bourgeois elements" in a clear reference to his ministers (Musemwa 1995). This no-holds barred meeting was a glimpse into the deteriorating relationship between the war veterans and the ruling party. The work of the ZNLWVA led to vital policy formulations, including the War Victims Compensation Act of 1993.

The compensation act resulted in the creation of the War Victims Compensation Fund to cater for financial compensation on a scale proportional to the severity of injuries suffered during the war of liberation. Chitiyo (2000) notes, however, that the fund was riddled with fraud and corruption which included falsifying injuries, and that it in large part benefitted politicians and politically connected individuals. This included cabinet ministers, then police commissioner Augustine Chihuri, members of parliament and then-president Robert Mugabe's brother in law, Reward Marufu— who was categorised as 95% injured and paid Z$821,668 (Mashike 2005). The ordinary ex-guerrillas were left in large part excluded from the benefits of the fund.

By the mid-1990s, over 25,000 ex-guerrillas were virtually destitute. The promises of the war were not being fulfilled, as noted in a *Motto* magazine headline: "'Son of the soil' during the armed struggle; 'squatter' after independence" (cited in Musemwa 1995: 31). In this respect, it should be highlighted that war veterans were not given any meaningful and sustained preferential treatment in the selection of beneficiaries under the land redistribution programme. There were no special quotas for war veterans beyond the farms provided for under the agricultural cooperative initiatives.

Further, both Makumbe (2003) and Herbert Moyo (2015) highlight that one of the key issues was the failure on the part of the ZANU-PF government to prioritise the psychological dimensions of ex-guerrillas during the demobilisation process. The government did not provide a holistic psychological, emotional, moral or mental

demobilisation of war veterans. This made reintegration into communities more difficult as "most veterans showed signs of mental disturbances, some are very violent and
in fact politicians have used them in subsequent violent activities mainly around election times" (Moyo 2015: 4). The majority of war veterans thus faced multiple physical
and mental challenges to which the Zimbabwean government failed to adequately
respond. Some civil society organisations did though attempt to assist in the reintegration exercise. This included the Zimbabwe Project Trust, a Roman Catholic linked
organisation run by Judith Todd which provided support for various projects by war
veterans.

In 1997, musician Clive Malunga released what turned out to be biggest song of
the year—*Nesango* ("in the forest")—which portrayed the plight of war veterans. In
the song, he narrates one of the chasms that had developed between war veterans
and ruling party, singing: *tinongoti mberere nenyika vamwe vakagarika zvavo mudz-
imba umu* ("we are wandering the country whilst some are living lavish lives"). It
seemed, in fact, that the ruling party was progressively disowning the war veterans,
as Musemwa (1995: 43) noted:

> In October 1992 [ZANU-PF stalwart] Shamuyarira stressed that there were some right-wing
> forces who were manipulating the ex-combatants to gain 'political capital.' Shamuyarira's
> comment sums up the government's lukewarm attitude towards the latter. He declared that
> it was never the intention of the government to create a 'privileged class of ex-combatants'.

Shamuyarira actually went on to argue that going to war was a voluntary act.

Nevertheless, by the late 1990s, the now functional ZNLWVA under the leadership of Chenjari Hitler Hunzvi was putting extensive pressure on then-president
Mugabe. The war veterans' association embarked on multiple protests in pursuance
of a specifically war veterans' agenda, including by disrupting National Heroes Day
celebrations in August 1997 where Mugabe was speaking. In this regard, Hunzvi
reorganised the association into a "highly effective lobbying force and carved out
a niche for his association in a ZANU-PF that was tired, corrupt and visionless"
(Nyathi 2004: 71). Their protests led to discussions between ZANU-PF politicians
and war veterans on 21 August 1997 where demands for a gratuity, pension and most
importantly land were put forward.

At a ZANU-PF summit in Mutare in September 1997, a financial package for war
veterans was set out. The package included the following:

> 52 000 war veterans received a one-off Z$50 000 (about US$4 500 at that time) gratuities
> and Z$2 000 monthly pensions each at an estimated initial cost of over Z$4,5 billion. The
> unbudgeted Z$4,5 billion payout triggered "Black Friday" on 14 November 1997 when the
> Zimbabwe dollar lost 71,5 per cent of its value from around Z$10 to below Z$30 to the US$
> over four hours of trading time and the stock market crashed 46%.[4]

This was the first clear sign that Mugabe was buckling under the increasing pressure
from the national war veterans' association, and that a possible fissure was arising
between the ruling party and war veterans. While the ruling party gave into the

---

[4]https://www.theindependent.co.zw/2017/11/25/war-veterans-love-hate-affair-mugabe-endgame/.

demands for gratuities and pensions, nothing was forthcoming with regard to land. The argument of the war veterans was:

> In order to resolve this issue peacefully, we demand that 50 percent of all ex-combatants needing settlement be given land by December 1997, the rest by July 1998. Failure to meet these deadlines will force war veterans to move in and settle themselves on farms that have been identified for resettlement. They will occupy white man's land because the white man did not buy that land. (quoted in McCandless 2005: 304)

The question of land hence remained a paramount grievance amongst war veterans, even twenty years after independence. Land was the primary grievance of war veterans who accused the ruling elite of engaging in wealth accumulation (Chitiyo 2000). As we discuss in the following chapter, this became abundantly clear in the context of changes to the constitution as put to a vote in a referendum in February 2000. In that month, a national referendum was held around revisions to the national constitution, including strengthening the powers of the executive president and more radical forms of land redistribution.

After seeing the final draft of the constitutional amendments without any reference to compulsory acquisition of land without compensation, the national war veterans' association demanded such a clause in the constitution before voting could take place. In expressing its extreme displeasure, the veterans' association thus vigorously and successfully lobbied ZANU-PF—even using "strong-arm tactics" (McCandless 2005: 404)—to ensure that a land expropriation without compensation clause was incorporated into the voted-on constitution. The ruling party called on its supporters to vote "yes" in the referendum while the new opposition party (the Movement for Democratic Change—MDC) campaigned for a "no" vote. In the end, the constitutional revision was rejected. This marked the first major defeat for the ruling party since independence.

Thus, by early 2000, there was significant tension between the war veterans and ruling party, with the former existing as a movement with some degree of autonomy from the latter (Moyo and Yeros 2005) and not simply subordinate to the dictates of ZANU-PF. This, we argue, is consistent with the claim made in chapter four about the conflictual relationship which existed between the nationalist and guerrilla movements during the war of liberation. It was demobilised ex-guerrillas who became involved in the sporadic occupations before the year 2000 and who then became central to the third *chimurenga* land occupations.

6.5   Conclusion

This chapter has focused on the intervening historical period between the second and third *zvimurenga*, examining in particular the intertwined issues of land reform, localised land occupation struggles and the development of an organised war veterans' association. As with the emergence of the second *chimurenga*, the third *chimurenga* emerged out of a complex array of struggles over an extended duration.

We do not claim that the dynamic interplay of these issues caused or led directly to the third *chimurenga* occupations; rather, they set the conditions for the possibility of their emergence. The next chapter is the first of three chapters on the third *chimurenga* occupations, with chapter seven considering the centrality of war veterans to the occupations. Some scholars, such as Chitiyo (2000) and Kriger (2003), suggest that the ZNLWVA and ZANU-PF were coalescing around a common nationalist agenda at the time of the occupations, a point which we in large part question. However, neither the war veterans as a whole nor ZANU-PF as a party were homogeneous entities, as we have tried to show in this chapter. Hence, the tensions which we describe in the next chapter, between ZNLWVA and ZANU-PF, is not an argument against the existence of some lines of communication and cooperation between elements across the war veteran/ruling party interface.

References

Alexander J (1994) State, peasantry and resettlement in Zimbabwe. Rev Afr Polit Econ 21(61):325–345

Alexander J (2003) 'Squatters', veterans and the state in Zimbabwe. In: Hammar A, Raftopoulos B, Jensen S (eds) Zimbabwe's unfinished business: rethinking land, state and nation in the context of crisis. Weaver Press, Harare, pp 83–118

Andrew N, Sadomba W (2006) Zimbabwe: land Hunger and the war veteran led occupations movement in 2000. Critique Int 31:126–144

Bratton M (1990) Ten years after: land redistribution in Zimbabwe: 1980–1990. In: Prosterman RL, Temple MN, Hanstad TM (eds) Agrarian reform and grassroots development: ten case studies. Lynne Rienner, Boulder, pp 265–291

Catholic Commission for Justice and Peace in Rhodesia (CCJPR) (1999) The man in the middle: torture, resettlement and eviction. Catholic Commission for Justice and Peace in Rhodesia, Harare

Chakona, L. (2012) Fast track land reform programme and women in Goromonzi District, Zimbabwe. Unpublished Masters Thesis, Rhodes University, South Africa

Chaumba J, Scoones I, Wolmer W (2003) From Jambanja to planning: the reassertion of technocracy in land reform in South-Eastern Zimbabwe? J Mod Afr Stud 41(4):533–554

Chavhunduka C, Bromley D (2013) Considering the multiple purposes of land in Zimbabwe's economic recovery. Land Use Policy 30(1):670–676

Chitiyo T (2000) Land violence and compensation: reconceptualising Zimbabwe's land & war veterans debate. Track Two Occasional Paper, Centre for Conflict Resolution, University of Cape Town

Chitiyo K, Rupiya M (2005) Tracking Zimbabwe's political history: the Zimbabwe defence force from 1980–2005. In: Rupiya M (ed) Evolutions and revolutions: a contemporary history of militaries in Southern Africa. Institute of Security Studies, Pretoria, pp 331–363

Cousins B (2003) The Zimbabwe crisis in its wider context: the politics of land, democracy and development in Southern Africa. In: Hammar A, Raftopoulos B, Jensen S (eds) Zimbabwe's unfinished business: rethinking land, state and nation in the context of crisis. Weaver Press, Harare, pp 263–316

De Villiers B (2003) Land reform: issues and challenges—a comparative overview of experiences in Zimbabwe, Namibia, South Africa and Australia. Konrad Adenauer Foundation Occasional Paper

Dombo S (2019) Zimbabwe's 'capturing a fading national memory project': an evaluation and reconsideration. South J Contemp Hist 44(2):55–73

Dzinesa AG (2000) Swords into ploughshares: disarmament, demobilisation and re-integration in Zimbabwe, Namibia and South Africa. Occasional Paper 120, Institute of Security Studies, http://www.iss.co.za/pubs/papers/120/Paper120.htm

Gaidzanwa RB (1994) Women's land rights in Zimbabwe. Issue J Opin 22(2):12–16

Gibbon P (1995) Structural adjustment and the working poor in Zimbabwe. Nordiska Afrikainstitutet, Uppsala

Goebel A (2005) Gender and land reform: the Zimbabwe experience. McGill Queens University Press, Montreal

Government of Zimbabwe (1981) Intensive resettlement: policies and procedures; growth with equity. Government Printers, Harare

Groves A (2007) The construction of 'liberation': gender and the 'National Liberation Movement' in Zimbabwe. http://www.e-ir.info/2007/12/13/the-construction-of-a-%E2%80%98liberation%E2%80%99-gender-and-the-%E2%80%98national-liberation-movement%E2%80%99-in-zimbabwe/

Helliker KD (2006) A sociological analysis of intermediary non-governmental organisations and land reform in contemporary Zimbabwe. Unpublished DPhil. Thesis, Rhodes University, South Africa

Herbst J (1990) State politics in Zimbabwe. University of California Press, Berkeley

Jackson P (2011) The civil war roots of military domination in Zimbabwe: the integration process following the Rhodesian war and the road to ZANLA dominance. Civil Wars 13(4):371–395

Jowah EV (2009) Rural livelihoods and food security in the aftermath of the fast track land reform in Zimbabwe. Unpublished MA Thesis, Rhodes University, South Africa

Kinsey BH (2004) Zimbabwe's land reform programme: underinvestment in post-conflict transformation. World Dev 32(10):1669–1696

Kriger N (2003) Guerrilla warfare in post war Zimbabwe: symbolic and violent politics, 1980–87. Cambridge University Press, Cambridge

Madhuku L (2004) Law, politics and the land reform process. In: Masiiwa M (ed) Post-independence land reform in Zimbabwe: controversies and impact on the economy. Friedrich Ebert Stiftung and Institute of Development Studies, Harare, pp 124–146

Makumbe JW (1999) The political dimension of the land reform process in Zimbabwe. Unpublished paper, University of Zimbabwe, Harare

Makumbe J (2003) ZANU PF: a party in transition. In: Conwell R (ed) Zimbabwe's turmoil: problems and prospects. Institute of Security Studies, Pretoria, pp 24–38

Makura-Paradza G (2010) Single women, land and livelihood vulnerability in a communal area in Zimbabwe. Wageningen Academic Publishers, Wageningen

Marongwe N (2002) Conflict over land and other natural resources in Zimbabwe. ZERO Regional Environment Organisation, Harare

Marongwe N (2008) Interrogating Zimbabwe's fast track land reform and resettlement programme: a focus on beneficiary selection. Unpublished PhD Thesis, University of the Western Cape, South Africa

Mashike JL (2005) Down-sizing and right-sizing: an analysis of the demobilization process in the South African National Defence Force. Unpublished PhD Thesis University of the Witwatersrand, South Africa

Masiiwa M (2005) The fast track resettlement programme in Zimbabwe: disparity between policy design and implementation. Round Table Commonwealth J Int Aff 94(379):217–224

Matondi P (2007) Institutional and policy issues in the context of land reform and in the context of land reform and resettlement programme in Zimbabwe. In: Khombe T (ed) The Livestock sector after the fast track land reforms in Zimbabwe. National University for Science and Technology, Bulawayo, pp 32–45

Mazarire G, Rupiya MR (2000) Two wrongs do not make a right: a critical assessment of Zimbabwe's demobilization and reintegration policies since 1980. J Peace, Conflict Mil Stud 1(1):69–80

Mbaya S (2001) Land reform in Zimbabwe: lessons and prospects from a poverty alleviation perspective. Paper presented at the conference on land reform and poverty alleviation in Southern Africa, The Southern African Regional Poverty Network (SARPN), June 4–5, 2001

McCandless E (2005) Zimbabwean forms of resistance: social movements, strategic dilemmas and transformative change. Unpublished PhD Thesis, American University

Moyo H (2015) Pastoral care in the healing of moral injury: a case of the Zimbabwe National Liberation War Veterans. HTS Teologiese Stud Theological Stud 71(2):1–11

Moyo S (1995) The land question in Zimbabwe. Sapes Trust, Harare

Moyo S (2000a) The political economy of land acquisition and redistribution in Zimbabwe, 1990–1999. J South Afr Stud 26(1):5–28

Moyo S (2000b) Land reform in Zimbabwe: key processes and issues. Unpublished Draft, August 2000

Moyo S (2000c) Land reform under structural adjustment in Zimbabwe: land use change in the Mashonaland provinces. Nordiska Afrika Institutet, Uppsala

Moyo S (2001) The land occupation movement and democratisation in Zimbabwe. Millennium J Int Stud 30(2):311–330

Moyo S (2004) The land question in Africa: research perspectives and questions: contradictions of neo liberalism. J Int Stud 30(2):311–330

Moyo S, Rutherford B, Amanor-Wilks D (2000) Land reform and changing social relations for farm workers in Zimbabwe. Rev Afr Polit Econ 27(84):181–202

Moyo S, Yeros P (2005) Land occupations and land reform in Zimbabwe: towards the national democratic revolution. In: Moyo S, Yeros P (eds) Reclaiming the land: the resurgence of rural movements in Africa, Asia and Latin America. Zed Books, London, pp 165–205

Musemwa M (1995) The ambiguities of democracy: the demobilisation of the Zimbabwean ex-combatants and the ordeal of rehabilitation 1980–93. Transformation 25:31–46

Musemwa M (2011) Zimbabwe's war veterans: from demobilisation to re-mobilisation. Transform Crit Perspect South Afr 75:122–131

Mushimbo C (2005) Land reform in post-independence Zimbabwe: a case of Britain's neo-colonial intransigence? Unpublished Master of Arts Thesis, Graduate College of Bowling Green State University, United States America

Muzondidya J, Ndlovu-Gatsheni S (2007) 'Echoing silences': ethnicity in post-colonial Zimbabwe, 1980–2007. Afr J Conflict Resolut 27(2):275–297

Nmoma V (2008) Son of the soil: reclaiming the land in Zimbabwe. J Asian Afr Stud 43(4):371–397

Nyathi PT (2004) Reintegration of ex combatants into Zimbabwean society. In: Raftopoulos B, Savage T (eds) Zimbabwe: injustice and political reconciliation. Institute of Justice and Reconciliation, Cape Town, pp 63–78

Palmer R (1977) Land and racial discrimination in Rhodesia. Heinemann, London

Palmer R (1990) Land reform in Zimbabwe 1980–1990. Afr Aff 89(335):163–181

Sachikonye L (2003) From "growth with equity" to "fast track" reform: Zimbabwe's land question. Rev Afr Polit Econ 30(96):227–240

Sadomba WZ (2008) War veterans in Zimbabwe's land occupations: complexities of a liberation movement in an African post-colonial settler society. Unpublished PhD Thesis, Wageningen University, Netherlands

Sadomba F, Dzinesa GA (2004) Identity and exclusion in the post-war era: Zimbabwe's women former freedom fighters. J Peacebuilding Dev 2(1):51–63

Schmidt E (2001) Review article: for better or worse? Women and Zanla in Zimbabwe's liberation struggle. Int J Afr Hist Stud 34(2):416–417

Sithole B, Campbell B, Dore D, Kozanayi W (2003) Narratives on land: state-peasant relations over fast track land reform in Zimbabwe. Afr Stud Q 7(2&3):81–95

Steen K (2011) Time to farm: a qualitative inquiry into the dynamics of the gender regime of land and labour rights in subsistence farming: an example of the Chiweshe communal area, Zimbabwe. Unpublished Doctoral Thesis, Lund University, Sweden

Thebe V (2017) Legacies of *Madiro?* Worker-peasantry, livelihood crisis and *'Siziphile'* land occupations in semi-arid north-western Zimbabwe. J Mod Afr Stud 55(2):202–224

Tshuma L (1997) A matter of (in) justice: law, state and the agrarian question in Zimbabwe. SAPES Books, Harare

Van Horn A (1994) Redefining "property": the constitutional battle over land redistribution. Zimbabwe J Afr Law 38(2):144–172

Von Blanckenburg P (1994) Land reform in Southern Africa: the case of Zimbabwe. In: Land reform; land settlement and cooperatives, FAO, Rome

Weiner D, Moyo S, Munslow B, O'Keefe P (1985) Land-use and agricultural productivity in Zimbabwe. J Mod Afr Stud 23(2):251–285

Wels H (2003) Private wildlife conservation in Zimbabwe: joint ventures and reciprocity. Brill Publishers, Leiden

Zimbabwe Human Rights Association (1982) Choosing the path to peace and development: coming to terms with the violations of the 1982–1987 conflict in Matabeleland and Midlands provinces. Zimbabwe Human Rights Association, Harare

Chapter 7
The Third *Chimurenga*: Party-State and War Veterans

Abstract This is the first of three chapters on the third *chimurenga*. This chapter focuses on the crucial question of the ruling party and state and their supposed involvement in the fast track land occupations (third *chimurenga*), raising the centrality of war veterans to the occupations. In showing that the ruling party and state were in large part peripheral to the occupations, the following chapter is then able to focus on the complex localised dynamics of the occupations as they unfolded across the Zimbabwean countryside. The chapter discusses the acrimonious debate which emerged soon after the occupations began, a debate about the very form of the third *chimurenga* occupations. This sets the stage for a detailed scrutiny of war veterans and the party-state vis-à-vis the occupations. The war veterans, even if articulating a nationalist agenda seemingly consistent with the ruling party, were neither acting at the behest of the ruling party nor acting in alliance with it. The chapter ends by indicating briefly the sheer diversity of the occupations nation-wide in order to demonstrate the decentralised character of the occupations, which becomes the central theme for the following chapter.

Keywords Third *chimurenga* · Fast track · Zimbabwean state · War veterans · Land occupations

7.1 Introduction

This is the first of three chapters on the third *chimurenga*. This chapter focuses on the crucial question of the ruling party (in conjunction with the state) and its supposed involvement in the fast track land occupations, raising simultaneously the centrality of war veterans to the occupations. In seeking to show that the ruling party and state were in large part peripheral to animating the occupations, we are able subsequently to concentrate in the following chapter on the complex localised dynamics of the occupations as they unfolded across the Zimbabwean countryside. The chapter is divided into three sections. The first section discusses the acrimonious debate which emerged soon after the occupations began, a debate which existed both within the public sphere and amongst Zimbabwean scholars about the very form of the third *chimurenga* occupations. This discussion sets the stage for a detailed scrutiny, in

© The Author(s), under exclusive license to Springer Nature Switzerland AG 2021   149
K. Helliker et al., *Fast Track Land Occupations in Zimbabwe*,
https://doi.org/10.1007/978-3-030-66348-3_7

the second section, of war veterans and the party-state vis-à-vis the occupations. We argue that the war veterans, even if articulating a nationalist agenda seemingly consistent with ZANU-PF, were neither acting at the behest of the ruling party nor acting in alliance with it. On this basis, in the third section, we briefly indicate the sheer diversity of the occupations nation-wide in order to bring to the fore the decentralised character of the occupations, which becomes the central theme for the following chapter.

7.2   Acrimonious Zimbabwean Debate

Moyo and Yeros (2005b) argue that the fast track land occupations represent, as of the turn of the century, "the most notable of rural movements in the world today"; that they obtained "the first major land reform since the end of the Cold War"; and that they have been "the most important challenge to the neocolonial state in Africa" under neo-liberal conditions (Moyo and Yeros 2005b: 165). Even more controversially, they claim that the occupations—as well as the subsequent fast track land reform programme—had a "fundamentally progressive nature" (Moyo and Yeros 2005b: 188). They argue that the occupations were in the main initiated and led by war veterans and had a decentralised character, and that the positive ramifications of fast track land reform—for example, in de-racialising the countryside and redistributing land assets—should not be underestimated.

This depiction of the third *chimurenga* (both occupations and fast track), in their early individual work (Moyo 2001; Yeros 2002a, b) and in their later more comprehensive collaborative projects (Moyo and Yeros 2005a, 2007a), became subject to significant criticism by a large grouping of Zimbabwean scholars. These scholars claimed (as many of them still do) that the land occupations were instigated and led by the ruling party (pure and simple) in a top-down manner and, along with the far-reaching negative implications of a deeply disorganised fast track reform programme, were marked first and foremost by significant authoritarian and violent restructuring by the Zimbabwean state (Raftopoulos and Phimister 2004; Moore 2004; Raftopoulos 2006). This led to an important but acrimonious debate about agrarian change, nationhood and state formation in postcolonial Zimbabwe. In considering this debate, we refer not only to the land occupations specifically, but also the broader reconfiguration of Zimbabwean state and society.

This controversy has tapered off in recent years and some scholars have sought to move beyond it, as we show later. But the debate's character and intensity became quite clear in the year 2008 with the publication of a piece on post-2000 Zimbabwe by acclaimed African scholar Mahmood Mamdani (Mamdani 2008). In drawing extensively if not exclusively on the arguments by Moyo, and in claiming that the occupations cannot be reduced simplistically and conspiratorially to state and ruling party actions (Mamdani 2008), Mamdani's interpretation of post-2000 restructuring was interpreted by a group of "concerned African scholars" as mere legitimation of the machinations of ZANU-PF (see Jacobs and Mundy 2009). In certain ways, it appeared that many of these Zimbabwean scholars were engaging not only intellectually, but also emotionally, as if Mamdani had the audacity to enter into a debate about which he had no real interest or expertise. These emotions have been regularly displayed by critics of Moyo as when, years earlier, Worby (2001: 476) asserted very crudely that the land occupations were "effectively carried out by gangs of hired youth armed and trucked into the countryside by the ruling party". At the same time, some critics such as Pilossof (2006: 635) claim that Moyo has been "radicalised by the emotive capacity" of the post-2000 reforms, despite the overwhelming evidence of a corrupt, chaotic and violent land reform.

In the same year as Mamdani's article, and in offering an overview of post-2000 Zimbabwean literature on land, Pilossof (2008: 273) identifies two positions within the critics of the occupations and fast track reform, namely "horrified reactions" within the "white and western world" (including white farmers), and "more considered responses" from academics. In referring to the "more considered responses", he goes on to argue that "[i]nitially those works that appeared in the first few years after 2000 were cautious in their assessments" (Pilossof 2008: 273). In fact, if the responses of the concerned scholars as late as 2008 are included, many academics critical of the third *chimurenga* were anything but cautious.

Amongst the "concerned African scholars", it was likely Hammar (in Jacobs and Mundy 2009) who articulated the character of the third *chimurenga* debate most succinctly when it comes specifically to the third *chimurenga* occupations. In seeking to counter the Moyo (now it seems Mamdani) thesis about the central role of war veterans in the occupations, as a somewhat if not completely autonomous formation vis-à-vis the ruling party and state, Hammar says that this thesis tends "to overestimate the capacities, resources and scale of the veterans' 'organisation', and to significantly under-estimate the overlapping (and persistent) projects of sovereignty and hegemony of the Zanu (PF) party-state" (Jacobs and Mundy 2009: 42). This debate, despite its problems, remains—at least for us—a crucial entry point in seeking to understand more fully the complexities of the fast track occupations.

In this context, critics of Moyo and Yeros claim that their statements about the third *chimurenga* entail—almost perverse—value judgements made by "patriotic agrarianists" (Moore 2004: 409) or "left-nationalists" (Bond and Manyanya 2003: 78) who fail to conceptualise analytically or even highlight empirically the increasingly repressive character of state nationalism in contemporary Zimbabwe, designated as an "exclusionary" nationalism (Hammar and Raftopoulos 2003), an "exhausted" nationalism (Bond and Manyanya 2003) or an "authoritarian populist anti-imperialism" (Moore 2003: 8). Raftopoulos and Phimister (2004) argue that this authoritarianism involves an "internal reconfiguration of Zimbabwean state politics" (Raftopoulos and Phimister 2004: 377) and now amounts to "domestic tyranny" (Raftopoulos and Phimister 2004: 356).

In discussing displacement and state-making more broadly in Zimbabwe since 1980, and with specific reference as well to the implications for white commercial farm labourers of fast track restructuring, Hammar (2008: 418) argues that practices of exclusion and violence by the Zimbabwean state have "strategically combined physical and symbolic removal of bodies (individuals, groups, institutions) from spaces [such as commercial farms] that the party-state and its allies have wanted to reclaim, purify and occupy" (2008: 421). These authoritarian acts of displacement, it is said, are ignored by a "number of African intellectuals on the Left" (notably Moyo) who have "leapt to the defence of ZANU PF" (Raftopoulos and Phimister 2004: 376) and its highly problematic re-distributive economic programme under fast track.

For their part, Moyo and Yeros (2005b) claim that their (supposedly) Movement for Democratic Change-aligned critics (who they call neo-liberal apologists for imperialism or "civic/post-nationalists") demote the significance of national self-determination and of resolving the land question in Zimbabwe—as expressed in the land occupation movement—by focusing on the movement's violence and the state's ill-planned and mismanaged fast track land reform. In this light, Moyo and Yeros insist on conceptualising the land occupations by reference to a re-radicalised (and revitalised) nationalism and an ongoing national democratic revolution under postcolonial conditions.

Clearly, this debate had pronounced political overtones and was initially linked to deep tensions (an almost chasm) within national party-politics of Zimbabwe with the rise of the opposition party MDC in 1999. Thus, just as ZANU-PF's unfolding *chimurenga* narrative about the occupations appeared consistent with the perspective of Moyo and Yeros, the critics of Moyo seemed to be reproducing the position of the MDC. In the case of ZANU-PF, President Mugabe is quoted as saying: "[W]hen the landless people *spontaneously* invaded white farmland to register their protest against this gross injustice [ongoing white control of land], Government then felt compelled to act. It thus embarked upon its fast track resettlement programme" (Zimbabwe Human Rights NGO Forum, 2001, as quoted in JAG and GAPWUZ 2008a: 11–12, our emphasis).

Like critics of the occupations, in his autobiography, former president of the MDC, Morgan Tsvangirai (2011), argues that the occupations were "aimed to smash, scatter and scuttle a potential MDC captive vote", that is, "the groundswell of support among the farm workers [on white commercial farms]" for the MDC (2011: 273). The occupiers themselves were in fact coerced by war veterans into moving onto white farms, as "villagers were forcibly driven from their [communal] homes to nearby commercial farms" (Tsvangirai 2011: 267). In its election manifesto from 2000 and even denying the significance of the land question to colonialism and the anti-colonial struggle in Zimbabwe, the MDC claimed that land reclamation was not central at all to the first two *zvimurenga* (Raftopoulos 2001: 19–20). Indeed, Tsvangirai (2011: 48) suggests that the term "chimurenga" is "a Shona word meaning revolutionary struggle for human rights" and not for land. In historical reflection, the war of liberation in the 1970s was seen, as pursued in particular by ZANU and ZANLA, as based exclusively on intimidation and coercion, with the rural population having to "toe the line ... or face gruesome consequences" (Tsvangarai 2011: 66) as it also experienced during the fast track occupations. Overall, then, the third *chimurenga* occupations were a misguided process devoid of any and all progressive merit.

Besides its overt political tone, the debate—as implied—involved "competing narratives of Zimbabwe's national liberation history" (Hammar and Raftopoulos

2003: 17) such that, post-2000, "history is at the centre of politics in Zimbabwe" (Ranger 2004: 234). This simultaneously involved fundamentally different conceptions of the crisis in Zimbabwe.

On the one hand, the "patriotic agrarianists" propounded a nationalist discourse which spoke of a land crisis and stressed national sovereignty and redistributive policies. In terms of this discourse, Raftopoulos (2006) says that land "became the sole central signifier of national redress, constructed through a series of discursive exclusions" (Raftopoulos 2006: 212). In an argument consistent with a historiography of nationalism*, this process of exclusion entails sidelining and undercutting counter-narratives about other themes (such as labour, democracy and identity) located in other spaces in Zimbabwe's past, including rural Matabeleland and the urban trade union movement (Alexander et al. 2000; Raftopoulos 2001). Likewise, Muzondidya (2007) highlights that, via the state's *chimurenga* monologue, the history of subject minorities falls outside notions of national belonging and citizenship, such as foreign workers on commercial farms and ethnic groups like Tonga.

On the other hand, the "civic nationalists" adopted a more liberal discourse, what Rutherford (2002: 1) refers to as a "managerial, modernising nationalism". This discourse speaks about a governance crisis in Zimbabwe and emphasises the importance of human rights and political democratisation in the struggle against colonialism and in resolving the contemporary crisis marked by state authoritarianism (Hammar et al. 2003; Sachikonye 2002).

Further, the first discourse, as exemplified in the work of Moyo, focuses on the external (imperialist) determinants of the crisis and the second discourse (as seen in Raftopoulos' work) on its internal (nation state) determinants (Freeman 2005). However, both discourses have roots in the notion of the national democratic struggle, with the former prioritising the "national" (in struggling against imperialism) and the latter the "democratic" (in struggling against an authoritarian state) (Moore 2004). For example, Ibbo Mandaza (who was linked to the ruling party for some time) said that, during the late 1990s, post-nationalist forces in alliance with foreign elements were engaged in a subterranean "social crisis strategy" that sought to make Zimbabwe ungovernable, and that the intellectual representatives of these forces sought to prioritise issues of governance and democracy "at the expense of addressing the National Question" (The "Scrutator" in *The Zimbabwe Mirror*, 28 April to 4 May 2000). Thus, the civic nationalism of these theorists (such as Raftopoulos) was portrayed as instigating civil society against the state, and as seeking to undermine economic (redistributive) nationalism rightly propagated (according to Mandaza) by a beleaguered nation state under the onslaught of imperialism in the global periphery.

For us, the key concern is that critics of the land occupations (as part of the broader third *chimurenga*) are amongst those scholars who readily argue for a historiography of nationalism with reference to the war of liberation (as set out in Chapters 4 and 5). However, there seems to be less readiness on their part to pursue this form of historiography when it comes to interpreting the third *chimurenga* occupations. In still hanging on to the claim that these occupations were driven by the state, as discussed in the next section, many scholars tend to adopt a version of nationalist historiography—or an account consistent with the *zvimurenga* monologue, though with negative moral connotations.

7.3 War Veterans and the Party-State in the Occupations

In terms of the debate, then, two grand narratives about the land occupations existed, and still exist to some extent. The first, and by far the most dominant, narrative was—and remains—that the occupations were simply an electoral ploy driven and directed by an authoritarian ruling party (ZANU-PF) to garner, bolster and consolidate rural support ahead of the national parliamentary elections which were to take place in June 2000, particularly given the emergence of a vibrant opposition party (MDC) with white commercial farmer backing (Hammar et al. 2003). Through this lens, the land "invasions" (as they were called from this perspective) were (and are) understood as an authoritarian state-led political campaign designed to crush opposition and bolster the ruling party's diminishing support.

In February 2000, as indicated earlier, a national referendum was held around revisions to the national constitution, including more radical forms of land redistribution—the constitutional changes were rejected, which went against the ruling party's stance on the matter. In response, it is typically argued that the ruling party called upon, and directed, the war veterans to occupy farms throughout the countryside in order to ensure certain electoral victory. In this respect, land occupations (as a short-term event) were triggered by the rejection of the draft constitution of 2000 (Masiiwa and Chipungu 2004; Hammar and Raftopoulos 2003). This triggering claim is accepted by Moyo but, for him, it was war veterans and mobilised communal area villagers who felt compelled to occupy farms and other landholdings and took it upon themselves to do so.

This dominated narrative was propagated quite forcibly by the MDC (as noted above) as well as white commercial farmers. A group of Zimbabwean farmers organised as Justice for Agriculture (JAG) arose in May 2002 as a splinter group in opposition to the well-established Commercial Farmers Union (CFU) because of the latter's willingness to negotiate with the government as well as its seeming unwillingness to resist strenuously the state's fast track programme. Jim Barker, who became a board member and vice-chairperson of JAG, recounted how "disgusted" he was "at the reluctance of the CFU to meet the Government head-on" (Barker 2007: 362). This splintering was part of a broader process of disunity emerging after the euphoria of the February 2000 referendum victory, with white farmers experiencing increasing isolation if not abandonment from the MDC-centred urban opposition coalition: "Farmers now saw themselves alone on the front lines of the battle against Mugabe and his corrupt ZANU-PF" (Pilossof 2012: 103). This was not unlike the experiences of white farmers during the war of liberation, who felt the full brunt of the war unlike their white counterparts in urban centres such as then-Salisbury. In a report on the occupations, the new commercial farmer organisation (JAG) argued that:

The invasions of white-owned farms were neither spontaneous, nor were they led by the landless poor [as claimed by ZANU-PF]. The groups of settlers consisted largely of ZANU PF youth led by one or two war veterans. The fact that this structure was so widespread shows that there must have been an organising entity [ZANU-PF] behind the invasions. (JAG and GAPWUZ 2008a: 62)

This report, which was co-published with the MDC-aligned farm workers' trade union (General Allied and Plantation Workers' Union of Zimbabwe—GAPWUZ), was based on interviews with white farmers, and throughout it highlights the state's complicity in the occupations. In defending what Rutherford (2004) refers to as the "interior frontier" of white farmers, namely the daily interface with their labour force and the ways in which (according to the report) farmers always prioritise the needs and interests of their labourers, JAG refused to acknowledge the existence of any injustices around land which could justify the occupations. In fact, any injustices were labelled as merely "perceived", with the "the invasions … carefully orchestrated by the Government to achieve its own narrow political ends" (JAG and GAPWUZ 2008a: 6).

These claims are supported by the existence on the occupied farms of "all night *pungwes* where farm workers were forced to stay awake, singing and shouting in support of ZANU PF and beating those accused of being 'sell-outs' or MDC supporters" (JAG and GAPWUZ 2008a: 6). This was quite similar to the *pungwes* at guerrilla bases held during the war of liberation, where singing and dancing took place, including the *kongonya* dance which "facilitated political mobilisation, moral boosting … [and] psychological anchoring" (Gonye 2013: 66) and was often followed by the identification and intimidation (if not death) of sell-outs.

In the end, the "war of attrition" (JAG and GAPWUZ 2008a: 14) by occupiers against white farmers was not about land reform at all, but about undercutting MDC support amongst farm workers. In a further report, JAG interprets the "invasions" as "operating in a non-random, systemic way, and thus as centrally organised and driven" (JAG and GAPWUZ 2008b: 33). Despite differences with JAG in terms of how best to counter the state's fast track programme, the CFU adopted a similar position on the occupations (Selby 2006). Overall, then, besides mobilising electoral support for ZANU-PF amongst communal farmers, the land occupations (from the perspective of white farmers) became a violent means of demobilising both commercial farmers and their workers in terms of supporting the rising MDC (Rutherford 2008).

This view has been reproduced repeatedly within the Zimbabwean literature (Raftopoulos 2006; Raftopoulos and Phimister 2004). Raftopoulos (2001: 17) for example claims that the ruling party "orchestrated" the land occupations in 2000 such that, in political terms, the occupations were a "frontal attack" on opposition politics, particularly the growing penetration of, and support for, the MDC in rural (including communal) areas. Marongwe (2008:133) argues that, in occupying farms, war veterans (who "in essence are ZANU P.F. party agents"—Marongwe, 2008:134) had "the full backing of the state". Though not disputing the centrality of war veterans' organisational structures from national level down to district level in the occupations, Alexander (2006: 186) likewise argues that:

A range of state and party actors were also directly involved in the organisation of the occupations. The Central Intelligence Organisation (CIO) and Zanu(PF) Headquarters directed veterans to specific farms, notably those where farmers had contested designations; veterans reported receiving lists of farms to occupy.

There were reports of people being bussed, particularly over weekends, from major urban centres to surrounding farms such as in Seke District (Marongwe 2003). Further, Angus Selby, who negotiated with occupiers as a son of a white farmer in the Concession area of Mashonaland Central Province and wrote his doctorate on the history of commercial farming in Zimbabwe, speaks about centrally coordinated occupations or the "orchestration of the land invasions using the state apparatus" (Selby 2006: 286). To Selby's credit, though, his claim is based on sustained fieldwork, as the literature embodying the dominant narrative is rarely based on in-depth evidence. Importantly, he also notes that dynamics at local level influenced the form of specific occupations. But this does not take him away from his broader point (Selby 2005).

Though providing astute and sophisticated analyses of relationships between white farmers and agricultural labourers, Rutherford also often makes bold claims similar to Marongwe, Selby and others, without arguing that the ruling party organised the occupations from the very beginning:

> [I]t is obvious that ZANU-PF politicians, leaders, members of Rural District Councils, Zimbabwean National Army, the police and the secret service (CIO) have played a role in leading and organising land invasions, though this has differed enormously from region to region. (Rutherford 2001: 636)

In focusing on the machinations of the ruling party and state, Chitiyo and Rupiya (2005) refer to the presence of a very structured organisational underpinning for the fast track occupations. They speak about the state's Operation *Tsuro* (Operation Rabbit) initiated in March 2000, a military operation used for political ends. Operation *Tsuro* had a command-and-control dimension, which involved in effect the reintroduction of the dreaded Rhodesian regime's Joint Operations Command (JOC) including the army, police and intelligence services but also the state-aligned war veterans' association, leading to "regular joint briefings and action plans" (Chitiyo and Rupiya 2005: 359). As well,

> Operational zones were established. The task was to identify 'loyal' and 'opposition' communities and individuals. ... The ultimate aim was that the rural areas in Mashonaland and Manicaland would be 'liberated' – that is, become pro-ZANU areas and 'no-go' areas for the opposition (Matabeleland was recognised as an opposition stronghold).... The methodology of operations included persuasion and violence. At first, the ground troops were landless villagers led by both genuine and nominal 'war veterans', with the state and ZNA [Zimbabwean army] operating as armourer, provider of logistics and enforcer. Attacks were initially on white farms, which were invaded or repossessed; however, as the June 2000 elections neared, the scale of violence increased, with auxiliary forces attacking known and suspected pro-MDC groups.

This involved mass politicisation in communal areas including rallies and *pungwes*. The MDC also stresses quite forcefully the existence of this operation (Tsvangarai

2011). But no significant local fieldwork evidence is provided to show the playing out of Operation *Truro*.

Those amongst the critics who have undertaken fieldwork on the occupations are more tempered in their conclusions. For example, Matondi (2001: 189), who did extensive research in Shamva, including during the occupations, initially adopted the dominant narrative in claiming that the Shamva occupations were "well orchestrated by the ruling party and the war veterans". His later research in Shamva led to conclusions which shifted away, though not entirely, from this narrative (Matondi 2012: 22). As well, Nelson Marongwe argues that "although Zanu (PF) and its structures, down to the district level, played a pivotal role in initiating and sustaining the land occupations, this does not suggest that it had complete control of what was happening on the ground" (Marongwe 2003: 165). This idea of somewhat autonomous practices (vis-à-vis the state and ruling party) in occupations at local levels, as well as the existence of some variation in the character of occupations across the countryside is, oddly, very common in the critical literature on the third *chimurenga* occupations. This also comes across when white farmers reflect upon their personal experiences of the occupations (Buckle 2001).

This is based on the same effort by first *chimurenga* scholars to address the "problem of scale". It leads to the general conclusion that, despite centralised direction and control by the ruling party, occupations simply became subject to local dynamics, as must invariably happen. From our perspective, as we seek to show later, this recognition should lead first and foremost to the conclusion that a decentralised character was embodied in the very anatomy of the occupations, and from the very beginning—in other words, that this spatial diversity across occupations prevailed because of the absence of centralised state and ruling party control, not despite it. This is consistent with the counter-narrative to the dominant one, as articulated initially by Sam Moyo.

The counter-narrative thus claims that the occupations, though not spontaneous, were in the main decentralised (all the way down to district, ward and farm level). Like the fast track critics, Moyo recognises—as shown in the previous chapter—that the character of land occupations in post-1980 Zimbabwe have shifted over time, including when the fast track occupations are taken into consideration. But this does not lead to a state-centric understanding of these occupations on the part of Moyo. For Moyo, the third *chimurenga* occupations took place in large part independently of ZANU-PF, and they were not centrally coordinated. Rather, they were subject to significant forms of decentralised mobilisation and organisation, particularly by war veterans whose relationship to the ruling party was characterised by overt tensions during the latter half of the 1990s, as discussed earlier. This position has been consistently argued by Moyo and Yeros (Moyo 2001; Yeros 2002b; Moyo and Yeros 2005a, 2007b, see also Sadomba 2008; Masuko 2013).

Though not focusing on the occupations specifically, Yeros (2002a: 244) undertook his doctoral research in the early 2000s and concluded that the "dismissal of the [occupation] movement as a merely 'orchestrated' affair is unfounded". Like Selby, but coming to different conclusions, Yeros speaks about local particularities or the importance of spatial diversities in understanding the land occupations. In his initial article about the occupations, Moyo (2001: 316) notes that ZANU-PF leaders "instigated as well as supported the land occupations" but he then goes on to argue that this point should not be overplayed. Ultimately, Moyo highlights the "organic and deep-seated local pressures for land reform" and claims that people were not "cajoled, paid or even forced to join occupations" (Moyo 2001: 323). The notion of ZANU-PF "instigating" the occupations is less prominent in his later work. But he always recognised, in certain instances, the prominent (personal) role of local ZANU-PF leaders in the occupations, such as Border Gezi, the then governor of Mashonaland Central Province and national political commissar for ZANU-PF. Overall, as Moyo argues, local ruling party and government structures slowly but surely became involved in the movement (or at least in specific occupations), including before the end of the year 2000.

Perhaps the most in-depth study of the occupations—and one conducted by an ex-guerrilla or war veteran engaged in the occupations (in Mashonaland Central)—is by Wilbert Sadomba. For Sadomba (2013), the Zimbabwean state, as a bourgeois neocolonial regime, promoted interests and values in opposition to those of marginalised war veterans, communal farmers, and rural and urban workers who—combined—were central to the land occupation movement. The ruling party and state negated the aspirations of the liberation struggle, as expressed symbolically and materially in terms of their failure to reverse a century-old grievance over unequal colonial land ownership structures (Sadomba 2013). In this light, Sadomba—in his doctoral thesis—concludes that the land occupations were not centrally initiated or organised (either by ZANU-PF or perhaps even the national war veterans' association). He thus argues that the occupations "were organised horizontally rather than vertically, with each local group employing its own tactics, determining its own boundaries of operation, and mobilising its own manpower and resources" (Sadomba 2008: 150). The fact that the war veterans' association and ZANU-PF were at loggerheads in the late 1990s suggests to Sadomba, as it does to Moyo, that the occupations represented a historical moment in which war veterans were struggling against the ruling party (in addition to the emerging opposition party). Other scholars have shown some sympathy to the arguments put forward by Moyo.

For example, McCandless (2005: 423) is somewhat ambivalent on the question of whether the ruling party led the occupations, but seems willing at least in part to accept the perspective as articulated by war veterans interviewed by her. She claims that there is "much evidence that government participated in the occupations in a variety of ways", including reports of payments to occupiers by Central Intelligence operatives and the use of army vehicles in transporting occupiers. This is despite all interviewed ZNLWVA members, from top-level to grassroots, insisting that they and not the ruling party led the occupations. Further, for the war veterans, the decentralised thrust of the occupations is "not seen to be in conflict with their close relationship with government" (McCandless 2005: 425). McCandless quotes—from 2003—a "former secretariat leadership representative" from the ZNLWVA, who articulates a point consistent with the claim by Sadomba:

The first farms to be occupied were government farms in Chiredzi and Masvingo [2000]. The war veterans took unilateral action and this was directed at the government rather than the white farm owners. It was meant to force the government to act, but ZANU PF took advantage of this to gain political mileage [through fast track]. (2005: 425)

As Moyo (2001) emphasises, war veterans broadly were dispersed throughout Zimbabwean state and society, including being embedded in state apparatuses (such as security and administration). This makes it difficult to compartmentalise war veterans and ruling party-state personnel, and even more difficult to say "who took more of the lead in the occupations" (McCandless 2005: 427). As well, that war veterans claim credit for the electoral victory of ZANU-PF in the year 2000 does not mean that "land occupations were conducted for the purpose of strengthening Zanu PF" (McCandless 2005: 528), rather than as a basis for localised land reclamation (McCandless 2012).

In a similar vein to Moyo and Sadomba, and consistent with our own position, Maruto (2010: 4) argues that "in practice, it was difficult to differentiate" between the occupations and the fast track programme, but "technically the farm/land invasions and the fast track land reform programme were two separate processes". In this light, Maruta is more willing than McCandless to accept the claim that:

[W]hile government elements were involved in the farm invasions, it is difficult to say that the government per se was involved, at least in its initial stages. The involvement of state elements, government vehicles and other equipment at this stage can be explained in other ways than state involvement. (Maruto 2010: 9–10)

This would include war veterans in state apparatuses supporting other veterans involved in the occupations by providing for instance food and other supplies. Mlambo (2010: 41) brings this to the fore, based on fieldwork in Mashonaland East Province, although seeking simultaneously to distinguish his position from that of Moyo:

[T]here was significant support for the farm invasions …; not everyone who participated in the land invasions was either duped or coerced. Interviews with resettled African farmers in the Upper Hwedza area revealed that villagers from the Hwedza Communal Areas were keen to participate in the farm invasions and were often ahead of local ZANU-PF structures in pushing for expropriation of white farms and their distribution to them… Clearly, while it is obviously absurd to claim that there a strong, well-organised and coherent land movement that systematically pushed the land invasions agenda [citing Moyo and Yeros in this regard] it is equally simplistic to dismiss the land reform exercise as an entirely ruling elite-driven exercise that had no grassroots support.

Hence, the occupations were meant to address the unresolved land question, including as a form of protest against the ZANU-PF government for the slow pace of land reform, with war veterans seeking to mobilise people on the nationalist basis of land dispossession and recovery. Contrary to what Mlambo implies, though, Moyo never argued for the existence of a "well-organised and coherent land movement" even if led by the war veterans' association.

As such, the land occupations were the culmination of a process that had been brewing and unfolding beneath the surface since 1980 to address a nagging political question (Masuko 2013; Moyo and Chambati 2013a, b). In considering earlier occupations, Chitiyo (2000) highlights that many of the "squatters" who occupied farms sporadically in the 1980s and 1990s were demobilised and disenfranchised war veterans who had deep and persistent land grievances against the ruling party. In the case of the fast track occupations, Chitiyo (2000: 2)—though not particularly supportive of the occupations—likewise highlights that "[m]any impoverished peasants are demobilised war veterans who have failed in various agrarian business ventures". In this respect, in their decentralised character, the fast track occupations represent a degree of continuity with the pre-2000 occupations.

Overall, we would argue that war veterans who led the occupations were not acting at the behest of the ruling party or even on its behalf. As Nelson Marongwe (2003: 178) is prepared to acknowledge, "it does not appear from available information that there was a final word 'from above' instructing war veterans to occupy the farms". The war veterans saw in the rejection of the referendum (in the year 2000) a repeat of the betrayal they had suffered at the pre-independence Lancaster House conference which had scuttled radical land reform by introducing market-led reform (Sadomba 2008). Subsequently, and mainly from the late 1990s, they became more militant by directly challenging settler capital, the state and ZANU-PF elites (Sadomba 2013). Additionally, the land occupiers did not only include war veterans but also diverse groupings with a range of motivations—discussed more fully later—which are simply irreducible to the nationalist machinations of the ruling party.

Overall, the two main narratives (of Moyo and of his critics) have different understandings of the relationship between marginalised war veterans and the ruling party-state. The dominant narrative sees war veterans as the storm troopers of the ZANU-PF party (depicted as directing the occupations) while the second, far less common, narrative perceives war veterans as initiating and leading decentralised occupations, and typically in tension with the ruling party. The dominant narrative inclines towards a nationalist historiography of the third *chimurenga* occupations, while the other one is more consistent with a historiography of nationalism. Nevertheless, both narratives recognise the centrality of war veterans in the occupations, and specific studies highlight this.

Nelson Marongwe (2008:188–189) notes for instance that by May 2000 on 20 farms in Goromonzi, war veterans were represented on all the farms, ranging from just one to fifteen, with a total of 103 war veterans and 405 people from communal areas on these farms. In February 2000, the Chiredzi District War Veterans' Association mobilised people largely drawn from local communal areas to occupy a number of properties in sequence across the district (Chaumba et al. 2003). In another example, the Nyabira-Mazowe War Veterans' Association was formed at the onset of the land occupations to organise the land occupation process in its area of operation close to Harare (Masuko 2013). In Chipinge, Zamchiya (2011) found that, during the land occupations, local war veterans entered into an alliance with war veterans from the adjacent Buhera district, as well as with local ZANU-PF leaders and traditional leaders.

There is no doubt that local ZANU-PF leaders and state functionaries were involved in the occupations in many cases. Thus Sadomba (2008: 120) says that, in the studied occupations, war veterans "interacted with forces within ZANU-PF" (at least at district level), and they at times "lobbied ZANU-PF politicians for financial contributions" (Sadomba 2008: 133). Further, in some areas, local state officials like District Administrators assisted the occupiers with for example transport. The general claim by Sadomba, however, is that ZANU-PF politicians and state personnel became involved in the occupations in their own individual and personal capacity and not as sanctioned by the party or state (see also Moyo 2001). As the occupations developed throughout the year 2000, at least after the introduction of the fast track programme (in July 2000), there was certainly more explicit institutional involvement by the party and state in seeking to regularise and subdue the occupations. Though part of ZIPA in the 1970s (like Mhanda), Sadomba remained part of ZNLWVA (until late 2019) and he claims a ZIPA legacy for the initial stages of the occupations as they were directed against Mugabe's party-state. As Hanlon et al. (2013: 3) argue:

> At first, the Zanu-PF leadership was opposed [to the occupations], but the occupations had party and government support at lower levels. Eventually Zanu-PF reversed itself, legalised 'fast track land reform,' and tried to take credit for it. But the veterans knew otherwise – they were challenging their own Zanu-PF leaders.

There were certainly instances where local state structures were more centrally involved in the occupations from the beginning, as in the case of the occupancy of Damvuri Conservancy in Mhondoro Ngezi in Mashonaland West Province which was "state-led" (Mkodzongi 2013: 33). As Mkodzongi (2013: 7) argues:

> [T]he land occupations in Mhondoro Ngezi did not take place a 'typical' … style. Instead, the occupations assumed a rather bureaucratic character from the start. Local state structures such as the District Administrator (DA) from the nearby Kadoma town and technocrats from key ministries … were present from the beginning, supervising the occupation.

Despite this, war veterans did first occupy the conservancy (sometime in 2000), with the owner then negotiating immediately and successfully with the DA for a "grace period in order for him to pack his property and vacant his land" (Mkodzongi 2013: 20), to which the veterans abided without enacting violence. It is not clear when in the year 2000 the conservancy was first occupied and, in many parts of the country, local state structures became involved in bringing order to the occupations from July onwards. As Chamunogwa (2019:76) notes in relation to his study of three farms in Mazowe District occupied in September 2001, longer after the start of fast track:

> The DA [District Administrator] played a far more active and direct role in occupations that took place in my three study sites in late 2001, compared to the first wave of occupations in the district in 2000 … The occupiers … highlighted that they were directed to occupy the three farms by the DA.

More broadly, Chamunogwa (2018, 2019) argues that the DA was enmeshed in negotiating clashes between war veterans and ZANU-PF elites over land access, which at times led the DA to evict war veterans from particular farms to facilitate their occupation by the ZANU-PF elites. The increasing subordination of the war veterans to the dictates of the ruling party and state (and to state planning rationalities), though taking place unevenly across the countryside, is a point we take up further in the concluding chapter.

Nevertheless, and despite claims to the contrary, there is no clear evidence (or no such evidence is yet known publicly) that the ruling party in any way directed war veterans in early 2000 to occupy farms nationally. In this respect, and against her seeming claims at times to the contrary, McCandless (2005:407) refuses to accept any denial by the ruling party of its culpability, as "government was involved from the grassroots to leadership levels [in the occupations] – despite some ... statements to the contrary in the first few weeks following the occupations." This argument is made with reference to statements, made soon after the occupations in February 2000, by the Home Affairs Minister Dumiso Dabengwa and Vice-President Joseph Msika who both ordered the occupiers (considered effectively as illegal squatters) to vacate the farms.

These statements were made, in April and May, soon after the gazetting by government of a constitutional clause on land expropriation (and its incorporation into the still-existing constitution) which had been rejected through the February referendum. A Land Acquisition Act was then passed on May 23. Both Dabengwa and Msika declared that the land "demonstrations" had had their necessary effect (i.e. the importance of land repossession) and hence the occupiers should be evicted. War veterans responded angrily and received Mugabe's backing in doing so (Alexander 2003). In August, Land Affairs Minister John Nkomo declared that the occupations had to stop. The fact that Mugabe took a somewhat hands-off approach does not imply that elements within the ruling party instigated the occupations. But, as Matondi (2012: 22) argues in his study of Mazowe, it is quite possible that "peasants, War Veterans and their supporters took to the land as a result of signals that the government would tolerate their actions".

This is not to deny direct war veteran engagement on occasion with the ruling party and Mugabe himself. The Zimbabwean army and Central Intelligence Organisation—along with war veterans—soon became involved in so-called "negotiations" with the CFU (including at meetings held at State House), where the CFU was warned to eschew any involvement in politics (Alexander 2003). In his manuscript on Zimbabwean history, Brigadier Felix Muchemwa (Muchemwa 2015: 217) recounts that Mugabe presided over a meeting of war veterans and white commercial farmers at State House on April 19, 2000, at which Mugabe declared to the director of CFU (Dave Hasluck) and the president of the Zimbabwe Tobacco Association (Richard Tate) that the police and army would not be deployed to farms to evict war veterans. An apparent agreement was reached between the CFU and war veterans (and specifically the then-leader of the ZNLWVA, Chenjerai Hunzvi) that veterans and other occupiers could stay on the farms and that there should be non-violent coexistence between the farmers and occupiers. Another meeting was held on April 28 with similar agreements. It was such agreements which led eventually to the breakaway of JAG from the CFU.

The mixed messages from ZANU-PF leadership reflected divisions within the ruling party pertaining to the occupations, with party divisions already existing in the late 1990s (Moyo 2001). Thus, the radical wing was prepared to support the repossession of land by physical seizure (and thus lent moral if not logistical support to the occupiers) while the more conservative wing was only willing to accept temporary and symbolic occupations in demonstrating the right of Zimbabweans to compulsorily acquire land. While Gezi (as national political commissar of the party) and others represented the radical wing, it seems that others like Msika and Dabengwa did not.

Further, certain groups of war veterans more firmly opposed to the ruling party stood aloof from the occupations, because they interpreted the latter as being driven by ZANU-PF. For example, the study by Zamchiya (2011) of occupations in Chipinge reveals significant tensions between war veterans. Some war veterans were simply reluctant to join the land occupations citing their allegiance to the ZANU-PF breakaway party known as the ZANU-Ndonga party, formed by ex-ZANU leader Nbabaningi Sithole (with its base in the Chipinge area). More broadly, there were some war veterans who denounced the land occupations in principle, notably those linked to the Zimbabwe Liberators' Platform, which was formed soon after the initial occupations in early 2000 and argued that the occupations were top-down rather than spontaneous and in effect "negated and betrayed the aims, objectives and values of the liberation struggle" (ZLP 2004: 39). As this organisation claimed:

> The ZANU PF leadership used the state apparatus to invade white-owned commercial farms, and later invited war veterans to participate in the exercise.... The war veterans associated with the invasions [under the leadership of Chenjerai 'Hitler' Hunzvi] did so in the name of the ZNLWVA, which now had close links with ZANU PF. (ZLP 2004:40)

One of the leaders of ZLP, and its founding director, was ex-ZIPA commander Dzinashe Machingura (i.e. Mhanda), whose critical views on Mugabe were highlighted in chapter four. In denying like the MDC that the second *chimurenga* was strictly about land, Mhanda (2011b) indicates that ZLP (which formed links with the MDC and associated civil society formations such as the National Constitutional Assembly) was launched to "salvage the honour of the former liberation war fighters and to help the nation refocus on the original ideals of the liberation struggle: freedom, democracy, social justice, respect for human dignity and peace" (Mhanda 2011b: 218). He claims that all the ex-guerrillas in ZNLVWA by the late 1990s were related to the post-ZIPA (or Mugabe) phase of the war of liberation, and thus they assumed their positions with Mugabe's blessing. He adds:

> For many of us, the idea of a so-called 'Third Chimurenga' was a ploy to legitimise the hero status of those who never participated in the war itself of the likes of Mugabe himself .... None of the genuine heroes of the war and senior commanders like Rex Nhongo, Dumiso Dabengwa and Josiah Tungamirai ever publicly associated themselves with the so-called 'Third Chimurenga' characterised by armed bandits murdering unarmed innocent and defenceless civilians. (quoted by Chung 2006: 20)

In the light of factional battles within the ZNLVWA in the 1990s, in which the Hunzvi faction became dominant, Chitiyo (2000: 21) notes that "[b]y 2000, Hunzvi's war veterans had effectively become the 'military wing' of ZANU-PF in the 'war' against white commercial farmers". In doing his bidding, Mugabe "designated the war veterans to spearhead the campaign for the party's re-election in the June 2000 general election" (Mhanda 2011b: 226), with the occupations led by "war veterans aligned to Hunzvi and Joseph Chinotimba, ostensibly in the defence of the legacy of the liberation struggle" (Mhanda 2011b: 227). Mhanda thus explicitly criticised the occupations and claimed that ZANU-PF's partisan war veterans involved in the occupations were criminally inclined "rogue war veterans" (Mhanda 2011a) and that their leaders (Hunzvi and Chinotimba) were likewise inauthentic and illegitimate. Because of this, the occupations cannot be ascribed to war veterans in general and did not involve "real" war veterans such as those in ZLP (Mhanda 2001).

It becomes important, hence, not to treat ZANU-PF and war veterans as undifferentiated and flattened political formations. It is intriguing, as well as problematic, that the dominant scholarly narrative on the occupations, which condemns patriotic history so strongly, sometimes tends to treat war veterans and ZANU-PF as homogenous groupings and fails to emphasise the divisions which existed within both. This is in addition to its tendency to interpret the fast track occupations as simply imposed from above, thereby denying the agency and aspirations of the diverse groupings involved in the occupations and the sheer diversity of localised occupations. Countering this perspective entails a deep sensitivity to heterogeneity, contingency and tension, and even to "struggles within the struggle", with reference to the third *chimurenga* occupations, as we make clear in the next chapter.

In this regard, and to reiterate, significant tensions existed between ZANU-PF and ZNLVWA-linked war veterans from the mid- to late 1990s which conditioned, and indeed animated, the occupations. In this context, and though writing about the assassination of Herbert Chitepo in the 1970s, White (2003) outlines what is at stake in the debate about the fast track occupations with particular reference to identifying the "founding moment" of the Zimbabwean nation. Up until the year 2000, the war of liberation or second *chimurenga* had been accepted unquestionably as the founding moment by war veterans and ruling party alike. But this became contested by the war veterans by way of the land occupations insofar as it is recognised and accepted that veterans in some way abandoned ZANU-PF and acted against it. White (2003) argues:

> The new entitlements of the war veterans and farm invasions scripted two new histories of the making of Zimbabwe. In one, the foundation of Zimbabwe was based entirely on the war, now recast as a unified and unflinching struggle for the land… In the other, the founding moment has been reduced to the agreement reached in the negotiations at Lancaster House in 1979. The negotiations … have been revived in political talk in Zimbabwe as an example of how British perfidy subverted the struggle. This particular history … claims that the cease-fire sold out guerrillas, denying them the land they were about to seize in battle. (White 2003: 95)

The history in which the founding moment is Lancaster House has been blamed for the failure of land reform, a history which is implicitly and often explicitly questioned by war veterans when they seek to explain their motivations for occupying land in the year 2000. This questioning is deeply critical of Mugabe and ZANU-PF, and disrupts any smooth transition between the second and third *zvimurenga*, as the Lancaster House Agreement undercut and reversed—and indeed sold out—the efforts, sacrifices and successes of the guerrillas during the war of liberation. In this way, it was an act of treachery by the nationalist politicians against the guerrilla armies.

In this way, the discourses and practices of war veterans were not reducible to the narratives and machinations of the party-state. Any attempt by ZANU-PF to subordinate and subdue the war veterans has never been effective in any totalising manner. This is seen in the manner in which the ruling party has sought to inscribe and embed patriotic history in local experiences even long after the fast track land occupations took off. As Chitukutuku (2017a) demonstrates, patriotic history is not simply about constructing a historical narrative (based on a set of crude nationalist ideas) and thereby legitimising the present rule of ZANU-PF, as "legacies of how and where the [liberation] war was fought, the landscapes of the struggle, ... [have been] ...invoked at district and ward level as ZANU-PF sought to instill loyalty, fear and discipline through its supporters and youth militia" (Chitukutuku 2017a: 133). In recent years, war veterans in the Zimbabwean National Army have returned to guerrilla bases (from the 1970s) and, by involving ex-*mujibas*, ex-*chibwidos* and ZANU-PF youth, have presented talks to local villagers about war battles while also performing liberation songs and detailing how spirit mediums assisted them during the war. Thus, besides textbooks and teachings, patriotic history as "ZANU-PF's memory work" is "also channeled through experientially engaging with the landscapes of liberation war violence to materialise these narratives" in trying to persuade people to be loyal to the party "through harnessing the affective qualities (unseen spirits and narratives of war) of the landscape" (Chitukutuku 2017a: 136, 146).

However, this memory work around the affective qualities of former guerrilla bases is not invariably successful in leading to "narrative closure" (Chitukutuku 2017a: 143), or at least it does not necessarily resonate with the experiences of marginalised war veterans and ordinary villagers in the communal areas. In invoking the spirits of the liberation war dead, including local guerrillas and villagers from Tribal Trust Lands who were killed during the war, stories contrary to patriotic history are brought forth. At one church vigil by the Apostolic Church near a guerrilla base, the spirits of the liberation war dead were experienced as restless and disturbed, which was interpreted as a betrayal of the liberation war on the part of the ruling party as well as pointing to the immense suffering for the living since the end of the war (Chitukutuku 2017b). This disrupts the ZANU-PF *chimurenga* discourse and brings into the open the disenchantment which otherwise remains hidden.

Fontein (2015) makes a similar point with specific reference to the fast track occupations. During these occupations, even if it appeared that war veterans were acting on behalf of the ruling party, they were not inherently co-opted into a pre-existing ZANU-PF political project as they (and indeed others) had their own set of local grievances which led to their involvement in the occupations. He develops this argument for instance by considering the relationship between war veterans and spirit mediums, based on a study in Masvingo Province (Fontein 2006). Fontein examines the ways in which war veterans have invoked legacies of war cooperation (from the 1970s) between themselves and spirit mediums, in the post-2000 period (including during the farm occupations). This, however, should not be seen as a crude attempt by war veterans to co-opt spirit mediums as part of a state-constructed and state-driven patriotic (*chimurenga*) history, as it also speaks to real shared experiences and perspectives dating back to the second *chimurenga* (as showed earlier). At the same time, during the war of liberation,—the aspirations of guerrillas did not derive pure and simple from nationalist propaganda, despite the political education they received during their training. This propaganda (tracing the liberation struggle back to Nehanda and the first *chimurenga*) may have sustained the nationalist movements (ZANU and ZAPU). But, for guerrillas and now for war veterans, the ideology of nationalism was synonymous with their lived experiences and practices, including living and fighting under the guidance of the ancestors.

Thus, the interaction between spirit mediums and guerrillas (and ex-guerrillas) "centres on shared understandings, memories and experiences of the guidance of the spirits during the first, second, and now third, *chimurenga*, which far exceeds the often hollow political rhetoric of 'patriotic history'" (Fontein 2006: 179). This shared legacy speaks to the significance of ancestral claims to particular tracts lands which were so central to fast track occupations (see Fontein 2009 for occupations around Lake Mutirikwi), and it became consolidated in the two decades prior to these occupations as it entails "a shared sense of alienation from the political processes of the [ZANU-PF] state" (2006:184). On this basis, Fontein (2006: 192–193) concludes that the practices of war veterans

> [R]epresent more than just a 'local re-shaping' and reworking of some of the themes of the 'patriotic history'… which now descends from the dominant circles of ruling party and government. Perhaps we can consider this process in reverse. Thus not only is some of the language and rhetoric of the dominant party located within the same episteme, as war veterans, chiefs and spirit mediums who are variously co-opted or not into the project of 'patriotic history', this shared conviviality also enables the conditions for the production of radical and alternative moral and historical narratives and practices; other imaginations of the way the state should be.

In the end, the shared legacies of spirit mediums and war veterans defies the cynical politicking of the ruling party. Overall, then, the works of White, Chitukutuku and Fontein all demonstrate (in different ways) that war veteran motivations, aspirations and practices are irreducible to the party-state.

For us, the (Moyo-influenced) second narrative's general claim about the de-centralised character of the third *chimurenga* occupations, along with the absence of a co-ordinated operation—let alone plan—by the party-state and even war veterans in pursuing these occupations nation-wide, seems plausible. Certainly, it is more consistent with historical trends, including the tension-riddled relationship which existed on a regular basis between the guerrilla and nationalist movements in the 1970s (see Chapter 4) and, more recently, between the war veterans and the party-state since 1980, including in the years immediately preceding the fast track occupations (see Chapter 6). Further, the existing fieldwork evidence tends to support, at least more fully, this narrative (see Chapter 8). Hence, the kind of nuanced claims about the complex and shifting relationships between guerrillas and nationalists which is central to the historiography of nationalism literature on the second *chimurenga* is missing from the third *chimurenga* literature about the relationships between war veterans and the party-state.

7.4   Fast Track Land Occupation Diversity

In considering the dynamics of the fast track land occupations, which we discuss in the following chapter, it is crucial to reiterate that there was significant spatial diversity across localities when it came to the character of the occupations and that, in certain places, occupations did not take place at all or began much later (Suzuki 2018). As Nelson Marongwe (2003: 167) notes, the "intensity of farm occupations varied across provinces and districts", with Masvingo and the three Mashonaland provinces possibly experiencing the greatest number of occupations. Occupations of white commercial farms have received the most media publicity and perhaps the most scholarly attention, including tobacco, cattle, citrus and maize farms as well as wildlife conservancies. But other forms of landholdings were occupied, such as forest estates, national parks, state-owned agricultural estates, black-owned commercial farms and NGO-owned farms. The sheer variation in the occupations across the countryside reflects in part the diversity of local agrarian histories and local contemporary politics, along with a range of other factors which become clear in the following chapter. Variation and diversity arose along a number of dimensions.

Rutherford (2005) for instance observes that, in certain places, the majority of the white farmers were displaced almost immediately through forced evictions but, elsewhere, it took much longer for displacement to occur as negotiated arrangements between land occupiers and white farmers emerged (see also Sadomba 2013). Sometimes land was occupied then abandoned, only to be reoccupied later. As well, at times, land was occupied for extended periods without any state intervention, formal planning or even state recognition; in other cases, the state was soon on the scene (Scoones et al. 2010). At the same time, this might reflect changes in the character of the occupations over time or the more interventionist approach of the state as fast track progressed, as we suggested earlier.

Nelson Marongwe (2008) also notes that, in relation to Goromonzi District, the closeness of commercial farms to communal areas and to the Mabvuku-Tafara high density area in Harare (where there was a high demand for land for residential purposes) shaped the dynamics of the occupations in the district. In many cases, people from communal areas (such as in Chiweshe) occupied farms adjacent to their communal areas and sometimes—mainly men—moved to more distant farms when an occupying force was required (Matondi 2012). Practical and logistical considerations (such as food supplies) were also of importance regarding which farms were occupied at least initially (Wolmer 2007), and the responses of white farmers sometimes determined how occupations unfolded (Zamchiya 2011). Thus comparative studies become quite significant in this regard (Scoones et al. 2010).

Besides spatial considerations, occupations—as noted—took on different forms across time (Wolmer et al. 2004; Wolmer 2007), with the state's fast track land reform programme in particular altering the conditions within which further occupations arose. However, even in the early months of the occupations (from February 2000), there was no "natural" progression of how occupations in general, or specific occupations, unfolded as if they had some sort of internal logic to them. Scoones et al. (2010: 22, 25) conclude regarding the occupations:

> Sometimes these were spontaneous efforts involving only local people, sometimes they were organised by networks of war veterans, and sometimes they involved the government and security forces. … [E]ach farm 'invasion' had a different character: different origins, different people involved and different forms of external support. … In most settings there were several phases, too, stretching over weeks or months. Spontaneous gatherings of people may have initiated an occupation of a farm, but subsequent organisation may have been facilitated by war veterans who established base camps and became base commanders, sometimes with support from the state, including transport, tents and food supplied by the security forces.

Similarly, Chaumba et al. (2003: 539) argue:

> Explanations of the farm occupations have tended to cast them as either a spontaneous rejection of bureaucratic process of land reform or a state-orchestrated process. But this dichotomy is not necessarily helpful. There were a variety of further motivations for the farm occupations, ranging from top-down directives to bolster support for ZANU(PF) in its rural heartlands, to localised desires for the restitution of ancestral land.

Our discussion in this chapter, and in the book as a whole, questions the existence of "top-down directives" cascading down from the national party-state level with reference to the emergence and unfolding of the occupations in the months immediately after February 2000. Though at times local party-state leaders were involved in these early occupations (in their personal capacities), the formal intrusion of party-state structures in the occupations became evident by mid-year. The importance of diverse motivations for the third *chimurenga* occupations, including "the restitution of ancestral land", is brought out in the following chapter.

7.5 Conclusion

This chapter has argued against the claim that the third *chimurenga* occupations were centrally initiated and orchestrated by the ruling party and state, and it thus goes against the dominant interpretation of the occupations within the Zimbabwean scholarly literature. The existing evidence, and certainly growing evidence based on original research including fieldwork-based studies, demonstrates that the occupations were in the main organised and led by war veterans in a decentralised manner. By the time of the emergence of the occupations in early 2000, significant tension existed between the ZNLWVA and ZANU-PF for a number of years, so that war veterans often framed the occupations as a political process in opposition to the ruling party. In line with this, though war veterans articulated—discursively—the occupations in terms of a nationalist agenda around the recovery of lost lands, this was not reducible to the nationalist claims of the ruling party. This was because their agenda was founded on an interpretation of post-colonial failure which was heavily linked to state failure dating back to the compromises made at the Lancaster House negotiations. The very diversity of the third *chimurenga* occupations, which we pursue fully in the next chapter, bears testimony to the manner in which war veterans and other occupiers pursued their own local nationalisms and indeed personal agendas—as took place during the second *chimurenga*. Likewise, this relates to Beach's understanding of the first *chimurenga*. The failure on the part of so many Zimbabwean scholars to recognise the decentralised character of the third *chimurenga* occupations rests uneasily aside their otherwise marked sensitivity to a historiography of nationalisms, as their claim posits the unfolding of a central state nationalism which engulfed the Zimbabwean countryside through the fast track occupations.

References

Alexander J (2003) 'Squatters', veterans and the state in Zimbabwe. In: Hammar A, Raftopoulos B, Jensen S (eds) Zimbabwe's unfinished business: rethinking land, state and nation in the context of crisis. Weaver Press, Harare, pp 83–118

Alexander J (2006) The unsettled land: state making and the politics of land in Zimbabwe 1893–2003. James Currey, Oxford

Alexander J, McGregor J, Ranger T (2000) Violence and memory: one hundred years in the 'dark forests' of Matabeleland. James Currey, Oxford

Barker J (2007) Paradise plundered: the story of a Zimbabwean farm. Jim Barker, Harare

Bond P, Manyanya M (2003) Zimbabwe's plunge: exhausted nationalism, neoliberalism and the search for social justice. Weaver Press, Harare

Buckle C (2001) African tears: The Zimbabwe land invasions. Covos Day, Johannesburg

Chamunogwa A (2018) Power at the margins of post-colonial states in Africa: remaking authority on fast track resettlement farms in Zimbabwe. Unpublished PhD Thesis, University of Oxford, Oxford

Chamunogwa A (2019) The negotiability of state legal and bureaucratic authority during land occupations in Zimbabwe. Rev Afr Polit Econo 46(159):71–85

Chaumba J, Scoones I, Wolmer W (2003) From Jambanja to planning: the reassertion of technocracy in land reform in south-eastern Zimbabwe? J Mod Afr Stud 41(4):533–554

Chitiyo K, Rupiya M (2005) Tracking Zimbabwe's political history: the Zimbabwe defence force from 1980–2005. In: Rupiya M (ed) Evolutions and revolutions: a contemporary history of militaries in Southern Africa. Institute of Security Studies, Pretoria, pp 331–363

Chitiyo T (2000) Land violence and compensation: reconceptualising Zimbabwe's land & war veterans debate. Track Two Occasional Paper, Centre for Conflict Resolution, University of Cape Town

Chitukutuku E (2017a) Rebuilding the liberation war base: materiality and landscapes of violence in Northern Zimbabwe. J Eastern Afr Stud 11(1):133–150

Chitukutuku E (2017b) Re-living liberation war militia bases: violence, history and the making of political subjectivities in Zimbabwe. Unpublished PhD Thesis. University of the Witwatersrand, South Africa

Chung F (2006) Re living the second Chimurenga: memories from the liberation struggle in Zimbabwe. Nordic Africa Institute, Uppsala

Fontein J (2006) Shared legacies of war: spirit mediums and war veterans in Southern Zimbabwe. J Relig Afr 36(2):167–199

Fontein J (2009) 'We want to belong to our roots and we want to be modern people': new farmers, old claims around lake Mutirikwi, Southern Zimbabwe. Afr Stud Q 10(4):2–35
Fontein J (2015) Remaking Mutirikwi: Landscape. Water and belonging in Southern Zimbabwe. James Currey, Oxford
Freeman L (2005) Contradictory constructions of the crisis in Zimbabwe. Carleton University. www.arts.yorku.ca/african_liberation/conference_papers
Gonye J (2013) Mobilizing dance/traumatizing dance: 'Kongonya' and the politics of Zimbabwe. Dance Res J 45(1):65–79
Hammar A (2008) In the name of sovereignty: displacement and state making in post-independence Zimbabwe. J Contemp Afr Stud 26(4):417–434
Hammar A, Raftopoulos B, Jensen S (2003) Zimbabwe's unfinished business: rethinking land, state and nation in the context of crisis. Weaver Press, Harare
Hammar A, Raftopoulos B (2003) Zimbabwe's unfinished business: rethinking land, state and nation. In: Hammar A, Raftopoulos B, Jensen S (eds) Zimbabwe's unfinished business: rethinking land, state and nation. Weaver Press, Harare, pp 1–48
Hanlon J, Manjengwa J, Smart T (2013) Zimbabwe takes back its land. Kumarian Press, Sterling
Jacobs S, Mundy J (2009) Reflections on Mahmood Mamdani's 'lessons of Zimbabwe'. Concerned Afr Sch, 82
Justice for Agriculture (JAG) and General Agricultural and Plantation Workers Union of Zimbabwe (GAPWUZ) (2008a) Destruction of Zimbabwe's backbone industry in pursuit of political power. JAG and GAPWUZ, Harare
Justice for Agriculture (JAG) and General Agricultural and Plantation Workers Union of Zimbabwe (GAPWUZ) (2008b) If something is wrong: the invisible suffering of commercial farm workers and their families due to "land reform". Research and Advocacy Unit, Harare
Mamdani M (2008) Lessons of Zimbabwe. Lond Rev Books 30(23):17–21
Marongwe N (2003) Farm occupations and the occupiers in the new politics of land in Zimbabwe. In: Hammar A, Raftopoulos B (eds) Zimbabwe's unfinished business: rethinking land, state and nation. Weaver Press, Harare, pp 155–190
Marongwe N (2008) Interrogating Zimbabwe's fast track land reform and resettlement programme: a focus on beneficiary selection. Unpublished PhD Thesis, University of the Western Cape, South Africa
Maruto S (2010) All culpable together: civil society and the fast-track land reform programme in Zimbabwe. Southern Institute of Peacebuilding and Development, Ruwa
Masiiwa M, Chipungu L (2004) Land reform programme in Zimbabwe: disparity between policy design and implementation. In: Masiiwa M (ed) Post-independence land reform in Zimbabwe: controversies and impact on the economy. Friedrich-Ebert-Stiftung, Harare, pp 1–24
Masuko L (2013) Nyabira-Mazowe war veterans' association: a microcosm of the national level occupation movement. In: Moyo S, Chambati W (eds) Land and agrarian reform in Zimbabwe: beyond white-settler capitalism. African Institute for Agrarian Studies, Harare, pp 123–155
Matondi P (2001) The struggle for access to land and water resources in Zimbabwe: the case of Shamva district. Unpublished PhD Thesis, Swedish University of Agricultural Sciences, Sweden
Matondi P (2012) Zimbabwe's fast track and reform. Zed Books, London
McCandless E (2005) Zimbabwean forms of resistance: social movements, strategic dilemmas and transformative change. Unpublished PhD Thesis, American University, United States of America
McCandless E (2012) Polarization and transformation in Zimbabwe: social movements, strategy dilemmas and change. University of Kwa-Zulu Natal Press, Durban
Mhanda W (2001) How Mugabe came to power: R.W. Johnson talks to Wilfred Mhanda. Lond Rev Books 23(4):26–27
Mhanda W (2011a) The role of war veterans in Zimbabwe's political and economic processes. http://solidaritypeacetrust.org/1063/the-role-of-war-veterans/
Mhanda W (2011b) *Dzino: memories of a freedom fighter*. Weaver Press, Harare.

Mkodzongi G (2013) Fast tracking land reform and rural livelihoods in Mashonaland west province of Zimbabwe: opportunities and constraints, 2000–2013. PhD Thesis, University of Edinburgh, United Kingdom

Mlambo A (2010) 'This is our land': The racialization of land in the context of the current Zimbabwe crisis. J Developing Soc 26(1):39–69

Moore D (2003) Zimbabwe's triple crisis: primitive accumulation, nation-state formation and democratization in the age of neo-liberal globalization. Afr Stud Q 7(2–3):1–19

Moore D (2004) Marxism and Marxist intellectuals in schizophrenic Zimbabwe: how many rights for Zimbabwe's left? A comment. Hist Mater 12(4):405–425

Moyo S (2001) The land occupation movement and democratisation in Zimbabwe. Millennium J Int Stud 30(2):311–330

Moyo S, Chambati W (eds) (2013a) Land and agrarian reform in Zimbabwe: beyond white-settler capitalism. African Institute for Agrarian Studies, Harare

Moyo S, Chambati W (eds) (2013b) Introduction: roots of the fast track land reform in Zimbabwe. In: Moyo S, Chambati W (eds). Land and agrarian reform in Zimbabwe: beyond white-settler capitalism. African Institute for Agrarian Studies, Harare, pp 1–27

Moyo S, Yeros P (eds) (2005a) Reclaiming the land: the resurgence of rural movements in Africa, Asia and Latin America. Zed Books, London.

Moyo S, Yeros P (2005b) Land occupations and land reform in Zimbabwe: towards the National Democratic Revolution. In: Moyo S, Yeros P (eds) Reclaiming the land: the resurgence of rural movements in Africa, Asia and Latin America. Zed Books, London, pp 165–205

Moyo S, Yeros P (2007a) The radicalised state: Zimbabwe's interrupted revolution. Rev Afr Polit Econ 34(111):103–121

Moyo S, Yeros P (2007b) The Zimbabwe question and the two lefts. Hist Mater 15(3):171–204

Muchemwa F (2015) The struggle for land in Zimbabwe: 1890–2010. Heritage Publishing House, Harare

Muzondidya J (2007) Jambanja: ideological ambiguities in the politics of land and resource ownership in Zimbabwe. J South Afr Stud 33(2):325–341

Pilossof R (2006) The unbearable whiteness of being: land, race and belonging in the memoirs of white Zimbabweans. South African Hist J 61(3):621–638

Pilossof R (2008) The land question (un)resolved: an essay review. Historia 53(2):270–279

Pilossof R (2012) The unbearable whiteness of being: farmer's voices from Zimbabwe. University of Cape Town Press, Cape Town

Raftopoulos B (2001) The labour movement and the emergence of opposition politics in Zimbabwe. In: Raftopoulos B, Sachikonye L (eds) Striking back: the labour movement and the post-colonial state in Zimbabwe 1980–2000. Weaver Press, Harare, pp 1–24

Raftopoulos B (2006) The Zimbabwean crisis and the challenges for the left. J South Afr Stud 32(2):203–219

Raftopoulos B, Phimister I (2004) Zimbabwe now: the political economy of crisis and coercion. Hist Mater 12(4):355–382

Ranger T (2004) Nationalist historiography, patriotic history and the history of the nation; the struggle over the past in Zimbabwe. J South Afr Stud 30(2):215–234

Rutherford B (2001) Commercial farm workers and the politics of (dis)placement in Zimbabwe: colonialism, liberation and democracy. J Agrarian Change 1(4):626–651

Rutherford B (2002) Zimbabwe: the politics of land and the political landscape. Green Left Weekly. Issue 487, 10 April

Rutherford B (2004) Settlers and Zimbabwe: politics, memory, and the anthropology of commercial farms during a time of crisis. Identities 11:543–562

Rutherford B (2005) The rough contours of land in Zimbabwe. Fletcher Forum World Aff 29(2):103–116

Rutherford B (2008) Conditional belonging: farm workers and the cultural politics of recognition in Zimbabwe. Dev Change 39(1):73–99

Sachikonye LM (2002) Whither Zimbabwe? Crisis and democratisation. Rev Afr Polit Econ 29(91):13–20

Sadomba WZ (2008) War veterans in Zimbabwe's land occupations: complexities of a liberation movement in an African post-colonial settler society. Unpublished PhD thesis, Wageningen University, Netherlands

Sadomba WZ (2013) A decade of Zimbabwe's land revolution: the politics of the war veteran vanguard. In: Moyo S, Chambati W (eds) Land and agrarian reform in Zimbabwe: beyond white-settler capitalism. African Institute for Agrarian Studies, Harare, pp 79–122

Scoones I, Marongwe N, Mavedzenge B, Mahenehene J, Murimbarimba F, Sukume C (2010) Zimbabwe's land reform: myths and realities. Weaver Press, Harare

Selby A (2005) Losing the plot: the strategic dismantling of white farming in Zimbabwe 2000–2005. QEH Working Paper Series-Working Paper Number 143

Selby A (2006) Commercial farmers and the state: interest group politics and land reform in Zimbabwe. Unpublished PhD Thesis, University of Oxford, United Kingdom

Suzuki Y (2018) The good farmer: morality, expertise, and articulations of whiteness in Zimbabwe. Anthropol Forum 28(1):74–88

Tsvangirai M (with Bango TW) (2011) At the deep end. Eye Books, London

White L (2003) The assassination of Herbert Chitepo: texts and politics in Zimbabwe. Indiana University Press, Bloomington

Wolmer W (2007) From wilderness vision to farm invasions: conservation and development in Zimbabwe's south-east Lowveld. James Currey, Oxford

Wolmer W, Chaumba J, Scoones I (2004) Wildlife management and land reform in southeastern Zimbabwe: a compatible pairing or a contradiction in terms? Geoforum 35(1):87–98

Worby E (2001) A redivided land? New agrarian conflicts and questions in Zimbabwe. J Agrarian Change 1(4):475–509

Yeros P (2002a) The political economy of civilization: peasant-workers in Zimbabwe and the neo-colonial world. Unpublished PhD Thesis, University of London, United Kingdom

Yeros P. (2002b) Zimbabwe and the dilemmas of the left. Hist Mater 10(2):3–15

Zamchiya P (2011) A synopsis of land and agrarian change in Chipinge district, Zimbabwe. J Peasant Stud 38(5):1093–1122

Zimbabwe Liberation Platform (ZLP) (2004) What happened to our dreams. In: Harold-Barry D (ed) Zimbabwe: the past is the future. Weaver Press, Harare, pp 31–42

Chapter 8
The Third *Chimurenga*: Land Occupation Dynamics

Abstract In developing the argument of the previous chapter, this current chapter demonstrates in diverse ways the manner in which the third *chimurenga* occupations embodied a decentralised character. This overall claim about the occupations is consistent with a historiography of nationalism and thus goes contrary to the dominant narrative about the third *chimurenga*. The chapter discusses the diverse memories, motivations and claims of different categories of occupiers which led to their participation in the occupations. It also considers issues around disorder and order during the occupations including the multiple ways in which occupations were organised locally and became crystallised in farm-level authority structures. In seeking to highlight the diversity more convincingly, the chapter examines the occupations in two areas of the countryside, namely, in Masvingo Province and in northern Matabeleland. Though a focus on farm labourers and women vis-à-vis the occupations shows that the fast track occupations were marked by forms of exclusion and subordination, it also evident that the agency of women and farm labourers signals a further need to question any understanding of the third *chimurenga* occupations rooted in party-state directives and machinations.

Keywords Third *chimurenga* · Fast track · Land occupations · Farm labourers · Women · Restitution

8.1 Introduction

In following on from the central argument in the previous chapter, this current chapter demonstrates in various ways the manner in which the third *chimurenga* occupations embodied a decentralised character. This overall claim about the occupations is consistent with a historiography of nationalism and thus goes contrary to the dominant narrative about the third *chimurenga*. In pursuing our argument, the chapter is divided into six sections. First of all, we discuss the diverse memories, motivations and claims of different categories of occupiers which led to their participation in the occupations. Secondly, we consider issues around disorder and order during the occupations, including the multiple ways in which occupations were organised locally and became crystallised in farm-level authority structures. In the following two sections,

© The Author(s), under exclusive license to Springer Nature Switzerland AG 2021
K. Helliker et al., *Fast Track Land Occupations in Zimbabwe*,
https://doi.org/10.1007/978-3-030-66348-3_8

in seeking to show the diversity more convincingly, we examine the occupations in two areas of the countryside, namely, in Masvingo Province and in northern Matabeleland. The sheer localised complexity of the occupations becomes even clearer when, in the last two sections, we focus on farmers and farm labourers, as well as women and patriarchy. A consideration of farm labourers and women vis-à-vis the occupations shows that the fast track occupations were marked by forms of exclusion and subordination. However, it is also evident that the agency of women and farm labourers (and the motivational basis for their uneven involvement in the occupations) signals a further need to question any understanding of the third *chimurenga* occupations rooted in party-state directives and machinations.

8.2   Occupiers: Motivations, Memories and Claims

Social groupings of a diverse range occupied farms to varying degrees. Besides war veterans, these included communal farmers, resettlement farmers from the 1980s, urban unemployed and workers, agricultural labourers, youth and women, and civil servants and ruling party elites. In fact, there were only a few cases where all the farm occupiers were war veterans, such as at Pangara Ranch, Burnaby Farm and Barwon Down Estate in Nyanga, and Janee Ranch in Gwanda (Marongwe 2002). As well, MDC supporters at times were involved in occupations, including those from high-density areas in cities and towns. In this respect, Mkodzongi (2018: 16) refers to the way in which occupiers at times "performed" their membership of ZANU-PF, or represented themselves as supporters of the ruling party, in order to establish their credentials as legitimate members of an occupying group. As with the composition of occupiers, there was great diversity in terms of what motivated people to occupy and why particular farms were occupied, including

> [P]roximity to resettlement and communal areas; social relations between farmers and bordering resettlement/communal areas; historical land claims by communities in relation to colonial land appropriations; multiple farm ownership and under-utilisation of land; political affiliation with the ruling party or the opposition; and urban demand for residential land. (Marongwe 2003: 171)

Unlike the dominant narrative pertaining to the occupations, the narrative as articulated by Moyo is by necessity more sensitive to the importance of identifying the diverse groupings occupying white commercial farms and other landholdings, as well as the multiplicity of motivations animating the occupations, including why a particular farm was occupied by a particular group of people. This is because, as we likewise argue, the dynamics of the fast track occupations are not reducible to occupiers being hoodwinked by the nationalist rhetoric of the ruling party. However, mild (though not strident) critics of the occupations increasingly recognise the need to identify these groupings and their motivations.

Thus, Scoones et al. (2010) relate stories of people who occupied farms because of problems in their communal lands, including the sheer lack of land, low agricultural productivity and the absence of good grazing land; as well as social problems such as disputes with neighbours and family problems including accusations of witchcraft. As James (2015: 13) rightly argues: "[T]here has been a tendency to fixate on the role of the state in the process of land takeovers, while ignoring that *jambanja* [or occupations] also presented an opportunity for the airing of frustrations and local grievances that might, in fact, have had little to do with the state or land reform per se". This is not to deny that many occupiers, and not just war veterans, seemingly echoed ruling party ideology by offering political arguments about the unfulfilled legacy of the liberation struggle around land reclamation.

Nationally, occupiers invoked historical grievances of both a broad nationalist and specific experiential kind. They brought forth memories of initial land loss and related colonial practices as well as ongoing postcolonial land dispossession. Thus Nelson Marongwe (2008: 171), in studying specifically Goromonzi district, notes that "land problems had been simmering in the district" for an extended period and that a "social basis for peasants to participate in the post 2000 land occupations existed". In this sense, occupiers were enacting their own agency arising from localised encounters characterised by racism, discrimination, abuse and aggression stretching back to colonial times. Such encounters included displacement and relocation into Native Reserves and then Tribal Trust Lands and the practice of *chibaro* (forced labour). In Gowe-Sanyati, the fast track occupations were a response to frustrations instigated by the 1950 Rhodesdale evictions, the Native Land Husbandry Act of the 1950s and the overall past disruption of livelihoods (Nyandoro 2012).

Besides the broad historical fact of white settlers' dispossession of African people of land in an extended colonial enclosure process, past and present relations with white commercial farmers clearly were of great importance in motivating occupiers. In the case of agricultural labourers, labour relations and the semi-despotic character of farm authority at times motivated these labourers to support or even initiate occupations. For nearby communal area villagers, commercial farmers regularly protected—and strenuously—their privatised space by, for instance, confiscating communal cattle straying into this space or punishing communal farmers for collecting firewood and other natural resources on their farms. For instance, a farmer from Marondera claimed that her sheep had died because Africans were relieving themselves on the farm, with 70 Svosve villagers occupying her farm (Marongwe 2003). Communal villagers also spoke about underutilisation of land on commercial farms and particular historical experiences with commercial farmers during the war of liberation, including a farm being used as a shooting range for the training of Rhodesian Army soldiers (Marongwe 2003). At the same time, cordial relations between commercial and communal farmers did not necessarily prevent a specific farm from being occupied.

Besides communal farmers, farmers from the early resettlement farms (dating back to the early 1980s) became involved in occupations. There are instances where the majority of occupiers came from resettlement areas, including occupiers of Charter Estate in Seke District who came from Masasa Resettlement Scheme, and those on Longfield Farm from Chinyika Resettlement Scheme (Marongwe 2003). Typically, the occupiers were young adults who were children in the early 1980s.

Thus, in his study in Mashonaland Central Province, James (2014, 2015) notes that, in these old resettlement areas, overcrowding was a problem so that accessing more or better-quality land, or having access to any land at all, was a key motivating factor for occupying white farms. In these resettlement areas, a young man often lived with his wife and children at his father's homestead, and shared his father's field, and hence the occupations became an opportunity for independent access to land.

Farm occupations by residents of high-density areas in cities and towns also took place, either of nearby farms or vacant city-owned land, in large part to access land for residential purposes. In the case of Harare, this included Stockade Farm and Mt Hampden Farm, involving occupiers from Kuwadzana, Dzivaresekwa and Mabvuku. Sometimes they just occupied land during the day and returned home at night, trying to stake their claim to land for residential purposes. Particularly intriguing is the involvement of urban residents of foreign origin and without a *kamusha* (rural communal home) in Zimbabwe: "In Chinhoyi and Banket, urbanites that had never had a rural home in Zimbabwe, particularly those of Malawian, Mozambican and Zambian origins, took part in the farm occupations" (Marongwe 2002: 49). At other times, urban residents were literally bussed into farms to ensure a sufficient occupying force on farms, such as Chitungwiza residents moving onto farms in Seke District.

In this context, in his recent study on landscapes in the Mutirikwi area in Masvingo Province, Fontein (2015) argues for diversity and complexity when it comes to understanding the motivations of those who occupied farms. As he seeks to show, the occupations were "far more complex" than the dominant narrative posits, as "[p]eople engaged with a myriad of different discourses and practices … The complex aspirations, motivations and subjectivities of people involved in land occupations matter" (Fontein 2015: 30).

The work by Mkodzongi (2011, 2016, 2018) on occupations by people from the Mhondoro Ngezi communal area demonstrate this complexity. These occupiers referred to the killing of a communal area resident by a white landowner on suspicion of poaching wildlife as initially triggering the occupations of the former Damvuri conservancy. In relation to their communal area, occupiers cited reasons such as overcrowding, poor soils and the need for better pasture for livestock. Mkodzongi (2018: 7) cites a Mrs. Changi who came from Chief Nyika's territory in Mhondoro Ngezi:

> We came from Bhururu in Mhondoro Ngezi in 2000. … When we came here we did not completely give up our home in the Mhondoro CA [communal area]. We wanted to be sure that this place was going to be secure enough before we completely gave up our home [thus, as a security precaution and hedging for land in the new area] … The soils in Mhondoro were bad and unproductive. We also had three sons who needed land to start their own families. Our land back home was too small for a growing family.

In terms of retaining their communal home, older occupiers in particular tended to have a strong sense of belonging to their communal area where graves of ancestors were located.

But many occupiers, in many parts of rural Zimbabwe, cited the recovery of (specific tracts of) ancestral lands lost, because of forced removals during the colonial era, as a major factor for occupations in a specific area. Many occupiers who came to Damvuri from Gokwe (over 200 kilometres away) claimed historical connections to the area, as they had been forcibly evicted from the former Rhodesdale estate to Gokwe in the 1960s and, after the decline of the cotton economy in Gokwe in the late 1990s, took advantage of the emerging occupations. In this respect, competing—including chieftainship—claims of autochthonous connections arose between those who came from Mhondoro Ngezi communal area (who claim to have been evicted from the area and dumped in the nearby Mondhoro reserve paving way for white commercial farmers in the 1940s) and those who had come from distant places like Gokwe (Mkodzongi 2011).

In this particular instance, and in the context of a revitalisation of chiefly authority by the Zimbabwean state in the late 1990s, chiefs became emboldened and participated in these occupations. In this light, "chiefs in collaboration with war veterans became dominant political actors in the countryside in leading land occupations" (Mkodzongi 2016: 102). The involvement of chiefs in specific fast track occupations, as in the occupations in the late 1990s, is of some significance.

More broadly, claims over particular pieces of land considered as ancestral, involving claims for restitution, characterised many of the occupations. Hence, specific claims to lost lands, and not merely racially based historical injustices around land, became a central issue. This at times led to the involvement of spirit mediums, though Scoones et al. (2010: 198) suggest that "[w]hile traditional religion played a critical role during the liberation struggle of the 1970s, it was perhaps not such an important consideration during the 2000s". Nevertheless, as noted by Nelson Marongwe (2003: 184) "restitution claims based on [colonial] dispossessions were a strong rallying point for participants in the occupations". For instance, those occupying the Eastern Highlands Plantation in Nyanga claimed it as the original home of the Tangwena, Zindi and Chavhanga people; and, in the case of Zvimba in Mashonaland West, the occupation of Machiori Farm amongst others was legitimised through narratives about recovering ancestral graves and honouring the spirits of the ancestors. Similarly, during the occupation of Wolfscrag farm in Chipinge, claims to ancestral lands were strong, and particularly amongst chiefs displaced in 1974 (Zamchiya 2011).

Mujere (2011) highlights that the significance of returning to ancestral lands explains why certain farms were occupied in Gutu in Masvingo Province. As with Mkodzongi, Mujere refers to the strident stance articulated by chiefs about specific farms based on ancestral claims of belonging, as well as competing ancestral claims by different chiefs over specific pieces of land. This was part of a broader dynamic in which occupying groups at times competed with each other in establishing their presence on a particular farm. In the case of Gutu, the village head of Mudziwaniswa Village in Headman Mawungwa's area left his communal home with other people from the village in 2001 in order to return to ancestral lands in the Harawe area under Chief Chikwanda's authority. They demarcated plots and pursued agricultural activities, simultaneously negotiating with the chief over re-establishing their original village there. However, they were told by police to vacate the farm because it had not yet been gazetted (under the fast track programme). Houses were burned and crops destroyed by the police. The village head nevertheless insisted: "[W]e still have an emotional attachment to the area because it is the land of our forefathers where they lay buried" (Mujere 2011: 1135).

Fontein (2006) seeks to show how spirit mediums (along with war veterans and other occupiers) engaged with the ruling party's nationalist rhetoric about land reform so as to substantiate their ancestral rights and the particular land claims of their clans. One example is the occupation of a particular piece of state land by members of the Haruzvivishe house of the Mugabe clan and the activities of a VaDuma spirit medium called Ambuya VaZarira. Ambuya was involved in reclaiming an area on Beza mountain on the northwest side of Lake Mutirikwi in Masvingo Province. This was not simply about lodging a historical claim, as it involved "a moral appeal to an imagination of the way things should be, in which the idea of 'tradition' or *chivanhu*, and ancestral ownership of land, demonstrated through the provision of rain and fertility, plays a central role" (Fontein 2006: 232). In the end, Ambuya was "involved in extensive negotiations with war veterans redistributing land, without being co-opted into their political agenda" (Fontein 2006: 242). Again, this speaks to the ways in which the state's *chimurenga* monologue history was subject to negotiation, translation and reconfiguration at local levels during the occupations. Just as war veterans were not merely co-opted into a ruling party agenda, other occupiers (including spirit mediums) were not simply co-opted into a war veteran agenda.

In Chiweshe, restitution-based claims regarding land occupations prevailed as well (Sadomba 2009). In 2000, peasants from the Hwata/Chiweshe and Zumba dynasties, then scattered around the country (in Mt. Darwin, Mupfure, Chiriseri and elsewhere) organised to occupy an area called Gomba on autochthonous and ancestral grounds. In this particular instance, no war veterans were involved as the occupations were "based on unique peasant traditions, history, and organisation" (Sadomba 2009: 437). One of the women occupiers noted: "We came from Madziwa. We left after receiving a radio announcement that Mbuya Nehanda was saying that the owners of the land should return, meaning Zumba people. This said, our elders came to see Mbuya Nehanda. ... [T]hey were told to go and collect the clan to come and occupy" (Sadomba 2009: 438–439).

Thus, for Sadomba (see also Andrew and Sadomba 2006), many third *chimurenga* occupations were peasant-led occupations organised through spirit mediums and chiefs, who convened *biras* and led groups of occupiers in reclaiming land based on autochthonous, cultural and historical grounds. Fontein (2009) shows, though, at least in relation to certain occupations near Lake Mutirikwi, that some occupiers made autochthonous claims ex post facto, or only after occupying a farm. This entailed reinventing their reasons for the occupation (which previously was simply the desire for land based on redistribution, not restitution). In this way, chiefs in particular began to assert control over occupied farms, claiming that these farms were the home of their ancestors and thus were religious sites where spirits resided (thus, emphasising a process of restitution) (Scoones et al. 2010).

8.3 Order and Disorder

There remain only minimal understandings of the complex dynamics which took place during a particular farm occupation, or on a number of conjoining farms. Overall, there was a complex and dynamic combination of order and disorder during each occupation, which varied across space and time. Undoubtedly, there was substantial district- and lower-level variation throughout the countryside in terms of the forms and levels of coordination and organisation, as well as of violence. Asserting that war veterans coerced, on all occasions, communal villagers into joining the occupations, as some argue characterised the relationship between guerrillas and Tribal Trust Land villagers during the war of liberation, is problematic. More significant, for this section, is the question of violence in the process of occupying farms.

Often, some degree of coordination of occupations took place across a limited number of farms within or across districts. This regularly entailed a fluid combination of war veterans, chiefs, village heads and spirit mediums in facilitating mobilisation processes. As well, informal networks and arrangements amongst groups of occupiers, including those coming from communal areas, facilitated movement onto particular farms, with war veterans mobilising communal area villagers, as guerrillas did during the second *chimurenga*. War veterans (mostly men) were particularly important once farms were occupied. For instance, in some areas, certain war veterans were selected to circulate between farms to give briefings and to monitor progress and identify challenges across occupied farms (Matondi 2012).

Evidence also exists, at least for some areas, of a clear sequence of farm occupations as if planned and coordinated in advance by local, district or even provincial war veteran structures, including the Masvingo and Harare provincial war veteran associations. Nelson Marongwe brings this to the fore: "During the mobilisation process, the war veterans and the communities [in communal and old resettlement areas] would agree on which farms to occupy, set the day and decide on the gathering point. Where necessary, requisite arrangements were made to ferry people to the farm" (Marongwe 2002: 45). He also refers to the presence and involvement of youth in mobilisation:

> In some cases, youths were used to recruit people into occupying farms. For example, evidence from Chiredzi has shown that youths frequented surrounding villages, targeting social gatherings such as traditional beer drinking ceremonies to mobilise communities into occupying farms. Some amount of coercion was also used to recruit more people into farm occupations, particularly the youth. In Chiredzi, the practice was later abandoned after it was discovered that some youths fled from the occupied farms as soon as they were brought in. In other areas, a weekly timetable was developed and specific days were set aside for the recruitment of more people into the process. (Marongwe 2002: 45)

The reference to youth (and particularly male youth) and their participation in various dimensions of the occupation process is telling, and it parallels the relationship between guerrillas and *mujibas* during the war of liberation.

The study by Chaumba et al. (2003a) of one of the first series of occupations, in Chiredzi, demonstrates some of these points about mobilisation. In February 2000, the already-existing Chiredzi district war veterans' association mobilised people largely from communal areas to occupy a number of landholdings in a reasonably clear sequence across the district. On each farm, a certain number of occupiers were left to remain on the farm, led typically by a war veteran (labelled as a base commander), who would report back to the district committee when necessary. At Fair Range ranch, a base commander spoke about several meetings at cell, ward, branch and district levels to mobilise communal area villagers to occupy farms, with villagers at times compelled to attend these meetings. More than 1000 villagers from Sangwe communal areas and elsewhere were mobilised.

Despite the "apparent disorder there was actually much organisation and a rapid emergence [on the occupied farms] of structures, plans and forms of authority" (Scoones et al. 2010: 191). Once on farms, formal structures of authority were generally put in place to coordinate the activities of occupiers (Chaumba et al. 2003a; Sadomba 2013) on what became known as base camps. The base camp authority structure varied considerably (Scoones et al. 2010) but its key members tended to consist, where available, of war veterans. Typically, the all-important position of base camp commander and/or chief of security was a male war veteran, so that farm-based structures were militarised and patriarchal. The authority structure, including perhaps a political commissar and treasurer as well, coordinated the activities and tasks on the farms such as food provisions, transport, pegging of plots, security and maintaining discipline (Chaumba et al. 2003a).

Occupiers were normally registered on each farm and regular roll calls were undertaken to monitor movement on and off the farm. Any and all movements onto an occupied farm by unknown persons were carefully regulated, as was the internal on-farm movement of the owner of the farm. In part, this was because some occupiers were wary about staying on the farm, if only because of fear of retaliation by well-armed farmers. On occasion, occupiers were attacked by opposition supporters, and young male (ZANU-affiliated) occupiers were sometimes drilled in liberation war-like training and deployed as sentries to monitor farm boundaries and possible attacks. Occupiers though became increasingly confident as their temporary presence turned into a sense of permanent belonging, especially given that support from government was becoming clearer.

Accommodation of the barest kind (such as makeshift shelters of plastic, canvas, grass and poles) was often constructed to house the occupiers (often within view of the main farm homestead) or existing structures on the farm were used or taken over for shelter. Food for the occupiers came from a variety of sources. Sometimes it was brought in from nearby communal areas. Occupiers close to their communal villages would replenish their food supplies daily while those further away might bring food supplies to last from one to four weeks; alternatively, arrangements were made with household members remaining in communal areas to bring supplies at regular intervals. As well, voluntarily or because of demands placed upon them, farmers or ranchers provided food to the occupiers such as meat and mealie meal. In other instances, district war veteran associations arranged food supplies for occupiers, even with the assistance of transport from the District Rural Council. In terms of accessing water, occupiers normally relied on untreated water from rivers and shallow wells, or were able to access borehole water. Some occupiers left on their own volition, and never returned, because of staying for an extended period in temporary structures and with only erratic food supplies.

The occupiers sought to make their presence visibly and sometimes forcefully known to the farmer and the state. At times, this merely involved the putting up of signs by war veterans in strategic locations on the farm indicating that, in effect, the farm was under new ownership. If not at night, then during the day, occupiers positioned themselves near the main farm homestead and along the entrance and exit roads. There were also incessant demonstrations at farm gates (involving singing and drum beating, often by women) and all-night vigils close to the farm homesteads. This was done to cause maximum discomfort for the (typically male) farmer and his family. As Catherine Buckle, a commercial farmer, wrote in her memoirs of the farm occupations in her area:

> [O]ccupiers stationed 100 metres below their house. There was almost nowhere we could go now without them watching us. …They watched when I went out to feed the chickens… They watched when we checked the cattle and sheep. … They watched every time we drove out of the gate. This was almost intolerable. (Buckle 2001: 16)

This tactic of intimidation, and rumours about widespread local occupations in the days ahead, led some male farmers to send their farming wives and children to the nearest city or town to stay with relatives, or to go away for a few days to recoup their emotional strength.

The manner in which occupiers sought to make use of land, or take over land, including the pegging of land to delineate plots, was particularly important in staking a claim to the land. In this respect, Matondi (2012: 24–25) notes an array of approaches by occupiers:

> Some placed themselves on underutilised land and started to 'utilise' it without interfering with the landowners; some immediately declared themselves the new owners and asked the owners to leave; others grabbed produce already on the land and declared it theirs; some went for the farmhouses and equipment and declared ownership.

In driving farmers from their fields (including burning their crops and preventing agricultural activities) or from their farms entirely, and then demarcating their own plots, occupiers were seeking to construct new agrarian relations on the farms in a reasonably orderly fashion. For instance, Chaumba et al. (2003a) reveal that—in Chiredzi district (and specifically Fair Range ranch)—the practice of technical land-use planning by occupiers consistent with both colonial and postcolonial state agrarian programmes was evident in the laying out of new homesteads and the establishment plots, thereby enhancing their legibility and recognition by the state (Scoones et al. 2010). Even riverbank cultivation bans, like during the colonial period, were enforced (Chaumba et al. 2003a). In the farms studied in Mazowe by Chamunogwa (2019), the occupiers (which included state security agents) negotiated and gained access to official land records in order to facilitate the eviction of white farmers and thereby legitimise their claim to the farms.

Chaumba et al. (2003a: 535) highlight that the use of such technocratic practices (and bureaucratic procedures), while enabling the "legitimacy of Zimbabwe's 'new settlers', ultimately was potentially inimical to their interests in the longer term". This was because, when the state's agricultural extension services for instance became involved (under the fast track programme) in allocating plots officially (by May 2001 at Fair Range), the state could more easily superimpose its standard bureaucratic and technical agrarian order on any existing informality. In this context, Chaumba et al. (2003a: 544) argue that the land-use methodologies of fast track occupiers went contrary to previous ways of asserting authority over land, including before and during the second *chimurenga*:

> This can be contrasted with the 'freedom farming' of the liberation war or the 'squatting' on vacant land immediately after independence, when colonial land-use plans and restrictions on where and how people lived and farmed were explicitly resisted. This is particularly ironic given the revival by ZANU(PF) and the war veterans of liberation era anti-imperialist rhetoric.

Nevertheless, this technocratic consolidation was part of the broader process by which the ruling party and state sought to normalise and subdue the occupations.

Despite this coordination and ordering of on-farm activities subsequent to the occupation of a farm, the general image of the occupations is one of chaos and anarchy, an image which the dominant perspective (analytically consistent with nationalist historiography) of the third *chimurenga* occupations has tended to foster. This claim is problematic, given that this perspective conceptualises the occupations as orchestrated and driven from above, mainly by and through the ruling party. In one sense, however, it seems that the chaos and anarchy is conflated with the diverse working out of the occupations as animated by fluctuating local dynamics.

The presence of anarchy and confusion appears to undercut the argument about ruling party coordination of the occupations as part of a grand master plan. Though claiming that the occupations were state-driven, in her memoirs, Buckle (2001) in fact makes constant reference to the absence of coherence, when she states for instance that farmers "assumed that sanity and legality would prevail"; "[f]rom area to area the situations were completely different"; "[w]e all began to wonder if in fact Chenjerai Hunzvi was still in control of his members and their supporters"; "everyone now wondered who, if anyone, was in control"; and "[h]ow many people had now claimed they were in charge of this farm? I had lost count" (Buckle 2001: 42, 75, 85, 96, 180).

In another sense, claims about chaos and anarchy are meant to highlight the levels of coercion and violence which marked the occupations broadly speaking. As the war veterans tended to label the land occupations as a reactivation of the war of liberation, and represented the occupations as in fact heralding the completion of the second *chimurenga*, the language, images and symbolism of the second *chimurenga* were invoked, thus making the possible use of violence as necessary and legitimate. In this way, war veterans were "quick to evoke a war-time inventory of names and labels" (Mlambo and Gwekwerere 2019: 138), including the sell-out label for those who opposed the occupations.

The systematic, robust and often violent move to reclaim land and oust white farmers is said to be clearly evident in the formulation and implementation of Operation *Tsuro*, as described previously. It is also plain to see, according to Pilossof

(2012), in a document circulated amongst local war veteran associations in some areas just ten days after the official launch of the fast track programme in July 2000, Titled "on the white farmers and opposition" and dated 25 July 2000, the document alludes to an Operation Give-Up-And-Leave and to the need to ensure that "farmers are systematically harassed and mentally tortured and their farms destabilised until they 'give in' and 'give up'" (Pilossof 2012: 52) in a kind of war of attrition.

The pervasiveness of violence by occupiers during the occupations is undoubtedly a point which the historiography of nationalism interpretation (as articulated by Moyo) of the occupations regularly underplays. Undoubtedly, there were a number of occupations which were devoid of any physical violence by occupiers. James (2014, 2015), in his study of North Star, Golden Star and Chiraramo farms in Mashonaland Central Province, thus speaks of generally peaceful occupations, at least mainly in the case of North Star. In his research at Damvuri Conservancy in Mhondoro Ngezi district, Mkodzongi (2018) likewise refers to relatively violence-free occupations. The forms and levels of violence, like other dimensions of the occupations, were open to significant variation over time and space, depending upon a range of local factors including the composition of the occupying force and the attitudes and responses of farmers. At times, violence (often against farm labourers) existed from the start in order to allow occupiers to assert their presence while, in other cases, initial negotiations with farmers led to coexistence for a short period.

Throughout the occupations, the Commercial Farmers Union gave daily reports about the violence taking place on particular farms in different provinces and districts, as farmers throughout the country sought to remain in constant contact with their local CFU representatives, updating them of the status of occupations on their farms. These daily situation reports speak of widespread violence. The autographical account by one farmer (Henry Jackson) (Jackson 2014: 111) concurs with this, as he talks of violence sweeping across the countryside "like a plague" or "cancer". In cases where there was limited or no violence, there were (as noted) tactics of ongoing intimidation, including the singing of liberation war songs and beating of drums at farm gates, which likely traumatised many farmers and their families.

At the same time, the perpetuators of—and reasons for—violence during the occupations appear more complicated than most analyses acknowledge. As Ngonidzashe Marongwe (2014) shows in his study of Shurugwi, any violence during the occupations is not reducible "in a simple manner" (Marongwe 2014: 77) to occupiers acting out the will of the ruling party. Like the history of nationalism understandings of the second *chimurenga*, it is necessary to move beyond merely asserting the significance of national-level politics by offering a refined understanding of local histories and dynamics in explaining violence. In the case of Shurugwi, the importance of "local politics and contradictory interests" and "petty jealousies" in fuelling violence came to the fore (Marongwe 2014: 77, 92).

Whatever the reasons, white farmers and particularly their agricultural labourers bore the brunt of significant violence and intimidation (Human Rights Watch 2002). Further, as the occupations progressed, tactics of intimidation and disruption widened to include for example local government offices and schools, as McGregor (2002) highlights with reference to areas of the country, such as northern Matabeleland, where the opposition party was particularly strong. As well, war veterans, even in the early period of the occupations, supplanted Rural District Councils as loci of district authority, including sacking officials in places like Zaka, Chivi and Mwenezi (Chaumba et al. 2003b).

It is notable though that some farmers were responsible for acts of violence as well. In addition, likely instigated by farmers, MDC supporters from Harare are known to have attacked occupiers on farms close to the city. At least at first, farmers regularly sought the assistance of the police and District Administrators, usually unsuccessfully, to minimise the extent of the occupations and encroachment on their lands (including the forced halting of their agricultural operations), and to ensure that acts of violence were not tolerated or kept to a minimum. Quite often, if a consensual outcome from these local negotiations arose, this was simply ignored by the occupiers subsequent to the departure of the state official. Farmers initially believed, during the early part of 2000, that the occupations were a temporary phenomenon, and some farmers even kept records of losses and damages on their farms to recover the costs later (presumably from the state). But they soon came to realise that they were on their own in defending their farms, and ended up relying quite extensively on the social networks and comradery that they had established over many years, in some cases dating back to the war of liberation. In fact, in doing so, they reverted to agri-alert security systems (Alexander 2003) used during the 1970s.

In the following chapter, we offer insights into many of the themes about the occupations discussed so far (in this and the previous chapter), by focusing on occupations in Shamva and Bindura in Mashonaland Central Province. But, in order to show more fully the diversities and complexities of the occupations throughout the country, we consider the dynamics of occupations in two outlying areas, namely in northern Matabeleland and in the Lowveld in Masvingo Province.

8.4 Lowveld in Masvingo Province

A number of writers (Wolmer et al. 2004; Wolmer 2007; Scoones et al. 2012) have considered occupations in the Lowveld of Masvingo Province, particularly in areas which were not initially incorporated into the fast track land reform programme (if at all), though they were subject to significant occupations. These include Nuanetsi ranch (owned by Development Trust of Zimbabwe), Save Valley and Chiredzi River conservancies, Gonarezhou National Park and numerous other game ranches. But they also included a small-scale irrigation scheme in Sangwe communal area (Chaumba et al. 2003b).

In this area at the margins (or borderland) or state power and authority, any notion of a well-established imposition of a centralised state power (including through the fast track programme) is questionable (Scoones et al. 2012). Because these occupations were not in all cases retrospectively sanctioned by the state, complex local power struggles took place for many years subsequent to the initial occupations. In this sense, in relation to the regularising, subduing and subordinating of the occupations by ZANU-PF and the central state, this took place unevenly over different time-spans across the countryside. This in itself speaks to diversity across the countryside regarding the dynamics of the occupations, and it tends to go against the undifferentiated claim by Moyo. As Scoones et al. (2012: 549) rightly argue, this helps to "qualify some of the Harare-centric analysis of Zimbabwean politics" in which it is assumed that "top-down imposition by the state takes place" unproblematically. Indeed, in terms of the occupations, there has tended to be a Mashonaland-centric understanding based on an examination of the three Mashonaland provinces.

At the same time, the argument by Scoones et al. (2012) might imply that ongoing contestations did not take place in areas where the fast track programme was implemented, which is also not the case. In this regard, Scoones et al. (2012: 549) clarify their argument as follows:

> In contrast to the 'fast-track' land reform areas elsewhere in the [Masvingo] province, a different political dynamic is observed in the Lowveld. While, in the 'fast-track' areas, the *jambanja* phase was characterized as disorderly and chaotic, a sense of control emerged relatively quickly, as plans were imposed and resettlements formalized. ... By contrast, in large areas of the Lowveld such formalization has not been evident, and people remained with the status of 'squatter' for over a decade.

However, any claim about absolute control occurring on fast track farms (even in the Mashonaland provinces) goes contrary to evidence of the existence of ongoing conflicts on many of these farms.

In the Lowveld, particularly from 1997 to 1999, there was a noticeable escalation of farm occupations in the area, including "squatters" moving onto Levanga and Angus ranches in Save, as well as Faversham and Ngwane Extension ranches near Chiredzi. Like these earlier "squatters", the occupiers from the year 2000 were mainly of Shangaan origin, and they had been moved from these areas (for example, Save) at various times in the past. However, historical memories and the importance of ancestral spirits lingered on. In referring to burial sites and to sacred places where rituals needed to be performed, there was—amongst the occupiers—"a sense of reparation – of re-asserting territorial identify claims by returning to 'our land' and 'our animals'" (Wolmer et al. 2004: 92–93). These local land claims were supported by war veterans and local politicians.

Occupations in the Lovweld were evident by April 2000 and, by mid-year, they were extensive, notably around Chiredzi. In the case though of the large-scale sugar cane estates, only minimal occupations took place. For Wolmer (2007: 194), even within the Lowveld, "different combinations of actors ... occupied different properties for different reasons in different ways". Overall, "[f]arm invasions in Mwenezi and Chiredzi Districts were ... a product of both [local] [party]-political orchestration and spontaneous opportunism" (Wolmer 2007: 198). Some occupations were peaceful and symbolic, with the pegging of plot boundaries and building of shelters but no land cultivation. Some farms were occupied, abandoned and then reoccupied. Other occupations were more aggressive, including maiming of cattle, cutting of trees, asset stripping, occupying of safari camps, fish poaching, clearing of fields and homestead construction. As well, boundary fences were taken down as "a symbolic attempt to erase the white signature in the landscape" (Wolmer 2007: 199). Also apparent was the deliberate use by occupiers of liberation wartime rhetoric, terminology and practices, while white farmers travelled in convoys as they did during the height of the war of liberation. As Chaumba et al. (2003b: 596) put it:

> The War Veterans Association and ZANU(PF) have deliberately echoed the language and symbols of the liberation war, including: reviving the former enemies (Rhodesians and imperialist, mainly British, aggressors); slogans, *pungwes*, *mujibas* (youth auxiliaries), *chimbwidos* (women supporters/cooks), 'sell-outs', and the creation of a new cadre of youth brigades. Even some of the guerrilla tactics, such as arson and stock theft and mutilation, were revived on the occupied farms.

In the Lowveld, significant land shortages in communal areas (such as in Bikita, Zaka and Chivi) animated the occupations. However, the occupiers were diverse, including the urban unemployed from Chiredzi town. Conflicts between occupiers from different administrative areas arose and became politicised (Wolmer 2007). For example, Jabula Ranch, despite bordering Matibi II communal area, falls within Mwenezi District, and it was occupied by people from Chiredzi District. This led to a series of quarrels between the occupiers themselves and between the Members of Parliament and District Administrators from Chiredzi and Mwenezi. Ranches in Triangle in Chiredzi was similarly contested by communal villagers from Zaka, Bikita and Chiredzi. At times, this entailed the involvement of competing factions of war veterans. There was an ethnic dimension at times as well, with Shona and Ndebele "immigrants" occupying Lowland land based on broad redistributive claims, while Shangaan people were motivated by ancestral and restitution claims. Local Members of Parliament, notably the (Shangaan) Member of Parliament for Chiredzi South, became involved in this contestation.

In the Save Valley conservancy, many occupations of the ranches likewise related to ancestral longings and thus revealed "long-term contests over the landscape" (Wolmer 2007: 210) (see also Wels 2003). For instance, the (ethnically) Ndau Gudo occupiers from Sangwe communal areas had burial and ritual sites and ritual pools on Levanga ranch. Before 1986, this was state-owned land and communal villagers had relatively easy access to it. When it became part of Save conservancy, conflicts over access arose, leading to fence-cutting and veld fires—before 2000. The case of Save also reveals considerable political involvement, including the Provincial Governor and ruling party Members of Parliament for Chiredzi North and Chiredzi South, with some politicians addressing rallies exhorting people to occupy farms. There were divisions as well within the state regarding the occupations, with the Ministry of Environment and Tourism and the Department of National Parks and Wildlife Management expressing concerns about the effects on wildlife. This shows that the state was "highly heterogeneous and fissured" (Wolmer 2007: 213), with "[i]n particular, the politicians and technocrats … working to different agendas rooted in different ways of seeing the landscape".

Gonarezhou National Park experienced, with time, a high-profile occupation by the Chitsa people (Chaumba et al. 2003b; Scoones et al. 2010). Headman Chitsa, together with a local Shangaan war veteran leader named Phikilele and supported by the local ZANU-PF Member of Parliament, led an occupation by Sangwe communal area villagers of a portion of the north-western part of the national park adjoining Sangwe. Because of past evictions, the Chitsa people had claimed rights to areas of Gonarezhou Park and, during the 1980s and 1990s, they regularly entered the park to poach and graze cattle. In the occupation in 2000, around 750 households (from Sangwe) moved to the park and fields were cleared, with tents supplied by the army" (Wolmer et al. 2004: 96). The Department of Parks was, in the end, able to remove most of the occupiers.

In the case of the Lowveld, then, it is possible to discern the marked importance of local political contestations. At the time when fast track land reform was being pursued and implemented in many other parts of the countryside, and war veterans and occupiers in large part were becoming subject increasingly to the dictates of the state, the Lowveld was characterised by greater fluidity, with ongoing occupations in which a complex array of local politicians and state officials became involved. As well, the significance of ancestral rights, and contested rights, to particular pieces of land prevailed. The return of ancestral lands would allow, for instance, rain-making and other ceremonies to take place, including at sacred sites at Chilunja and Gonakudzingwa hills in Gonarezhou Park. As Scoones et al. (2012: 544) argue:

> This landscape was one where their ancestors lived, where the graves were sited, where Shangaan circumcision and initiation ceremonies were practiced, and where the spirits lived. Within the new settlement there were reputedly three different spirit mediums, each with powerful arguments about how the spirits owned the land. The spirits, as guardians of the land on behalf of the ancestors, got intensely involved in the land dispute through their mediums ... These were powerful, symbolic stories asserting authority over the land and its use.

Once an occupation took place, the war veterans and other occupiers were at pains to appease ancestral spirits. Chiefly authority was only called on later, as if (like the war of liberation) spirit mediums and "war veterans have usurped chiefly authority" (Chaumba et al. 2003b: 599). More broadly, in the Lowveld at least, "[n]either the motives for the violent occupations of land ..., nor the shrill emotive response to it, can be properly understood without an understanding of the competing ways in which this landscape is and has been physically managed and invested with symbolic meaning" (Wolmer 2007: 1).

8.5 Northern Matabeleland

Jocelyn Alexander and Joann McGregor, as part of their broader work on Matabeleland, have examined some of the dynamics of fast track occupations in that area. Like the Lowveld, many parts of northern Matabeleland have been on the margins of state presence and power, with the postcolonial state failing to prioritise the area in terms of socio-economic development. As well, like the examination of the Lowveld shows, occupations in northern Matabeleland had their own particularities.

In discussing occupations until July 2000, Alexander and McGregor offer comments about the occupations countrywide, claiming—like so many others—that occupiers were provided with systematic and pronounced support from the ruling party, army personnel and Central Intelligence Organisation operatives, who were amongst the occupiers and played leadership and coordinating roles during the occupations. They do highlight, however, the central role of war veterans and, in particular, the marginalised status of those war veterans involved in occupations in Matabeleland and elsewhere: "Everywhere, there was a class element to active support for the farm occupations: those veterans in secure and well-paid jobs tended to oppose occupations, while the unemployed and those languishing in communal areas were more likely to favour them" (Alexander and McGregor 2001: 515).

The ZNLWVA provided various district-level war veterans' associations with lists of farms in their respective districts to be prioritised for occupation. Though, as we argue as well, "[o]ccupations brought together a variety of agendas and unstable alliances" (Alexander and McGregor 2001: 551), they tend to conclude—in large measure—that the primary purpose of the nation-wide occupations was to bolster the ruling party's rural support and undermine that of the MDC. Given the strong support base of the MDC in Matabeleland, as manifested in the June 2000 election, there is likely some truth in the idea that political-party contestations became intertwined with the occupations in the case of this region, at least more so than in other regions. But localised agendas of war veterans and other occupiers also come to the fore.

In understanding the occupations in Matabeleland North, Alexander and McGregor (2001) rightly argue that the political history of the region becomes important. Occupations in Matabeleland did not take off immediately and only gained momentum in late March, and then only "only half-heartedly": "Leading politicians from the region such as Joseph Msika and Dumiso Dabengwa did not share [ZANU-PF Political Commissar] Gezi's enthusiasm for occupations, and many veterans in Matabeleland were unwilling to engage in violence in this period" (Alexander 2003: 101). Undoubtedly, war veteran-ruling party dynamics in Matabeleland had been marked by a degree of political distance for an extended period, and this was one legacy of the *Gukurahundi* during the 1980s, with ex-ZIPRA guerrillas feeling more marginalised than ex-ZANLA guerrillas and disinclined to support ZANU-PF.

In resonating with the inter-guerrilla army dynamics of the 1970s, some senior ex-ZIPRA commanders criticised the occupations. Others supported them but, at the same time, condemned the violence underpinning them as well as their partisan character vis-à-vis ZANU-PF. In reflection, these veterans spoke about negotiating with white farmers and sharing the occupied land, though farmers recounted compulsion in relenting to war veteran demands and threats. Where significant violence did take place, this was due to veterans being bussed in from other areas (Alexander and McGregor 2001). For instance, close to Nkayi communal area, national war veteran leader Hunzvi entered Gourlay Ranch on March 29, 2000 with 70 veterans (from Mutoko and Murehwa) along with ZANU-PF youth. This occurred on the request of local veterans, who had been beaten by farm workers when they first sought to occupy the farm. By 2001, with ZANU-PF-linked war veterans becoming more assertive, there emerged a "political assault on the local state" in northern Matabeleland (McGregor 2002: 23), as these veterans sought to push fast track redistribution forward.

In the case of local war veterans occupying and setting up base camps on farms (at least by July 2000), they did not normally present their action in terms of the militant language of ZANU-PF national leaders, or like war veterans elsewhere: "Occupations here were … not the popular repossessions of ZANU-PF rhetoric" (Alexander and McGregor 2001: 520). Rather than articulating any version of the ruling party's *chimurenga* discourse, occupiers generally spoke angrily about those farms handed out in "payment" to white settlers for service in the Pioneer Column (or as ex-soldiers from the Second World War), and about the need to secure land for those threatened with eviction by the Gwaai/Shangani Dam.

In contrast to many other parts of the country, there were few occupiers from communal areas on the largely cattle and game ranches prevailing in the region. In terms of agrarian history, local villagers tended to focus on demands for additional grazing land to be added to their communal areas, rather than on movement of households to new agricultural lands. The game and cattle lands were marginal arable areas unsuitable for small-scale agriculture, notably in Hwange and Lupane districts. As well, many communal area villagers did have not have historical-ancestral claims to neighbouring commercial ranches and farms in Matabeleland, as they had been moved hundreds of kilometres from their former homes during the colonial era. There were, however, exceptions to this trend. In the case of Gourlay Ranch (and nearby farms), there was significant occupations by Nyaki communal area villagers, in part because these farms are located on more fertile and well-watered soils. Nyaki villagers had deep memories of evictions from these farms and had not received any redistributed land under land reform since 1980.

Lastly, the relevance of local histories and concerns, including their irreducibility to ZANU-PF's (and even war veterans') nationalist discourses and practices, comes out clearly in the case specifically of the Tonga people. McGregor (2009) indicates that—in the national elections of June 2000—Binga District delivered the highest opposition vote for the MDC of any rural constituency. In this context, ZANU-PF tried to use former ZIPRA guerrillas as proxies against the MDC in this area, including by way of occupations. However, ZANU-PF's discourse on land (and race) did not fully resonate with the memories of local Tonga people, as those in Binga did not lose land to white farmers but to state conservation bodies. Hence, "they did not want to be resettled on former commercial farms but wanted access to state resources and reparations for unsettled grievances relating to the [Kariba] dam and displacement" (McGregor 2009: 170) dating back many decades. Thus, the nationalist rhetoric of ZANU-PF did not displace local ethnic Tonga claims and aspirations.

This discussions of occupations in the Lowveld in Masvingo Province and in northern Matabeleland bring out the fundamental importance of local histories and contingencies in understanding the diverse character of the third *chimurenga* occupations. While both cases demonstrate some involvement, at local level, of ruling party personnel (and even state officials), this does not imply that the occupations were initiated and organised by the ruling party. It might be argued that a centralised directive from the ruling party to occupy land is not inconsistent with local differentiation and dynamics regarding occupations as, in cascading down, top-down directives are shaped by local conditions. However, in our case studies in Mashonaland Central Province (in chapter nine), we demonstrate the decentralised character of the occupations, and how they were conditioned and constituted by, and entangled in, a web of specific and fluctuating local dynamics, and how occupations unfolded in an indeterminate manner on this basis.

8.6   Farmers and Farm Labourers

Agricultural labourers found themselves in an in-between and ambiguous location vis-à-vis the occupiers and white farmers, a situation with some parallels to their situation during the war of liberation. Though farm labourers had engaged, in the post-1980 period, in their own everyday forms of resistance and at times overt struggle (notably in the late 1990s) (Tandon 2001; Rutherford 2017), by the year 2000 they were still positioned structurally in a highly subservient and dependent relationship under the command of white farmers in the quasi-feudal "domestic government" (Rutherford 2001a). At the same time, a social, cultural and spatial distance often existed between farm labourers and the occupiers (who mainly came from communal areas), as the two groupings had diverging social and personal histories and experiences in different spaces within the Zimbabwean countryside.

In fact, many farm labourers (particularly those of foreign origin) had no discernible links with communal areas and had no alternative place to call home (beyond the farm). This gave labourers a sense of belonging to the farm, as Sinclair-Bright (2016: 58) argues: "[T]he longevity of farmworkers' lives on farms, their labour on the land, their knowledge of the area and, in many cases, the burial of their dead in the [farm] land all stood as a counter to claims over the land on the basis of having fought in the liberation war". This should not imply that agricultural workers, during the second *chimurenga*, played no role whatsoever in supporting the guerrilla armies. It does indicate, however, that claims over farmland, and notions of belonging, were open to contestation between farm workers and land-short communal area villagers.

There were cases, though, in which farm labourers had certain connections to communal areas. In the Lowveld in Masvingo Province, farm workers on many of the large farms had ancestral roots in nearby communal areas, unlike in the tobacco-growing zones of the Mashonaland Provinces where workers often originated from Malawi and Mozambique. This meant that "they had social connections with nearby communal lands, allowing linkages with the land invasions to be forged" (Scoones et al. 2010: 128), as we discuss later.

As well, as indicated earlier, there was a close relationship established between white farmers and the main opposition party, with farmers declaring their support for the MDC and becoming local MDC representatives, transporting their workers to MDC rallies (which led to workers at times becoming MDC members) and generously bank-rolling the party. All this was done in a very public manner, contributing to party-politics becoming embroiled in the land contestations during the third *chimurenga* occupations. In relation to Mashonaland East Province, the MDC "assumed that farm workers could be reached only through the lineaments of domestic government" (Rutherford 2013: 860). In this way, farmers mediated the relationship between the MDC and their farm labourers, which in effect compromised the labourers as they "were made even more vulnerable to the nationalist *poritikisi* unleashed by ZANU(PF) as [MDC] 'sell-outs' to the postcolonial nation" (Rutherford 2013: 861). Hence, an "intertwinement of electoral politics in these land occupations" (Rutherford 2014: 229) existed and became more pronounced as the June 2000 parliamentary elections approached. Indeed, with the occupations, the "erstwhile impermeability of commercial farms to *poritikisi* was permanently breached" (Rutherford 2013: 857). The white farmer-MDC alliance, and the very subordination of agricultural labourers to white farmers, was bound to lead to suspicion by occupiers about the loyalties and sympathies of farm workers (Rutherford 2013).

At the same time, white farmers assumed—often on dubious grounds—that farm labourers were inherently loyal to them, based on farmers' discursive construction of domestic government as a caring and intimate relationship along the "interior frontier", which was meant to bolster their private and public identity and image. In the context of the occupations, ZANU-PF criticised implicitly the "interior frontier" by claiming that white farmers compelled their farmers to vote "no" in the constitutional referendum (Rutherford 2004b). Further, and again if only implicitly, occupiers challenged the "interior frontier" by seeking to undercut any farm labourer loyalty to white farmers so as to gain the support (or at least to minimise the resistance) of the labourers as the occupations unfolded.

Nevertheless, collaboration between farm workers and occupiers was often difficult, and this led at times to violence against labourers by occupiers (Sachikonye 2003; Chadya and Mayawo 2002). Overall, the widespread exclusion of farm labourers from the occupations was not entirely deliberate on the part of the occupiers, as it also arose from the intertwining of labourers' lives (and their security and belonging) with the lives and politics of white farm owners. Occupiers in many areas came to realise the importance of gaining some level of support from labourers. For instance, "[i]n Mazowe Valley and Mvurwi, occupiers were repelled by farm workers and MDC supporters from February 2000 and it became clear to them that that they could not win the 'war' without mobilizing agricultural labourers to embrace the movement" (Sadomba and Helliker 2010: 214).

Like the guerrillas during the second *chimurenga*, war veterans and other occupiers felt compelled to hold all-night rallies and vigils (or *pungwes*), in this case with farm labourers. They did so in trying to propagate and justify the reasoning behind the occupations, with these rallies regularly entailing liberation and nationalist messages as well as tactics of intimidation. There was, however, some variation between farms in terms of relationships between labourers and occupiers. Magaramombe (2010) argues that most farm labourers opposed and even resisted the land occupations (as the occupations threatened their farm-based livelihoods). Overall, though, there was "a mixed reaction" to the occupations by farm workers:

[S]ome opposed it and resisted farm takeovers, while others welcomed it and were active in farm invasions … in their respective districts … Those workers who supported the invasions saw them as an opportunity to access land in their own right, whereas those who opposed tended to see them as a threat to their jobs, housing and other benefits. (Magaramombe 2010: 366)

Those labourers who opposed occupations were not necessarily against land reform. Rather, no matter how subordinate they were to white farmers, they were likely concerned about maintaining their current position on the farm, particularly given that the outcome of the occupations remained uncertain (Rutherford 2004a). In fact, white farmers often assured their workers that the occupations were a temporary phenomenon. In this sense, labourers were hedging their bet.

In the case of Brylee farm on the outskirts of Harare, there was some resistance by farm workers against occupiers and "[p]ro-ZANU PF elements" (Hartnack 2005: 178) who had come from Harare. At times, though, labourers welcomed the occupiers and were active in the occupations as a possible basis for accessing land. In fact, there is evidence that workers sometimes led the occupations on the farms on which they laboured (Scoones et al. 2010). The work by Chambati (2013) also highlights that some farm labourers engaged in occupations while others fought against occupiers (see also Rutherford 2001b). Sadomba and Helliker (2010: 214) also note the role of casual workers, on occasion, in occupations:

In Matepatepa, seasonal farm workers who lived in Chiweshe communal lands provided information to peasant-led occupations. These farm workers recommended which farms to occupy and they gave details to occupiers of farm ownership, farmers' dispositions and behaviour, and farm security systems. They also fed the occupying peasants when supplies from the communal lands were depleted.

White farmers tend to deny the significance of labourers' involvement in occupations and even acts of violence by labourers in pursuing occupations. For instance, based on fieldwork in Manicaland and Mashonaland provinces, JAG (JAG and GAPWUZ 2008b) notes that fifteen per cent of farm labourers interviewed indicated that fellow workers had committed acts of violence against them, and labourers engaging in such acts against farmers as well. A very simplistic explanation is then offered, which cleanses farmers of any responsibility (see also JAG & GAPWUZ 2008c):

[T]he war veterans and youth militia who invaded the farm would immediately set one up [a ZANU-PF committee] and recruit from farm workers sympathetic to ZANU-PF. This minority of workers, often tempted by promises of personal gain, thereafter perpetrated violations against their employers and their fellow workers (JAG & GAPWUZ 2008b: 29–30).

Undoubtedly, significant research is still required to understand more fully the different responses to the occupations by farm labourers. Rutherford (2014: 230) for example quite rightly argues that it is simply not possible to explain "those [different farm labourer] actions [supporting or opposing occupations] through the lens of a nationalist version of the agrarian question (farmworkers are either for independence and autonomy of having their own land or are inured within the ties of racist dependencies)". Rather, it is necessary to "attend to the particular cultural politics of the location, the discursively constituted power relations through which access to resources are channelled" (Rutherford 2014: 230). In this regard, Rutherford (2012: 152) notes elsewhere that "[f]arm workers who resisted" the occupations "were said to be 'very loyal to their employers' ... and not, say, trying to grapple with uncertain and ambiguous power relations and threats from all sides". Rutherford's argument has relevance to the second *chimurenga* and is consistent with our claim that villagers in Tribal Trust Lands had a diverse array of reasons for why they engaged with guerrilla units in the manner that they did. This is part of our broader argument about not reducing the complexities of the *zvimurenga*, including the third *chimurenga*, to grand narratives and centralised impositions.

Like farm labourers, white farmers themselves were not fully united in their responses to the occupations. At national level, as noted with regard to the formation of JAG, splits within the white farming community arose because of the occupations. There were divisions at local levels as well. Some farmers in Mashonaland Central, who were eventually forced to leave their farms, spoke derogatively about other farmers (a minority) who negotiated staying put, seemingly at all cost and without principles (Kalaora 2011). At least in the Mashonaland provinces, some form of farm-level negotiation with occupiers was initiated by many farmers, so as to prevent immediate eviction (Laurie 2016). These entailed either "short-term extortion practices" or "long-term protection agreements" (Laurie 2016: 205, 206). In the case of the former, which were quite common, occupiers would minimise intimidation and violence in exchange for money and assets. For the latter, powerful local ZANU-PF members would request protection money from the farmer to prevent further occupations (and thereby facilitate ongoing agricultural production).

With regard to those farmers engaged in long-term agreements, evicted farmers would refer to them as corrupt, mad and contaminated (if not sell-outs to the cause of white farmer solidarity), as they violated the basic tenets of domestic government on which white farmer authority rested (Kalaora 2011). This was evident in the Shamva district, at least in relation to Doug Bean who owned Douglyn Farm and was one of the largest pig farmers in the country. He entered into a business arrangement with then Minister of National Security (Nicholas Goche), who owned a farm in the area. Bean allowed occupiers in the area to use an old pick-up truck to carry occupiers around and his sons were often seen in Bindura town wearing ZANU-PF caps. Other Shamva farmers treated Bean with great disdain.

There is now a large body of literature on white farmers in post-2000 Zimbabwe (Suzuki 2018), including scholarly writings as well as memoirs and fictional writings (Manase 2016), which is termed as "white writings" (Pilossof 2006). The scholarly work includes Kalaora (2011), Pilossof (2006, 2012), and Hughes (2010) and the extensive work by Rutherford since the late 1990s, which all offer useful framings for making sense of white farmers' stances on the occupations.

Though applying his arguments to whites more broadly, Kalaora (2011: 750) argues that the farmers' "power and social identity were continually defined by, and activated in, constantly renewed efforts to enforce and delimit ... [geographical, social and moral] 'boundaries'"; including along the "interior frontier". In defending these boundaries, they were not only protecting their economic power but also their "identity, 'manners', values and way of life" (Kalaora 2011: 750). In this respect, the farm occupations severely undercut their way of life by disrupting the socio-spatial borders in which they seemed protected and which ordered their very existence: this entailed a political, moral, social and at times literal spatial displacement. The occupations became a traumatic existential threat like no other, entailing a loss of meaning and comprehension for farmers, even more so than during the war of liberation because in part, during the occupations, they were left completely to their own devices and had no government support. Further, during the second *chimurenga*, farmers—though in danger—

Felt that they understood the logic of what was happening. Now the white farmers ... say they no longer understand what is happening and find themselves thrust into an uncertain zone where they have no protection at all... [F]armers have lost control over the manner and timing of contact with Africans. (Kalaora 2011: 751)

As we noted in citing the work of ex-farmer Catherine Buckle, rather than a regime of law and order, all that existed for white farmers were "unclear, anarchic modalities of negotiation" (Kalaora 2011: 752) in which they were "forced to acknowledge the presence of the other [the occupier] on the farm" (Kalaora 2011: 753). This was particularly troublesome for the hypermasculinity of the male white farmer, who saw his domestic government regime whittled away and who felt unable to defend his wife and children against intruders.

Kalaora (2011) discusses this with reference to farmers' long-held conceptions of landscape, which was subjected to rupture and ruin because of the occupations, with the post-occupation landscape diametrically opposite to manicured landscape of the past. As Pilossof (2006: 629) notes, this conception entailed a romanticised vision of the farm setting and of life before the land occupations, or "an idealised view of race and labour relations" on the farms:

[T]he way white farmers have talked, and continue to talk, about the land reforms and their evictions, demands a great deal of scrutiny because it contains much more than a mere description of confusing and chaotic events. The language and description used so often has a very real and deep connection to pre-independence tropes of land, belong and race. In addition, much of the evidence used to justify place and history, relies on highly problematic readings of the past. (Pilossof 2012: 3)

Typically, this reading of the past involved the absolute legitimacy of "domestic government", understood from the perspective of white farmers as paternalistic and caring labour relations devoid of all exploitation.

Farmers' post-eviction memories of the farm revolve around land and nature, entailing "a project of belonging" (Hughes 2010: 2) without almost spiritual connotations. In doing so, they connected with the natural and built landscape of the country and not with the African workers who toiled for them. More broadly, when farmers' connections to Africans are made, it comes across as paternalistic or, as Hart-nack (2016) emphasises, maternalistic in the case of farmers' wives. Of course, this landscape vision differs considerably from the vision of the occupiers, as discussed earlier.

White farmers' landscape vision and attachment to the land (and to Zimbabwe as well as Africa more generally) was devoid of (or not mediated by) African subjects including their labourers. Farmers, at least thus studied by Hughes (2010) in Virginia east of Harare "*managed* them [black workers] but did not construct an identity around them" (Hughes 2010: 74 his emphasis). As Hughes (2010: 126) argues more fully:

> Before 1980, they [farmers] thought about black society almost entirely in the context of what was known as the 'native problem': the administrative project of locating and disci-plining black labor. The project of belonging stood apart, inflected toward the landscape. Independence, if it suggested to whites that they try to belong among blacks, did so only briefly and indeterminately. Even the farm invasions initially provoked as much colonial-style paternalism as humility.

For African occupiers to forcefully enter the space of white farmers and, to break all boundaries, was to tear the latter's landscape vision asunder. This is because "by writing themselves intensely into landscape they [white farmers] wrote themselves out of society" (Hughes 2010: 25). Even in post-independent Zimbabwe, until the late 1990s, white farmers remained as decentralised despots in their freehold silos, with only minimal intrusions by the state, and they never sought to integrate themselves publicly into the postcolony. In this sense, postcolonial Zimbabwe was invisible and of no serious consequence to white farmers. The land occupiers turned invisibility into strident visibility and undercut farmers' imaginations around landscape.

A crucial point that arises from the views of Hughes, Kalaora and others is that the motivations and practices of white farmers, in responding to the third *chimurenga* occupations, are not explainable (at least not in full) in terms of farmers seeking to counter the authoritarian nationalism of the ruling party. Insofar as they did seem to be doing this, it was grounded in (and mediated by) by questions of boundary-setting and identity-formation.

The parallels between white farmers during the second and third *zvimurenga* are of some importance. Just as war veterans returned to the days of the war of liberation in terms of their perspectives and practices, so did white farmers (as if they were at war once again). The days of reconciliation, over the first two decades, was replaced with the language of war on both sides. As Pilossof (2012: 197) argues: "Twenty years after the end of the Liberation War, the white farming community was able to seamlessly fit the 'land invasions' into the paradigm of war". Likely, this re-emerged "bush war" narrative was facilitated by the significant presence of war veterans (ex-guerrillas) in the occupations, and vice versa.

During the war of liberation, the mouthpiece of the Rhodesian farmer's union—the *Rhodesian Farmer*—rarely reported on attacks on white farmers, so that fear would not grip the farmers and thereby leave their lands. With respect to the fast track occupations, perhaps in order to declare and claim the status of victimhood in the face of what they perceived to be a violent state, the CFU reported daily on attacks taking place on farms throughout the country. In defending their farms as best they could, and in defending their country as only true patriots would, they would stand tall (and if need be, alone) against the postcolonial state, just as they stood alone in remaining faithful to the colonial state. Just as they felt an absence of support and sympathy amongst white urban Rhodesians during the war of liberation, they felt abandoned by white urban Zimbabweans when the fast track occupations unfolded. In the process, white farmers sought to undermine the legitimacy of the occupiers by referring to them as *mujibas*, that is, as mere youth without liberation credentials who were not involved in the any past (let alone present) legitimate struggle for land (Pilossof 2012). Simultaneously, they projected themselves as defending the livelihoods and security of farm labourers.

One noticeable practice reactivated during the occupations was the Agric-Alert radio system which became a crucial part of farmer's lives during the war of liberation. District-wide inter-farm radio communication systems did exist before the occupations but these served agricultural and social purposes (such as the Omnicom system in parts of Mashonaland Central Province). However, at times in the late 1990s, these radio systems were used to mobilise local farmers when sporadic occupations (including plot pegging) took place. During the third *chimurenga* occupations, the agri-alert radio system was

> [A]times a mixed blessing for farmers. While it made it easy and efficient to communicate important information, call for help, have evening roll calls, and so on, it also contributed to the climate of fear and uncertainty, as people in serious situations could be constantly heard on the radio calling for assistance. (JAG and GAPWUZ 2008a: 60)

This communication system was part of a broader effort to support each other, as action plans were put in place (often at district and sub-district levels) based on then-existing neighbourhood-watch programmes and involving negotiation (or talk) and reaction teams. Overall, "[t]hese groups would provide both mental and logistical support to farmers who found themselves in difficult situations. Volunteers were found in each area and they became either the Talk Team [those who negotiated with the war veterans because of their disposition] or the Reaction Team" (Buckle 2001: 40). Reaction Teams, with certain members on constant call as per a formal register, were expected to move around and make their presence felt at a farm where a particular farmer was under distress or even in danger. This was meant to give the farmer moral support and in effect to intimidate the occupiers so that they would think twice before furthering their activities. These reaction teams were reminiscent of the "reaction sticks" of farmers during the war of liberation (Pilossof 2012).

8.6.1 Women and Patriarchy

Ranger, in his foreword to the book by Nhongo-Simbanegavi (2000: xvi–xvii), notes:

> Many people have asked, as they read reports of attacks on opposition supporters or of farm occupiers, 'Where are the women?'... [H]ardly any female ex-combatants have been

involved at all. The women one sees on television have been bussed into dance and sing; none of them has taken a decision; ... Why is this? ... One answer is that women were never allowed to make policy during the liberation war nor to take command of men. Another answer is that most women were 'demobilised' after the war, returning to civilian life and the 'command' of their husbands.

Ranger's key argument, which has significant validity, is that the fundamental basis for systems of patriarchy remained intact after 1980. Kesby (1996: 581) makes a similar argument: "The post-independence restructuring of society removed the arenas in which local agendas [involving challenges to local patriarchs, as occurred during the second *chimurenga*] could readily be pursued and opened the way for patriarchal agents to reassert their masculinity in their homes and communities". Thus, there was "a return to normality" (Kesby 1996: 582) for female guerrillas and women in general after the war of liberation, with any openings for women to reshape gendered practices during the war eventually undermined by post-1980 restructuring (Kesby 1999).

As noted earlier, there is a vast literature on the fast track land reform programme in Zimbabwe but significantly less on the occupations themselves. Further, in the case of the limited literature on occupations, few studies focus on women and the occupations. In considering women and patriarchy, and despite some exceptions, we consider the ways in which women were incorporated into the occupations in a subservient manner, just as farm labourers—as a general trend—tended to be excluded from the occupations.

Nelson Marongwe (2008: 49) argues that the participation of women was varied with "some farms, particularly those close to communal and resettlement areas" incorporating "balanced numbers of male and female occupiers". Moyo (2001) as well acknowledges an active role for women in land occupations which took place in Svosve, Goromonzi, Murewa and Insiza. However, social reproduction and caring responsibilities inhibited women's active involvement in the occupations. In the occupation of Nuanetsi ranch in Mwenezi district in Masvingo province, women were not able to engage fully in occupying Merrivale because of family responsibilities in the nearby communal areas from which they came and, further, they had to continue with farming activities in their communal homesteads (Mutopo 2011a, 2014).

In certain instances, women occupied farms on their own in the absence of men. This occurred in the case of women now living in Nyabamba A1 resettlement area (formerly part of a wattle company's land) in Chimanimani district. Women's prominent position in the occupations in large part arose because most men in the local communal areas were employed elsewhere at the time of the occupations (Chingarande 2010). In other cases, women as war veterans or with important ruling party affiliations spurred on certain occupations, such as in Mazowe district (Chiweshe 2011). In Mazowe, women seemed to play a leading role in the mobilisation campaigns during the occupations when compared to other districts (Sadomba 2009).

The continuation of gendered relations once on occupied farms, including with respect to social reproduction, was very pronounced, as women were in large part confined to the domestic sphere. In certain occupations, women and men were separated after 7 pm and not allowed to see each other (Chaumba et al. 2003b). In relation to Masvingo Province, and indeed elsewhere, women "often took on highly gendered roles in the base camps [on the occupied farms] (including cooking, collecting firewood and water)" (Scoones et al. 2010: 55). Women were directly affected by water access challenges which regularly existed on occupied farms, as it is their duty in rural communities to collect water. The problems experienced by women are vividly portrayed in the following comment from a woman at Dunstan Farm in Goromonzi district: "As women we always bear the burden of walking long distances in search of water… We don't have rest. If we are to wash clothes then we have to dedicate the whole day to that particular activity since the [distance to the] place is long (quoted in Chakona 2012: 101).

As during the war of liberation, women also sang and danced during the *pungwes* on the farms. As Ngonidzashe Marongwe indicates with regard to specifically Shurugwi District, the dancing and singing of women "served only to glorify … the country's male political leaders" (Marongwe 2013: 466). The land occupations hence did not entail any reconfiguring of the kinds of social relations (including patriarchal relations) existing at the time in communal areas (Marongwe 2012). Though patriarchally based chiefly authorities were very rarely in place on the occupied farms, hierarchies at district and farm level existed in which male war veterans were particularly prominent while women war veterans were in large part absent.

Despite these patriarchal challenges, certain categories of women used the openings provided by the occupations to manoeuvre, as women, for purposes of advancing their social security. As Mazhawidza and Manjengwa (2011: 7) argue about women during the occupations, "old and new actors" negotiated "the path, producing trade-offs, as the process unfold[ed]". While some women, especially young unmarried women, joined land occupations in their own right, most married women did so on behalf of their spouses, who were away at work or attending to other commitments. The position of unmarried women—divorced, widowed or single—is thus particularly important.

Many single, divorced and widowed women, who often have difficulties in accessing land in communal areas—did occupy farms (Scoones et al. 2010: 52). These unmarried women from communal areas took tactical advantage of the movements onto commercial farms during the year 2000 to gain access to land which was never available to them (or which they had lost) in communal areas because of their unmarried status. In this respect, it is common for widows and divorcees in communal areas to be accused of witchcraft and causing the death of husbands by in-laws (particularly in HIV and AIDS cases), and they are frequently even chased away by their in-laws. Although Mutopo (2011b: 1028) tends to overstate the extent to which the occupations "subverted formal forms of patriarchal traditional or administrative authority", it was certainly the case that opportunities arose for women, including widows and divorcees ostracised from their communal area villages, to access land which simply was not available in their place of origin. In the case of Merrivale farm (and likely in other cases), there was also significant tracts of land on occupied farms which facilitated the enactment of specific cultural practices. This was particularly pertinent to polygamous households in communal areas in which each wife (after giving birth to a second child) would be given her own homestead accompanied by a *tseu* field, which was exceedingly difficult to ensure in overcrowded communal areas.

Hence, for some women at least, embedded in the occupations were the "emancipatory potentials of joining a new community" (Scoones et al. 2010: 52) in which some degree of independence from the strictures of local patriarchies might be attained. In this regard, although war veterans used various means (and notably nationalist discourses around land) to mobilise communal area villagers to occupy farms, women (perhaps more than men) had their own very personal reasons for occupying farms, which were not tied directly or invariably to the broader historical grievance over land.

8.7 Conclusion

This chapter sought to further the argument about the decentralised and variegated character of the third *chimurenga* occupations by going beyond a mere focus on war veterans and the party-state and the convoluted relationships between them. To consider exclusively the significance of war veterans to the occupations might lead to the conclusion that they, and not ZANU-PF, organised and led the occupations in a centralised fashion. This chapter, in a variety of ways, has sought to show otherwise. Local histories, memories, claims and aspirations alongside the diverse and contingent combinations of different categories of people involved, gave specific character to particular occupations or a series of occupations. The forms of mobilisation and organisation were open to significant variation. Though we argue, broadly, that farm labourers were excluded from the occupations and that women occupied subordinate positions within them, there are many examples to the contrary. As well, despite the centrality of accessing land from the perspective of nearly all occupiers, the reasoning behind accessing land varied considerably, with many reasons irreducible to nationalist narratives about recovering land lost through colonialism. By way of two case studies about northern Matabeleland and the Lowveld in Masvingo Province, which involved detailing the specific character of the occupations and occupiers, we would also suggest that no particular series of occupations (anywhere in the countryside) is necessarily representative of the third *chimurenga* occupations more generally. This is brought out more forcefully in the next chapter when we consider occupations in two districts in Mashonaland Central Province.

The decentralised form of the third *chimurenga* occupations, the importance of local histories and circumstances, the tensions between disgruntled war veterans and the ruling party, the non-nationalist and often personalised motivations for engaging in occupations, along with other trends, all point to certain parallels with both the first and second *zvimurenga*. This is not to claim, as ZANU-PF's patriotic history-based *chimurenga* monologue does, that the fast track occupations in any way represent a seamless—though interrupted—temporal continuity with the earlier *zvimurenga*.

At the same time, we argue that the main scholarly perspective about the third *chimurenga* occupations adopts a kind of nationalist historiography when it comes to analysing the occupations. In a troubling manner, then, the third *chimurenga* literature has not undergone a discernible and significant shift from a nationalist historiography to a historiography of nationalism, as transpired in the second *chimurenga* literature.

References

Alexander J (2003) 'Squatters', Veterans and the state in Zimbabwe. In: Hammar A, Raftopoulos B, Jensen S (eds) Zimbabwe's unfinished business: rethinking land, state and nation in the context of crisis. Weaver Press, Harare, pp 83–118

Alexander J, McGregor J (2001) Elections, land and the politics of opposition in Matabeleland. J Agrar Chang 1(4):510–533

Andrew N, Sadomba W (2006) Zimbabwe: Land hunger and the war veteran led occupations movement in 2000. Critique Internationale 31:126–144

Buckle C (2001) African tears: the Zimbabwe land invasions. Covos Day, Johannesburg

Chadya JM, Mayawo P (2002) The curse of old age: elderly workers on Zimbabwe's large scale commercial farms, with particular reference to "foreign" farm laborers up to 2000. Zambezia XXIX(i):12–26

Chakona L (2012) Fast track land reform programme and women in Goromonzi district, Zimbabwe. Unpublished Master Thesis, Rhodes University, South Africa

Chambati W (2013) Changing agrarian labour relations after land reform in Zimbabwe. In: Moyo S, Chambati W (eds) Land and agrarian reform in Zimbabwe: beyond white-settler capitalism. African Institute for Agrarian Studies, Harare, pp 157–194

Chamunogwa A (2019) The negotiability of state legal and bureaucratic authority during land occupations in Zimbabwe. Rev Afr Polit Econ 46(159):71–85

Chaumba J, Scoones I, Wolmer W (2003a) From Jambanja to planning: the reassertion of technocracy in land reform in South-Eastern Zimbabwe? J Mod Afr Stud 41(4):533–554

Chaumba J, Scoones I, Wolmer W (2003b) New politics, new livelihoods: agrarian change in Zimbabwe. Rev Afr Polit Econ 30(98):585–608

Chingarande S (2010) Gender and livelihoods in Nyabamba A1 resstlement area, Chimanimani distict of Manicaland province in Zimbabwe. Institute of Development Studies, Harare

Chiweshe MK (2011) Farm level institutions in emergent communities in post-fast track Zimbabwe: case of Mazowe district. Unpublished PhD Thesis, Rhodes University, South Africa

Fontein J (2006) Languages of land, water and 'tradition' around Lake Mutirikwi in Southern Zimbabwe. J Mod Afr Stud 44(2):223–249

Fontein J (2009) 'We want to belong to our roots and we want to be modern people': new farmers, old claims around Lake Mutirikwi, Southern Zimbabwe. Afr Stud Q 10(4):2–35

Fontein J (2015) Remaking Mutirikwi: landscape—water and belonging in Southern Zimbabwe. James Currey, Oxford

Hartnack A (2005) 'My life got lost': farm workers and displacement in Zimbabwe. J Contemp Afr Stud 23(2):173–192

Hartnack A (2016) Ordered estates: welfare, power and maternalism on Zimbabwe's (once white) Highveld. Weaver Press, Harare

Hughes D (2010) Whiteness in Zimbabwe: race, landscape, and the problem of belonging. Palgrave Macmillan, New York

Human Rights Watch (2002) Fast track land reform in Zimbabwe. Human Rights Watch, New York

Jackson D (2014) Another farm in Africa. Litera Publications, Pretoria

James GD (2014) Zimbabwe's 'new' smallholders: who got land and where did they come from? Rev Afr Polit Econ 41(141):424–440

James GD (2015) Transforming rural livelihoods in Zimbabwe: experiences of fast track land reform, 2000–2012. Unpublished PhD Thesis, University of Edinburgh, UK

Justice for Agriculture (JAG) and General Agricultural and Plantation Workers Union of Zimbabwe (GAPWUZ) (2008a) *Destruction of Zimbabwe's Backbone Industry in Pursuit of Political Power*. JAG and GAPWUZ, Harare

Justice for Agriculture (JAG) and General Agricultural and Plantation Workers Union of Zimbabwe (GAPWUZ) (2008b) If something is wrong: the invisible suffering of commercial farm workers and their families due to "land reform". Research and Advocacy Unit, Harare

Justice for Agriculture (JAG) and General Agricultural and Plantation Workers Union of Zimbabwe (GAPWUZ) (2008c) Reckless tragedy: irreversible?—a survey of human rights violations and losses suffered by commercial farmers and farm workers from 2000–2008. Research and Advocacy Unit, Harare

Kalaora L (2011) Madness, corruption and exile: on Zimbabwe's remaining white commercial farmers. J South Afr Stud 37(4):747–762

Kesby M (1996) Arenas for control, terrains of gender contestation: Guerrilla struggle and counter-insurgency warfare in Zimbabwe 1972–1980. J South Afr Stud 22(4):561–584

Kesby M (1999) Locating and dislocating gender in rural Zimbabwe: the making of space and the texturing of bodies. Gend, Place & Cult 6(1):27–47

Laurie C (2016) The land reform deception: political opportunism in Zimbabwe's land seizure era. Oxford University Press, New York

Magaramombe G (2010) 'Displaced in place': agrarian displacements, replacements and resettlement among farm workers in Mazowe district. J South Afr Stud 36(2):361–375

Manase I (2016) White narratives: the depiction of post-2000 land invasions in Zimbabwe. University of South Africa, Pretoria

Marongwe N (2002) Conflict over land and other natural resources in Zimbabwe. ZERO Regional Environment Organisation, Harare

Marongwe N (2003) Farm occupations and the occupiers in the new politics of land in Zimbabwe. In: Hammar A, Raftopoulos B (eds) Zimbabwe's unfinished business: rethinking land, state and nation. Weaver Press, Harare, pp 155–190

Marongwe N (2008) Interrogating Zimbabwe's fast track land reform and resettlement programme: a focus on beneficiary selection. Unpublished PhD Thesis, University of the Western Cape, South Africa

Marongwe N (2012) Rural women as the invisible victims of militarized political violence: the case of Shurugwi district, Zimbabwe, 2000–2008. Unpublished PhD Thesis, University of Western Cape, Cape Town

Marongwe N (2013) Political aesthetics, the third Chimurenga, and the ZANU-PF mobilization in Shurugwi district in Zimbabwe. J Dev Soc 29(4):457–485

Marongwe N (2014) Localised politics, conflicting interests and Third Chimurenga violence, 2000–2008: reflections from Shurugwi district, Zimbabwe. Int J Dev Confl 4:44–92

Matondi P (2012) Zimbabwe's fast track and reform. Zed Books, London

Mazhawidza P Manjengwa J (2011) The social, political and economic transformative impact of the fast track land reform programme on the lives of women farmers in Goromonzi and Vungu-Gweru Districts of Zimbabwe, Women Farmers Land and Agriculture Trust, Harare

McGregor J (2002) The politics of disruption: war veterans and the local state in Zimbabwe. Afr Aff 101:9–37

McGregor J (2009) Crossing the Zambezi: the politics of landscape on a central African frontier. James Currey, Suffolk

Mkodzongi G (2011) Land grabbers or climate experts? farm occupations and the quest for livelihoods in Zimbabwe. Paper presented at the 4th European Conference on African Studies, The Nordic Africa Institute Uppsala, Sweden

Mkodzongi G (2016) 'I am a paramount chief, this land belongs to my ancestors': the reconfiguration of rural authority after Zimbabwe's land reforms. Rev Afr Polit Econ 43(S1):99–114

Mkodzongi G (2018) Peasant agency in a changing agrarian situation in central Zimbabwe: the case of Mhondoro Ngezi. Agrar South 7(2):1–23

Mlambo OB, Gwekwerere T (2019) Names, labels, the Zimbabwean liberation war veteran and the third *Chimurenga*: the language and politics of entitlement in post-2000 Zimbabwe. Afr Identities 17(2):130–146

Moyo S (2001) The land occupation movement and democratisation in Zimbabwe. Millenn: J Int Stud 30(2):311–330

Mujere J (2011) Land, graves and belonging: land reform and the politics of belonging in newly resettled farms in Gutu, 2000–2009. J Peasant Stud 38(5):1123–1144

Mutopo P (2011a) Women's struggles to access and control land and livelihoods after fast track land reform in Mwenezi district, Zimbabwe. J Peasant Stud 38(5):1021–1046

Mutopo P (2011b) Gendered dimensions of land and rural livelihoods: the case of new settler farmer displacement at Nuanetsi Ranch, Mwenezi District, Zimbabwe. International Conference on Global Land Grabbing, Institute of Development Studies, University of Sussex, UK

Mutopo P (2014) Women, mobility and rural livelihoods in Zimbabwe: experiences of fast track land reform. Afr-Stud Ser 32. Koninklijke Brill NV

Nhongo-Simbanegavi J (2000) For better of worse: women and ZANLA in Zimbabwe's liberation struggle. Weaver Press, Harare

Nyandoro M (2012) Zimbabwe's land struggles and land rights in historical perspective: the case of Gowe-Sanyati irrigation (1950–2000). Historia 57(2):298–349

Pilossof R (2006) The unbearable whiteness of being: land, race and belonging in the memoirs of white Zimbabweans. S Afr 61(3):621–638

Pilossof R (2012) The unbearable whiteness of being: farmer's voices from Zimbabwe. University of Cape Town Press, Cape Town

Rutherford B (2001a) Working on the margins—black workers, white farmers in postcolonial Zimbabwe. Weaver Press, Harare

Rutherford B (2001b) Commercial farm workers and the politics of (dis)placement in Zimbabwe: colonialism, liberation and democracy. J Agrar Change 1(4):626–651

Rutherford B (2004a) Desired publics, domestic government, and entangled fears: on the anthropology of civil society, farm workers, and white farmers in Zimbabwe. Cult Anthropol 19(1):122–153

Rutherford B (2004b) Settlers and Zimbabwe: politics, memory, and the anthropology of commercial farms during a time of crisis. Identities 11:543–562

Rutherford B (2012) Shifting the debate on land reform, poverty and inequality in Zimbabwe, an engagement with Zimbabwe's land reform: myths and realities. J Contemp Afr Stud 30(1):147–157

Rutherford B (2013) Electoral politics and a farm workers' struggle in Zimbabwe (1999–2000). J South Afr Stud 39(4):845–862

Rutherford B (2014) Organisation and de(mobilisation) of farmworkers in Zimbabwe: reflections on trade unions, NGOs and political parties. J Agrar Chang 14(2):214–239

Rutherford B (2017) Farm labor struggles in Zimbabwe: the ground of politics. Indiana University Press, Bloomington

Sachikonye LM (2003) The situation of commercial farm workers after land reform in Zimbabwe. A Report Prepared for the Farm Community Trust of Zimbabwe

Sadomba WZ (2009) Peasant occupations within war veterans-led land occupations: grassroots conflicts and state reaction in Zimbabwe. J Peasant Stud 36(2):436–443

Sadomba WZ (2013) A decade of Zimbabwe's land revolution: the politics of the war veteran Vanguard. In: Moyo S, Chambati W (eds) Land and agrarian reform in Zimbabwe: beyond white-settler capitalism. African Institute for Agrarian Studies, Harare, pp 79–122

Sadomba W, Helliker K (2010) Transcending objectifications and dualisms: farm workers and civil society in contemporary Zimbabwe. J Asian Afr Stud 45(2):209–225

Sinclair-Bright L (2016) This land: politics, authority and morality after land reform in Zimbabwe. Unpublished PhD Thesis, University of Edinburgh, UK

Scoones I, Marongwe N, Mavedzenge B, Mahenehene J, Murimbarimba F, Sukume C (2010) Zimbabwe's land reform: myths and realities. Weaver Press, Harare

Scoones I, Chaumba J, Mavedzenge B, Wolmer W (2012) The new politics of Zimbabwe's Lowveld: struggles over land at the margins. Afr Aff 111(445):527–550

Suzuki Y (2018) The good farmer: morality, expertise, and articulations of whiteness in Zimbabwe. Anthropol Forum 28(1):74–88

Tandon Y (2001) Trade unions and labour in the agricultural sector in Zimbabwe. In: Raftopoulos B, Sachikonye L (eds) Striking back: the labour movement and the post-colonial state in Zimbabwe 1980–2000. Weaver Press, Harare, pp 221–250

Wels H (2003) Private wildlife conservation in Zimbabwe: joint ventures and reciprocity. Brill, Leiden

Wolmer W (2007) From wilderness vision to farm invasions: conservation and development in Zimbabwe's South-East Lowveld. James Currey, Oxford

Wolmer W, Chaumba J, Scoones I (2004) Wildlife management and land reform in Southeastern Zimbabwe: a compatible pairing or a contradiction in terms? Geoforum 35(1):87–98

Zamchiya P (2011) A synopsis of land and agrarian change in Chipinge district, Zimbabwe. J Peasant Studies 38(5):1093–1122

Chapter 9
Local Fast Track Occupations: The Cases of Bindura and Shamva

Abstract This chapter examines the third *chimurenga* occupations in Bindura and Shamva, two districts in Mashonaland Central Province. Consistent with the claims made throughout the book, the chapter seeks to show that the occupations in both districts are not understandable in terms of top-down orchestration by the ruling party and state. In this sense, the chapter contributes to restoring the presence of occupiers, including ordinary men and women, onto the historical stage through a case study of occupations in these two districts. This involves considering the decentralised forms of organisation and mobilisation which existed during these occupations, along with the grievances and motivations which animated the occupiers. Central to the occupations were war veterans. The chapter does not claim that the occupations in Shamva and Bindura were in any way organised at district level, or that there was no inter-district coordination. Further, while parallels exist across the two districts, there were also differences, in particular the more pronounced organisational foundation of the occupations in Shamva.

Keywords Third *chimurenga* · Land occupations · Bindura · Shamva · War veterans

9.1 Introduction

This chapter examines the third *chimurenga* occupations in Bindura and Shamva, two districts in Mashonaland Central Province. Consistent with our claims throughout the book, we seek to show that the occupations in both districts are not understandable in terms of top-down orchestration by the ruling party and state. In this sense, we seek to contribute to restoring the presence of occupiers, including ordinary men and women, onto the historical stage through a case study of occupations in these two districts. This involves considering the differentiated forms of organisation and mobilisation existing during these occupations, along with the grievances and motivations which animated the occupiers. We do not claim that the occupations in Shamva and Bindura

This chapter draws on two published articles from: *African Studies Quarterly* 18(1) (2018) and *Journal of Asian and African Studies* 53(1) (2018), for which permission to use has been granted by the journals.

© The Author(s), under exclusive license to Springer Nature Switzerland AG 2021
K. Helliker et al., *Fast Track Land Occupations in Zimbabwe*,
https://doi.org/10.1007/978-3-030-66348-3_9

were organised at district level, or that there was no inter-district coordination at times. Further, while parallels exist across the two districts, there are also differences, in particular the more pronounced organisational foundation of the occupations in Shamva.

9.2   Bindura and Shamva Fieldwork

This chapter is based on qualitative research carried out in Bindura and Shamva Districts in Mashonaland Central Province between May 2015 and June 2016. The explicit focus of the research, as articulated during the fieldwork, was the fast track land occupations and not lives and livelihoods on fast track farms. There was no attempt to hide this from the research participants, and hence access had to be negotiated in a very sensitive manner. Shamva was chosen as the initial research site because one of the authors lived on a white commercial farm in the district during the occupations and was sensitive to some of the key local dynamics requiring investigation.

Shamva District is located north-east of the capital city of Harare in Mashonaland Central Province. It currently consists of communal areas (Bushu and Madziwa), old resettlement areas (dating back to the early 1980s), fast track farms, black commercial farmers and, at the time of the research, a few remaining white commercial farmers. In the late 1990s, before the occupations, Shamva district had numerous "marginal lands and extreme land pressure and shortages" (Moyo 2000: 41) with "growing 'squatter' trends in the district" (Matondi 2001: 71). Prior to fast track, there were 74 commercial farms in the district, mostly owned by white farmers. The most recent comprehensive statistical overview (from the late 2000s) indicated that there were, post-fast track, 12,400 communal farmers, 1,406 old resettlement farmers, 1,737 A1 small-scale farmers (on about 34 fast track farms), 92 A2 commercial farmers (on about 13 fast track farms) and fifteen black and four white commercial farmers (Sukume n.d.: 4, 110). The remaining white commercial farmers engaged in farming on a smaller scale than they did prior to fast track. In large part, the communal areas and the old resettlement farms are located in Shamva North and the former commercial farms (now fast track farms) in Shamva South.

Entry into Shamva district was first negotiated with the Provincial Administrator's office in Bindura, the capital of Mashonaland Central. At the district level, access was pursued through the chairperson (named Manduna) of the District War Veterans' Association. The researchers collected information from war veterans who commanded the land occupations, fast track farmers (mainly from nearby communal areas) who participated in the occupations (or who were present at the time of the occupations) and government officials. Although no individual interviews were conducted with former farm workers, they participated in the group discussions. Seeking individual access to these former workers was viewed with suspicion by "gate-keepers" because of the ongoing politicised character of local land issues as well as the ZANU-PF factional politics during the time of the fieldwork.

In terms of sampling, the research adopted a purposive sampling method. To locate the specific war veterans who led the land occupations in Shamva, the researchers relied on referrals and snowballing. Some of the key war veterans who participated in the 2000 occupations had died. The research was primarily qualitative and evidence was collected through life histories, key informant interviews and focus group discussions. The purpose of the research and the questions that it sought to address could only be qualitatively addressed. Some of the war veterans enthusiastically provided their liberation war autobiographies and these were crucial in understanding their motives and strategies during the land occupations. All in all, individually, 14 war veterans, six fast track farmers, the Chief Executive Officer of the local Chaminuka Rural District Council, one communal area chief and three officers from the Ministry of Agriculture were interviewed. The three focus group discussions conducted varied in size, from seven to twenty-six. These were conducted in Chiraramo, Mushambanyama and Zvataida villages on fast track farms.

The ongoing political sensitivity of the fast track land reform programme was readily apparent during the fieldwork in Shamva. Some war veterans were suspicious of the whole research endeavour, stating for example that the research appeared to be some kind of spying mission. This was particularly the case because of ruling party factional politics and purges taking place within local party structures. As well, government officials repeatedly emphasised the politicised character of land and many were reluctant to be interviewed. Additionally, access to female participants (as war veterans and occupiers) proved to be difficult. It was not only any initial access to women which proved difficult, as their voices in focus group discussions were muted and they mostly supported and validated male views. This was not entirely surprising considering the patriarchal character of Zimbabwean society which, besides domesticating women, also curtails their voices in the public domain. Methodologically, it would have been more appropriate to have separate focus group discussions with women to obtain their views more freely. However, according to the councillor of Ward 20, where all the focus group discussions were conducted, it was not possible to organise further groups because people were becoming increasingly wary of the research.

The second case study of the land occupations was conducted in 2016 in Bindura district, which is also situated north-east of Harare. The district had 21 wards including nine communal wards, with the rest consisting of old resettlement areas from the 1980s, fast track resettlement areas and a few remaining (pre-fast track) commercial farms. Bindura District shares boundaries with Mazowe, Mt. Darwin and Shamva districts. During our earlier fieldwork on land occupations in Shamva in 2015, consistent reference was made to farms in Bindura (for example, in Matepatepa) by the Shamva occupiers. A number of villagers from Bushu communal lands and other areas in Shamva had occupied farms in Bindura, specifically those close to the boundary shared by the two districts. This realisation of inter-district occupations informed the selection of Bindura District as a second study site. In addition, Bindura is the administrative capital of Mashonaland Central Province and provincial leaders of the national war veterans' association (who actively participated in district level occupations) were based there.

The fieldwork for the Bindura occupations took place in five wards: Ward 2 borders Madziwa communal area and Mt. Darwin District; Ward 5 is situated in Bindura North; Ward 7 runs along the Shamva District boundary; Ward 19 borders Mazowe District and Chiweshe communal lands; and Ward 21 is located in Bindura South and borders Mazowe District. Access to the district was negotiated with the Provincial Administrator for Mashonaland Central Province in Bindura town and the Rural District Council at Manhenga.

Regarding sampling, the research adopted a purposive sampling technique. Using this technique, we selected wards bordering communal areas and wards with people who had moved from Shamva District to occupy land in Bindura. We also sought to ensure that research sites were selected in diverse parts (or wards) of the district. The availability of the local councillor also conditioned the particular research sites selected. The research again was qualitative in character in line with the importance of hearing the voices and collective experiences of those who participated in the occupations. The fieldwork was based on extended—in some cases, life history—interviews as well as focus group discussions with war veterans, farm occupiers and former farm workers (who lived through the occupations).

Similar to the Shamva study, the fieldwork relied on referrals and snowballing to select participants. Participants for focus group discussions were selected with the assistance of ward councillors who knew (former) communal farmers, war veterans and farm workers who had participated in the occupations. Because women's voices are sidelined in patriarchal settings, as noted in the Shamva fieldwork, focus group discussions for men and women were conducted separately. The focus group discussions were facilitated by a female Zimbabwean researcher (with women) and a male Zimbabwean researcher (with men).

More specifically, the study involved key informant interviews with two ward councillors and life history interviews with five war veterans and four (former) farm workers. As well, six focus group discussions which varied in size from 7 to 15 were conducted, with mostly farm occupiers. The focus group discussions consisted of people of various ages, occupational backgrounds, marital statuses and origins (such as ethnicity, clan and pre-fast track place of residence). Most of the occupiers originated from communal lands within and beyond the district. At Saimoona Estates in Ward 21, a number of farm occupiers in the focus group discussions were based in Bindura town before the occupations. Some occupiers were also former members of the civil service such as soldiers, teachers and police officers based in towns. They were not full-time occupiers as they would come to the farms over weekends to ensure that their names were maintained on the registers kept by war veterans at the occupied farms. Just like absentee occupiers in the study by Masuko (2013) in Mazowe-Nyabira, these occupiers in Bindura showed their loyalty to the occupations through contributions in cash and kind.

At Chenenga farm in Ward 2, the wife of a former farm manager, the wife of a former soldier and former female farmers from Bushu communal area participated in the female focus group discussion. At Katanya farm in Matepatepa, the male focus group discussion comprised the ward councillor, war veterans, a former member of the ZANU-PF Central Committee, a former police officer and former farm workers (such as a mechanic and driver). At Chipadze farm, the female focus group discussion was composed of the female ward councillor, married women, divorcees, unmarried women (with children) and widows.

9.3 Land Occupations in Bindura

There are no focused studies on the land occupations in Bindura District in Mashonaland Central Province. Perhaps the work by Sadomba (2008, 2013) is the most relevant, as he touches on Matepatepa in Bindura District along with Mazowe (which borders Bindura) and Nyabira (bordering Mazowe) in Mashonaland West Province. By selecting a number of wards in different parts of Bindura District for our study (including Ward 2 in Matepatepa), we are not claiming that the occupations were in any way or necessarily organised or coordinated within wards or at ward level. In fact, the delineation of the wards at the time of the research were different than they were in the year 2000. Overall, the forms and levels of organisation of the occupations varied considerably within wards and the district more broadly, such that any coordination of occupations was quite localised.

9.3.1 Organisation and Mobilisation

There were cases of significant levels of organisation just beyond the borders of Bindura District. In Mazowe, for example, the well-structured Nyabira-Mazowe War Veterans' Association (NMWVA)—formed in February 2000—performed a key coordinating function for land occupations in Nyabira and Mazowe. In Mashonaland West, the NMWVA also collaborated with the Hunyani War Veteran's Association (Masuko 2013). Both Nyabira and Mazowe are closer to Harare than to most communal areas (where many war veterans resided), so that war veterans from Harare led the occupations there. Harare-based war veterans were crucial to the occupations around the capital city, and certainly in places reasonably far from communal areas, because of the sheer absence historically of war veterans on commercial farm areas. Many of the Harare war veterans would only come to the farms on weekends because they worked full-time, but they would ensure that there was a sufficient occupying force. The same was the case for some war veterans based in Bindura town with reference to the occupations in Bindura District (Interview with war veteran Muronda).

Matepatepa was one of the six zones of operation in Mashonaland Central identified by war veterans in Harare, with the other zones being Mvurwi, Concession, Mazowe South, Nyabira and Glendale. This clearly implies at times cross-zone coordination and interaction. Based on his study of Mazowe, Nyabira and Matepatepa, Sadomba (2008: 149) concludes:

The land occupations can ... be characterised as relatively isolated and disconnected, with variation reflecting the nature and characteristics of the occupying group and its linkages with surrounding institutions. Boundaries of groups and their overlaps were determined by mutual negotiation between and amongst the War Veterans groups. The boundaries kept shifting as the groups adjusted in order to achieve effectiveness, considering terrain, manpower and level of resistance to the occupations.

Our research in different parts of Bindura District tends to validate this conclusion. Though there is evidence of coordination at local levels (or cross-farm coordination), this was fluid and adaptable and not highly structured as in the case of Nyabira-Mazowe. Inter-district coordination of farm occupations cannot be clearly demonstrated, but the presence of war veterans from Shamva on occupied farms in Bindura was repeatedly noted. This was particularly evident in the Matepatepa area and Ward 7. The now councillor of Ward 2 in Matepatepa recalled working with Comrade Makasha and Comrade Muropa, who were also heavily involved in the Shamva occupations. As well, some occupiers in Ward 7 indicated that they first went to Shamva were then brought to farms in Bindura by the war veterans leading the occupations in Shamva. Some occupiers had actually participated in the land occupations in Shamva but eventually ended up settling at farms such as Beacon Hill and Koodo Vlei in Bindura as the occupations progressed. In Ward 19, war veterans located there went to Mvurwi and Concession was well.

In the main, the occupations in Bindura District were led by war veterans. The role of war veterans included providing leadership and mobilising people (often in communal areas), especially by drawing lessons from their involvement in the guerrilla struggle and explicitly making it clear that the land issue forced them to join the war and still needed to be resolved. As noted by Councillor Zanamwe:

> As you may ... be aware, the war of liberation was all about repossessing our stolen land and, in 2000, the time was now ripe for us to take over the farms.

A war veteran added that:

> We joined the occupations in fulfillment of our desire to take over the land that had been taken by our enemies over the past years. (Interview with war veteran Chingwaru)

At times, war veterans were sent to particular areas where a shortage of veterans existed. Wilbert Sadomba, as a war veteran and land occupier, speaks of attending a meeting in Glendale in early 2000 in which a request was made for war veteran reinforcements in Matepatepa, and war veterans from Mt. Darwin soon moved into the area (Sadomba 2008: 124). But some farms in Matepatepa, close to Chiweshe communal lands, were occupied by communal farmers without any war veteran involvement. Chief Negomo in Chiweshe apparently supported these occupations, if only as a way of decongesting the communal area. In reflecting upon this in 2016, Sadomba sought to draw a parallel between the guerrilla struggle in the 1970s and the occupations in the year 2000:

> It is interesting to analyse the [liberation] war time strategy of Bindura area and see how the land occupations used a similar strategy. For example, Matepatepa and Bindura farms were surrounded by TTLs [Tribal Trust Lands] (Chiweshe, Masembura, Mt Darwin etc.). For the

war, the operational zones of TTLs [now called communal areas] were used to launch attacks on farms and quickly retreat back to TTLs. This is because in TTLs you had [the] lowest infrastructure development for settler needs. They were areas that attracted least attention, neglected or abandoned. Farms were heavily protected militarily. During occupations, the now CLs [Communal Lands] became centres for organising occupations and launching attacks on farmers. So you see people coming from Chiweshe, Mt Darwin/Rushinga/Chesa and Madziwa to occupy Matepatepa. (email to one of the authors, 27 June 2016)

This parallel between the two *zvimurenga* resonates with the ways in which farmers sought to defend their private space against forays by guerrillas during the 1970s and against "invading" intruders during the fast track occupations.

In Bindura District, occupations took on different forms, with occupations relatively spontaneous or more organised depending on locality. In Ward 7, occupiers appeared to have settled on the following pre-set order for occupying farms: Chipadze, Koodoo Vlei, Beacon Hill, Chiwaridzo and then eventually Avillion farm. Avillion was occupied last because part of it was farmed already by black farmers dating back to the 1980s. In Ward 5, the first farm to be occupied in February 2000 was Kanjinga. After this farm, there was no particular order followed. New occupations emerged in the context of prevailing circumstances such as the existence of too many occupiers on a particular farm, and hence some occupiers needed to move elsewhere. For instance, in Ward 5, when war veterans heading the base camp at Kanjinga realised that there were more people present than the farm could accommodate, some people were moved to occupy Chenenga farm.

Occupiers admitted that there was confusion, disorder and even conflicts and displacements during the occupations, with war veterans and other occupiers displacing each other. Women occupiers at Chenenga therefore indicated that they were once displaced to Munzi farm and, when they thought they had firmly settled, they were forced to move back to Chenenga. War veteran Musangomuneyi also talked about his conflict with war veteran Muropa, who wanted him to move from Kanjinga farm in Bindura to Mhokore in Shamva. In Ward 7, some of the occupiers admitted that they attacked fellow occupiers and burnt their temporary shelters so that they could displace them, as the latter were aligned to former state vice-president, now opposition party leader, Joice Mujuru. The fluid character of the occupations in Bindura District was somewhat different to the case in Shamva District, where occupations involved a district-wide coordinating committee led by war veterans and tended to follow a clearer sequence.

This is not to argue of course that farm occupiers in Bindura district had no reason for occupying certain farms as opposed to others. We have already indicated that closeness to communal areas meant a greater likelihood of occupation. Kanjinga in Ward 5 for example was occupied for this reason. But the cruelty of particular white farmers was also a factor. One female occupier at Chenenga farm, who previously lived in a communal area, thus said:

He [the farmer] was not a good person at all. For example, if one of your cattle strayed into his farm, you were made to pay a fine. He never allowed us to pick dry wood; if you were caught you had to pay a fine.

Therefore, farmer reputation comes into play when making sense of the logic of the occupations. In many instances, farmers' relationships with locals (both nearby communal farmers and farm labourers) configured the occupations in a particular locality, and these relationships even explain why a particular farm was not occupied immediately. As one occupier observed, and not unlike the case of some farmers during the war of liberation:

> From my understanding, the reason why the white farmer at Piedmont farm remained unoc-cupied was because he quickly accepted the occupations. He voluntarily donated inputs and supported us by donating food. (Focus group discussion with Chenhenga women founders, Ward 5)

In Ward 7, as referred to later, Chipadze farm was occupied by people from Chipadze clan because they claimed it was their ancestral land. In this way, *autochthonous claims were of some significance. But war veterans still played a part in these cases, such as* Comrade Nhunzva and Comrade Tito Pela at the Chipadze farm base

Regarding how the occupations began, it was generally argued that several meetings were held by war veterans, normally in communal areas, where people were informed about the intention to occupy land. One occupier narrated how the occupations started:

> It all started in meetings that we attended. Usually before we attend meetings, we needed to know the purpose of the meeting. … Once the purpose of the meeting is satisfactory to us, we all agree to attend. This is how we got to know about the land occupation programme. The war veterans led discussions at these meetings and we supported them. Our leader at these meetings was Comrade Jones. (Focus group with Chenhenga women, Ward 5)

Another occupier also expressed how the occupations began:

> The war veterans and all other stakeholders … sat down and discussed the way forward. It was then agreed to approach the white farmers and advise them verbally, without any violence, that we were going to take over our land. People were then deployed to the farms. (Male focus group discussion, Matepatepa Country Club)

At the same time, there was some discouragement and lack of support from many villagers. In the case of Bushu communal lands, for instance, there were some village headmen who did not heed the call by war veterans to support the land occupations.

While the occupations were largely initiated and mobilised—by war veterans—through meetings, word about the occupations was spread by way of the radio, church meetings and other informal gatherings and announcements. The case of Sirewo Chipadze from the Chipadze clan stands out as he mobilised his kinspeople when occupying Chipadze farm. He moved to and around several places (such as Bushu, Madziwa and Murewa) where Chipadze kinspeople had settled after they had been displaced originally from their ancestral lands, and where they had only limited access to land. Sirewo Chipadze ferried people using his lorry to his farm called North Star and then people subsequently moved to Chipadze Farm:

> When they got to Comrade Sirewo's farm, they proceeded to a mountain in Ivrin and tradi-tional rituals were performed; those of a Christian background prayed. After these rituals, they came across a dead animal. They took the animal with them and proceeded to walk

and join others at an already set up base. It was at this base that there was the 'jambanja' that mobilised people to occupy Chipadze, Kodoo Vlei, Chiwaridzo and Muponda's farms. (Women focus group discussion, Chipadze Farm, Ward 7)

This series of occupations were seemingly guided by ancestral spirits, just like what occurred in other places, including, the occupations in Hunyani Hills led by sekuru Mushowe (Masuko 2013).

The spiritual dimension is particularly significant for the Bindura District and surrounding districts. The medium of the spirit of the original Nehanda during the first *chimurenga* was a woman named Charwe, who was born in the Mazowe Valley. The discursive construction of Charwe as a warrior against colonial rule, as noted previously, became important during the guerrilla-nationalist struggles against Rhodesian rule in the 1970s. Other mediums arose in and around Mazowe after independence in 1980, such that the memory of Nehanda remained alive "among her people in the Mazowe" (Charumbira 2015: 211). The occupations, as an awakening moment to repossess land expropriated under colonial rule, were seen as fulfilling Nehanda's clarion call of "my bones shall arise", as manifested in the messages of local spirit mediums. These include the spirit mediums in Madziwa and the spirit of Chapo in Ward 5.

Various rituals were conducted by mediums in the midst of the occupations, supposedly to strengthen the occupiers. For instance:

When we embarked on the journey to this farm, we had traditional rituals that we performed; we knelt down and clapped our hands. We asked our forefathers to lead and guide us on our journey to repossess our land. We also asked the spirits of that area to guide us. After the ritual, we continued with our journey. (Female focus group discussion, Chipadze Farm)

Sadomba mentions meeting mediums of Charwe in Chiweshe at the time of the occupations and, as a war veteran occupier, indicated that: "We followed guidelines from these mediums, particularly prohibition of sex and spilling blood in the occupied farms" (email to one of the authors, 27 June 2016). He goes on to argue more fully elsewhere that:

Chiefs and spirit mediums organised meetings to prepare the people. I managed to attend some of these all-night ceremonies [*biras*] in this part of the Chiweshe Communal Lands… The chief and other old men and women … explained to the possessed medium that the community was sending people to go and 'reclaim the land from the usurpers'. The spirit, through the medium, set rules for doing this and gave instructions on what ceremonies and actions had to be performed before occupying. (Sadomba 2008: 138–139)

Snuff was to be given to the occupiers to weaken the resolve of white farmers. In this way, many Bindura On 27 June 2016, at 04:25, Sandra Bhatasara <sandrabhatasara@gmail.com> wrote: Thank you Cde for taking time to respond to my questions. I really appreciate. Am really fascinated by everything you said. One of the reasons why I decided to go to Bindura particularly Matepatepa was to follow up from my Shamva research. There was a lot of reference given to the area by occupiers in Shamva. Besides that, I did it for to see if there is room for comparison with what emerged from Shamva. Shamva was really illuminating and interesting for me. I have checked your thesis online several times but I can not locate it. Do you mind sharing the copy? And any other material that you are free to share. Again, thank you very much. Occupiers spoke of being imbued with, and guided by, the same ancestral spirit that the two earlier *zvimurenga*.

9.3.2   On-Farm Dynamics

While responses of white farmers to the occupations varied across the district, there was significant resistance by many farmers, as an occupier from Chenenga farm highlights:

> The white farmers did not take that [the occupations] well and they tried to fight us off the farms. The MDC supporters also joined in the fighting [supporting the farmers] and urged us to go back to our rural homes. They accused us of occupying the farm unlawfully. We lost a lot of our property during that time, and some of our belongings were burnt. (Focus group with Chenenga men, Ward 5)

Others spoke about how farmers would fire their guns into the air, or dump rotten fruit near the base camp of the occupiers, to discourage their ongoing stay. Another occupier made specific reference to one farmer:

> There was also a thick-headed white farmer called Greamer. He would derogatively tell us that he would uplift the farm from where it was presently located to another place. He went on to tell us that we were being naïve in thinking that he would voluntarily leave the farm. Interestingly, on one of those days, he left in the dead of the night after realising that we were not relenting in our quest to take over the farm. (Women focus group discussion, Chipadze Farm, Ward 7)

There were therefore various unpleasant encounters with white farmers during the occupations. Accusations were made against some white farmers that they would "cook up" stories (such as occupiers beating up school teachers) so that the occupiers would be arrested by the police for criminal acts.

At first, white farmers expected the occupations to be a passing phase and were quite adamant in resisting the occupations without fear of reprisals from the ruling party and state, and they would often call the police to evict the occupiers, though unsuccessfully. However, intimidation by white farmers (and their labourers) led to some occupiers in Bindura District returning to their communal homes. Further, when the occupations began, the initial response from the ruling party was to label the occupiers as "invaders" and have them evicted and even arrested. For instance, war veteran Jones Musangomunei emphasised:

> Some of our [state] ministers supported the white farmers when the occupations started... Minister Msika is one of those who sent police to the farms so as to force us out of the farms.

But, with support for the occupations (particularly from Mugabe) becoming increasingly clear, even before the announcement of fast track (in July), white farmers soon realised and concluded that the occupiers were there to stay, including with the tacit support of local state and ruling party personnel. In the case of Bindura District, the presence of the ruling party's national political commissar (Border Gezi) in the area, as governor for Mashonaland Central Province, was significant. In his personal capacity, he often provided material support to the occupiers. Gezi was also a public figure in the Mazowe Apostles weChishanu church, which became publicly linked to ZANU-PF during this period (Mukonyora 2011).

Violence by occupiers certainly did take place, though they claim that this was a reaction to farmer-instigated violence. Otherwise, the sheer presence of occupiers and their ongoing tactics of intimidation were the weapons deployed in trying to force farmers off their land. For instance, occupiers were involved in singing, dancing and beating drums on the farms, and normally just outside farmers' main homestead, day and night. War veteran Jones Musangomuneyi said:

> We would … camp at his gate and start singing and playing our drums. The wives of these farmers got irritated by the noise and we knew they would be the first to leave. Indeed, that's what happened. They were the first to leave. The men braved it and tried to stay on until they eventually gave up and left.

This loud presence was intended to constantly annoy farmers, to strike fear into their families, and to ensure that farmers left the farm "voluntarily".

In the case of specifically Matepatepa (in the district), Kufandirori (2015) speaks about significant levels of violence directed at both farmers and farm workers. Farm workers in her study highlighted that war veterans in particular were the main perpetrators of violence. Farm workers were beaten, their houses burnt and they were forced to buy ZANU-PF party cards and attend rallies and *pungwes*); as well, they were intimidated and threatened with disappearances. Some farm workers admitted that violence was committed against farm workers and farmers by fellow farm workers. Women occupiers at Rotchford farm in Matepatepa revealed that they were victims of sexual violence in base camps and war veterans forced women to cook for them.

As a general trend in Bindura district, when a farm was initially occupied, war veterans would immediately go and talk to the farmer and often request (or demand) a map of the farm. This was done in part to understand the layout of the land and to identify a location for their base camp in order to signal their visible presence on the farm. One occupier indicated:

> When the war veterans got to the farms, they produced their war veterans' cards and informed the white farmer that they were there to get land, the same way they had taken land from our forefathers. (Male focus group discussion, Chipadze Farm)

It was claimed that the war veterans were able to discern if the map was the original one or an altered and fabricated photocopy. As the war veterans talked to the farmer, the occupiers would always remain at a distance. This was followed by the immediate establishment of a base, or base camp, on the now occupied farm. For Masuko (2013), the establishment of a base camp signalled the transition from a land invasion movement to a land occupation movement.

The bases were led by a base commander (normally a male war veteran) who was supported by a team of mainly war veterans, with positions such as chief of security, secretary and treasurer. Registers were usually kept at the bases to indicate the names of the occupiers and their interest in acquiring land. There were also *pungwes*, or large meetings, that were conducted to boost the morale of the occupiers, as well as to mobilise farm workers who might have resisted the occupations.

In the meantime, occupiers had to organise accommodation and to survive. The living conditions in the bases were at times challenging, including living in tents during heavy rains or even sleeping on the ground without any accommodation. An occupier noted that:

> The challenges we faced included the time when it was raining. We used tents as shelter, but they were blown away by wind during the night. But we did not give up, we soldiered on. (Male focus group discussion, Chipadze Farm)

Councilor Zanamwe, a *mujiba* (male war collaborator) during the 1970s, highlighted:

> During this exercise, we did not stay in the farm compounds, we were based in the bush and this is where we constructed temporary bases … We never sought permission to construct these bases at the farms.

Occupiers sometimes grew their own vegetables, received cabbage and mealie meal donations from sympathetic black farmers in the area, or went back to their communal area to replenish their food supply. Most occupiers, for the June 2000 parliamentary election, were not registered to vote in the area of the farm they occupied, so they had to return home temporarily to cast their vote. War veterans ensured though that there was a minimal and sufficient occupying force at all times. Indeed, it was normal for occupiers who were on farms close to their communal area to spend time between the two places. This was particular the case with women occupiers, who at times left their children at home in the communal area in the care of relatives and needed to check on their well-being on occasion.

The occupations in Bindura District were in large part decentralised with no clear evidence of pre-planned and ongoing coordination by war veterans across a range of farms. This resonates with the claim by Sadomba (2013) that, with reference to Mazowe and Matepatepa, no centralised organising took place. Rather, occupations were horizontally structured by way of locally organised units with no central command. War veterans of course did mobilise communal area farmers to occupy farms and they ensured that there was a base camp established on each farm. But there was also considerable self-initiative on the part of occupiers in moving onto a particular farm.

9.3.3   Memories and Motivations

The land occupiers in Bindura district were diverse in terms of their socio-historical backgrounds. Male war veterans were present on all the farms studied. Many of these war veterans, in recounting their guerrilla days, were clearly familiar with the area as it was their zone of military operation during the 1970s, with some meeting for the first time again during the occupations. Ordinary men and women as well as youth also participated and occupied farms. Significantly, some farm workers facilitated and supported the land occupations. The presence of such diverse actors echoes the argument by Sadomba (2008) that the land occupations in the district became quite inclusive. Certainly, from our study, both rural and urban people were present in the Bindura occupations, with urban occupiers more prominent at farms, such as Saimoona Estates, close to Bindura town. Occupiers came from areas as far away as Masvingo, Bulawayo, Murewa, Seke, Dotito and Uzumba. Most occupiers, however, came from surrounding communal areas such as Chiweshe, Madziwa and Bushu, mainly due to close proximity to the farms and strong ancestral ties to land dispossessed under colonial rule.

One particularly contentious matter about the occupations, as we have noted, is why the land occupations took place and at the time they did. In contrast to the dominant scholarly narrative on this matter, the occupiers in Bindura District deny that they acted at the behest of the ruling party. War veteran Musangomunei for example highlighted the political distance which existed between the ruling party and war veterans leading up to the constitutional referendum and then occupations. In intimating as well the close connections between war veterans in Shamva and Bindura districts, he clearly stated that:

> We as war veterans continued to have our meetings here in Shamva. We constantly met to check on each other's well-being, discussed our grievances and also wanted to know what our leaders were thinking of our plight. ...During the war of liberation, our ZANU PF leaders had promised us office jobs, a decent way of living, with plenty of food for us and our families. Sadly, all these promises were not fulfilled. So, we would meet to get feedback on these promises. We had held on to that promise and we expected it to materialise. We had anticipated that our leaders would still remember the promises that they had made. Sadly, it seemed they had forgotten all about us as they were now comfortable and in power.

War veterans referred to the rejection of the new constitution in the referendum as the spark which ignited the occupations, without being ordered by the President or the ruling party to occupy land. In the end, the "no" vote in the referendum meant the denial of more radical land reform—including land expropriation without compensation—as stipulated in the revised draft constitution. From the perspective of the war veterans, they had lost once again, in the same manner that they had lost during the independence negotiations at the Lancaster House Conference in 1979.

Land was central to the national liberation struggle, including the guerrilla war in the 1970s. The ongoing unresolved character of the land question resonated with the experiences not only of marginalised war veterans, but of ordinary people as well. Most occupiers were aware of how the Lancaster House Constitution made it difficult for the ZANU-PF government to recover and redistribute land from 1980. The process of giving black people land was very slow under the agreement:

> After the discussions at Lancaster House in Britain, the war ended. Margaret Thatcher was the Prime Minister then. It was agreed to give back the land to the blacks. This arrangement went on well from 1980 to 1982. In 1984 to 1985 the British government started to renege on their promise to give us back our land. (Male focus group discussion, Chipadze Farm)

But the stories of ordinary occupiers in the district speak to multiple grievances and motivations for joining the land occupations, with current maladies hopefully being addressed and rectified through the occupations. Occupiers were aggrieved by various encounters with the colonial system and nearby white farmers (before and after independence), including worker exploitation, racism and denial of personal liberties. War incarcerations during the 1970s, loss of land, cattle and other wealth, long working hours, *chibaro* (forced labour), and curtailed freedom of movement were all part of local memories. Placement in "keeps" (the protected villages during the war of liberation) was also cited. In Matepatepa, occupiers referred to the famous "Keep 10" in Madziwa where black people faced severe ill-treatment. For young men, growing up in protected villages was one of the most alienating and inhumane experiences under colonialism. One male occupier at Chipadze farm described "being harassed to unimaginable proportions when we were at these keeps". Other occupiers from nearby communal areas raised concerns about overcrowding in the communal lands, land fragmentation, poor quality soil and land conflicts. They also bemoaned the hilly mountainous areas in communal areas where they farmed, with the land more suitable to cattle ranching than crop farming.

Narratives of displacement from ancestral land, as noted earlier, were evident. As such, the land occupations were an opportune moment to reclaim lost ancestral land. The case of Chipadze clan clearly illustrates this:

> I am married in the Chipadze family. The Chipadze family is originally from here; they were however forcibly moved from this place and relocated at Bushu. They were given a small area to occupy by Chief Bushu. We were so happy when we heard about the land occupation programme. To us, it meant going back to our original place of residence. In Bushu, we did not have any space to carry out our farming activities and so we did not hesitate to move back to Chipadze so as to acquire land for farming. (Female focus group discussion, Chipadze Farm)

The Chipadze people (from the Zenda, Hungwe ethnic group) lived in areas consisting of Chipadze and Chiwaridzo farms before dispossession by white settlers. One Chipadze person lamented:

> My grandfather told me that our forefathers were chased away from Bindura by the white people, in 1918. We were forcibly moved to Madziwa, on a very small piece of land. Others were relocated to the Bushu area, Murewa, Chiweshe and Musana. (Didymus Zaranyika, Chipadze Farm)

When they were forcibly moved from their land by white settlers, they moved to different places. The Wembo family settled in Chigunde, but others went to the Murewa, Rushinga, Bushu and Madziwa areas. The original Chipadze chieftainship settled in the Bushu area after expropriation by the settlers, and they resided under the authority of chief Bushu. When in Bushu after independence, the Chipadze chieftainship and clan members tried to follow up on reclaiming their lands. They assumed that, after independence, they would return to their ancestral lands but this did not happen. At one time, some were told that they could be resettled in Muzarabani but they refused that offer. When the land occupations began in early 2000, they readily sought a return to their original lands. Besides this *autochthonous* motivation,

Hence, whatever the ostensible objective of the land occupations, deeply felt historical memories and contemporary grievances were pervasive amongst the occupiers. These highly localised memories and motives made communal area villagers susceptible to the national call by the ZNLWVA to occupy farms and to the mobilisation activities of war veterans in Bindura District. However, with reference to farm workers and women, the land occupations were mainly characterised by forms of exclusion and subordination.

9.3.4 Farm Workers and Women

Farm workers occupied an intercalary position during the occupations. In nearby Concession, Selby (2005) indicated that farm workers opposed occupations and, in the case of Collingwood farm, they marched to evict Ngwenya (linked to Grace Mugabe), though they did not succeed. Kufandirori (2015) likewise highlights how farm workers in Matepatepa aligned with their farmer employers to protect their jobs and future livelihoods. Similarly, occupiers in our Bindura study reported that farm workers were generally against the occupations, though some labourers did participate in occupations because of their marginalised and exploited status on white commercial farms. In this respect, in Matepatepa, farm workers indicated how white farmers only wanted them to work but did not really care about their livelihood security (Kufandirori 2015).

In this context, some farm labourers did participate. One farm worker (from our study), in reflecting upon his role in the occupations, highlighted that, after being a farm labourer for 35 years, he was unable even to save sufficient money for a bicycle. One Ashcot farm worker in Matepatepa spoke about his experience:

> My boss did not allow us to participate in the land invasions, so I would sneak out at night and cycle to join the group led by war veteran Muchenje.

Timothy Derek, a former farm worker, said:

> We supported the land occupation because as farm workers it was difficult to access land. We would go around looking for land so as to better our livelihoods, but it was not possible to get the land. To make matters worse, once our kids were grown up, they were not allowed to stay with us in the farms that we were working at. The white farmers did not allow our grown-up children to stay on. We even went on to suggest that the white farmers could perhaps build us lodgings in the locations in the cities so that our children could stay there and attend school, but the white farmers refused to entertain all this. Our relatives were also not allowed to visit us.

Similar sentiments were expressed by Kaisi at Katanya farm in Matepatepa, who worked as a mechanic when the land occupations started. Kaisi had first-hand experience of oppression under the white farmer and was paid a paltry wage. He did not hesitate in joining the occupations as he "knew that this was the only way to improve our lives".

Conflict between occupiers and farm labourers though was prevalent, with white farmers sometimes instigating their labourers against the occupiers. This reflected a tactical weakness on the part of the Bindura occupiers. Unlike in Nyabira, war veterans in Mazowe and Matepatepa only realised months into the occupations the need to mobilise farm labourers in making the occupations more effective (Sadomba 2008). In the main, the evidence from Bindura district indicates that occupiers engaged in significant forms of intimidation against assumed MDC-linked farm labourers (including through *pungwes*) who were regarded as anti-land reform traitors, and this took place increasingly as the June 2000 national elections approached. At Chipadze farm, female occupiers stated that:

> On one of the evenings, some of our [MDC] deserters were beaten up and urged to come back the following day and rejoin the ruling party and surrender their MDC regalia.

There was an unusual arrangement in Matepatepa along the border with Chiweshe communal area, in which farm labourers with roots in this area worked undercover by assisting the occupiers in providing information about farm layouts and farmer presence as well as necessities such as food (Sadomba 2008). At Rosedale farm in Matepatepa, one senior farm worker manager was an active ZANU-PF supporter, offered his house as a base camp, undertook surveillance of activities on the farm and supplied information to war veterans (Kufandirori 2015). Farm workers who had permanent homesteads in bordering communal lands such as Madziwa were also co-opted by communal area occupiers, and they supported the occupations more so than those labourers of Malawian descent who resided on the farms (Kufandirori 2015).

Motivations for occupying land had gender connotations as well, which highlights the importance of local identity-based experiences in making sense of the occupations. Polygamy was a key issue in this regard:

> As women within the clan, we faced a lot of problems, and these forced us to come here. It was difficult to be women married in a big [polygamous] family. If there were many sons, it meant many daughters-in-law and all of us in one homestead. (Focus group discussion with Chipadze women)

In wanting to lead a different kind of life, a female occupier (at Chipadze farm) stressed that she "wanted freedom and independence from my family so as to have a home I call my own where people can visit me". The issue of *chipari* (polygamy) in communal areas, as intimated in these two quotations, spurred on many women to occupy farms. Those women married into polygamous families had small plots of land in communal areas to call their own, and the emerging occupations provided a basis of escape from very troublesome and stressful arrangements.

In occupying farms, some men from communal areas left their wives behind to care for the communal homestead. Women at times though took the initiative in joining the land occupations. Many of these women had husbands working elsewhere when the occupations first took place. As one woman put it:

> When we first heard about the land occupations in Chipadze, my husband was still working. I am the one who came and registered for the land. My husband only got to know about this when he came back home at the end of the month. (Female occupier, Chipadze Farm)

Even when the husband was present, some women seized on the possible opportunities arising from the occupations:

> My husband was initially doubtful of the occupations. I had to push and convince him that the land was going to benefit us and our children. I told him if the occupations failed, we would have at least tried. (Female occupier, Chenenga Farm)

Once on farms, there was evidence of women partially breaking out of the patriarchal mould. Thus, at some occupied farms in Matepatepa, women participated in security patrols and in pegging land plots. Further, the base camp secretary at Koodoo Vlei farm was a woman, and she was involved in integrating new occupiers into base-life and ensuring food supplies. Another woman was later selected as the treasurer and, with the secretary, she collected and registered cash donations. From a gendered viewpoint, women (as care-givers and domestic "managers") were seen at times as more responsible than men as treasurers, as men would often use the cash donations for their own pleasures. Yet, patriarchal discourses about women, including their moral character and sexuality, prevailed on the occupied farms. Women, particularly widows and divorcees, were sometimes accused of being prostitutes or seeking illicit love affairs when they joined the occupations. Married women, if their husbands were away or did not join them, were accused likewise. Women tended to be subordinated to the dictates of patriarchy on the occupied, including undertaking all the domestic chores at the base camps. Even the singing and dancing was done primarily by women and, besides intimidating white farmers, it was designed to boost the morale and militancy of the male occupiers.

Broadly, the failure of the occupiers to incorporate farm labourers and their families into the occupation process, as well as the subordinate incorporation of women, raises serious doubts about the inclusive character of the occupations. Unlike Moyo (Moyo and Yeros 2005), we do not see the challenges undercutting any progressive impulse in the occupations as arising exclusively from the Zimbabwean state's increasing intrusion into the occupations via fast track land reform.

9.4 Land Occupations in Shamva

In Shamva district, there had been localised and diverse conflicts around land, water and natural resources ever since independence in 1980, including within communal areas, so that the land occupations from the year 2000 were "an epitome of a long gestation period of conflict building up" (Matondi 2001: 175). The exact relationship between the pre-2000 land conflicts and the later occupations though remains unclear. The 2000 land occupations mostly took place in Shamva South with the occupiers coming from diverse places. But "[m]ost...were from communal areas and old resettlement schemes both within and outside Shamva. Shamva district contributed two-thirds" of the occupiers. Another 25% came from neighbouring districts such as Mount Darwin and Bindura, and the rest "from districts further afar". For those emanating from within Shamva, 30% came from old resettlement areas (in Mufurudzi Valley) and 50% from Madziwa and Bushu communal areas (Sukume n.d.: 7, 11,111).

9.4.1   Invoking Historical Grievances

In his study of fast track farms in Shamva, James (2015) notes that reasons for occupations were varied, some of which were reinvented over time as land occupiers deployed old and new narratives to legitimise their claims to land. In this regard, our study shows that occupiers invoked historical grievances of both a broad nationalist and specific experiential kind. Thus, as in the case of Bindura district, in considering the motivations behind the occupations in Shamva, it is necessary to focus on colonial dispossession of land and subsequent developments under white rule (as well as postcolonial developments). This is because of the long and vivid memory amongst occupiers of initial land loss and related colonial practices as well as ongoing postcolonial dispossession. The Shamva occupiers lost their ancestral lands through imperial conquest and then concessions made after the Second World War (which saw a number of white ex-soldiers such as Captain Mobil acquiring land in Shamva). As well, after the Unilateral Declaration of Independence by the Rhodesian government in 1965, a significant number of white people accessed further land in Shamva. When people in Shamva moved onto farms in 2000, they were reclaiming what had been taken from them. As one war veteran noted, "we took them [farms] for the sole purpose of us repossessing what belonged to us" (Focus Group Discussion, Mushambanyama Village).

Memory does not simply refer to a sweeping nationalist-inspired memory as it is also based on localised personal experiences as well. Like in Bindura, the occupiers had direct experience of the colonial system. They worked for white employers and experienced racism, discrimination, abuse and aggression. They remembered how they lost their sense of belonging, as *mwana wehvu* (a child of the soil), through land alienation. The loss of land intertwined with the loss of other entitlements which a real child of the soil ought to have. They were forcibly relocated to reserves or Tribal Trust Lands. Dating as far back as the 1930s, there was a "deliberate state sanctioned process that involved the systematic relocation of people from areas such as Mazowe-Bindura to Bushu …. [and] … Madziwa" (in Shamva), with people then "dumped in the present communal areas without due regard to basic means of survival" (Matondi 2001: 117). As Matondi (2001: 179) highlights, older people in Shamva have "vivid memories" of this "painful experience" of forced removals and other events from the past. Villagers were also restricted in terms of how much livestock they could keep or the crops they could grow.

Stories of *chibaro* (forced labour), involving no remuneration, were narrated by many occupiers. People in the pre-1980 communal areas would be rounded up and brought to the farms by white farmers for periods of up to four months. Those who resisted working on the farms under such conditions were forced to make *madunduru* (contour ridges) which today are scattered all over the farms. Thus, dispossessed blacks, living in overcrowded reserves such as Bushu marked by *pfukarushesha* (sandy, unproductive soils), became sources of cheap labour. As well, villagers were placed in "protected villages" during the war of liberation in the 1970s, including the ten villages constructed in Madziwa in 1974 under Operation Stronghold. Others villagers fled the area during the war or were forcefully removed (as far as Beitbridge along the South African border) for supporting the guerrillas. Another war experience was provided in a focus group discussion (Mushambanyama Village):

> I recall sometime in 1975, could have been 1976 in the Madziwa area, they put poison on a small plane and sprayed all our crops. All the crops wilted and died and all our domestic animals including cattle perished. We were left with nothing.

These memories are reiterated in an empirical study of three villages on now fast track farms in Shamva (North Star, Golden Star and Chiraramo) (James 2014, 2015). On these farms, the deep memories of the occupiers and their experiences of the war—or the experiences of others retold—drove them to "take the land", as well as the failure of the post-colonial government to fulfil the promises of the liberation war.

Overall, the core values of the war of liberation remained etched in the minds of communal area villagers in Shamva. Most war veterans in Shamva indicated that the idea propagated during the war was "from bush to office" (war veteran Kashiri), meaning that when the war (bush) was over and independence (office) attained, black people would take over everything, including land. But the guerrilla struggle culminated in the Lancaster House Agreement. The war veterans in Shamva felt betrayed by this and were effectively claiming that they did not spend years fighting on harsh conditions, only to be given terms by others for their own independence. For the Shamva occupiers more broadly, the agreement symbolised a betrayal of the ideologies of the war of liberation and marked the continuity of colonial domination. What made it more treacherous was that the agreement was an alliance between representatives of white settler capital and black nationalists.

In our study, the Shamva occupiers reported that the terms of the agreement (i.e. willing buyer/willing seller for purposes of land distribution) privileged white farmers. As well, white farmers had every intention of holding onto their prime land. White farmers were seen as cunning and clever as they sold land that was infertile and unproductive to the government for resettlement of blacks, including land exhausted from years of monoculture, particularly in growing cash crops such as tobacco. The land they gave up for resettlement was in undeveloped areas, with no significant infrastructure such as roads, schools and clinics.

It was unanimous amongst war veterans and ordinary land occupiers in Shamva that there was no directive given by the ruling ZANU-PF party to people to occupy land in the district. Most interviewees claimed that the rise of the MDC and the ensuing rejection of constitutional changes in the February referendum were non-events in leading to the emergence of the occupations. One occupier argued:

> The fact that a new opposition party had been formed in Zimbabwe was not a factor in us taking over the farms. The war that we waged against our former colonisers was about land, and it is just the land that we wanted. (Focus Group Discussion, Chiraramo Village)

However, not all occupiers argued this stance so forcefully.

Hence, Matondi (2001: 165) refers to a war veteran interviewed in Shamva district in August 2000, who indicated that a key priority during the occupations was to "uproot" MDC supporters who had "infested" the area. White farmers in Shamva had publicly aligned themselves with the MDC soon after the latter's formation, including during a well-attended meeting of famers at Insingisi farm just outside Bindura town at which the president of the MDC spoke and received donations from farmers. Some farmers in Shamva became actively involved in local MDC party structures and they sought to mobilise their farm workers in support of the MDC. This included transporting them *en masse* in the back of 8-tonne farm trucks to MDC rallies in Bindura town. But, for us, the intricate entanglement of the MDC and the white commercial farmers in Shamva and elsewhere indicates that the uprooting of the MDC was primarily about isolating and weakening the political power of the farmers in the struggle to take over land.

Memory was not only crucial for spurring on the land occupations in 2000, as it also provides insights into the character of the occupations in Shamva, including why particular farmers were targeted. Further, providing an overview of the character of the farm occupations in Shamva and their organisational dynamics contributes to understanding the extent to which, if at all, the occupations were in any way directed (or orchestrated) by the ruling party, either initially or throughout the occupation process.

9.4.2 Overview of Farm Occupations

Farm occupations, or at least squatting, in Shamva and indeed elsewhere existed prior to the year 2000. In 2001, Yeros (2002) did fieldwork in Shamva, demonstrating that land occupations had been occurring with reference to gold panning and contested settlement, including the unauthorised use of land by so-called squatters. The squatters (including communal area villagers) had encroached onto commercial farms and game reserves, as well as state and other lands. Gold panners were present at Lion's Den estate, at John White's farm and also in Pote valley. Nelson Marongwe (2002) likewise speaks about land occupations and illegal settlements in Shamva.

Despite the often-made claim that the ruling party initiated and organised the occupations, the land occupations—as noted previously—are viewed regularly as chaotic and anarchic. At first sight this might seem true. But, certainly in the case of our Shamva case study (and unlike the occupations in Bindura), it is possible to discern some semblance of order regarding for instance the sequence of farm occupations. From his fieldwork in Shamva, James (2015) also reports that, in some places in Shamva, the unfolding land occupations appeared to be coordinated and organised in a coherent fashion.

When the occupations began, as our study shows, war veterans and others occupied Shamva farms in a systematic and tactical manner though with some flexibility. Most of the occupiers' recollections regarding the farms which were occupied first (and why) centred on overall relations with specific white farmers and their treatment of blacks (including farm labourers), including their previous encounters with current white farmers or even previous farm owners.

By early March 2000, at least four farms were occupied in Shamva (including Robin Hood, Douglyn and Lion's Den) at least on a temporary basis, with claims that some occupiers on one farm came from Mashonaland East Province (*Zimbabwe Situation*, March 3rd 2000). The first farm to be occupied in February 2000 was Robin Hood farm, followed by Douglyn and then Lion's Den. The sequence of the occupations is aptly summarised as follows by one war veteran:

> The first farms we occupied were for David Hastings [Robin Hood]. We occupied two farms there. Then we moved to Magobo farm [Douglyn farm owned by Doug Bean] and stayed there thinking we got the farm but he was still there. Then we moved to Keith's farm [Lion's Den] and then proceeded to Mupfurudzi. From there we went to Bata and next to Francis's farm at Kanjinga. Then we moved southwards to Rutherdale, [then] proceeded to Soma [owned by Burnleigh] and then to Harmish Logan who had two farms. Our intention was just to say we have taken this land and deploy people. (war veteran Kajauta)

The war veterans in Shamva also extended their occupations to nearby Matepatepa, Mazowe and Bindura, and for two main reasons. First of all, some were of the view that no boundaries were set in terms of where people should or should not occupy. Secondly, war veterans from Shamva had to deploy elsewhere because war veterans leading occupations in these other areas were either too few or they were not creating base camps on the occupied farms (as was happening in Shamva) to ensure a firm presence on the farms. The war veterans in Shamva also sometimes communicated with war veterans in those districts without significant numbers of commercial farms, so that they could send occupiers to Shamva. At the same time, there was no clear inter-district coordination by the war veterans in Shamva.

Broadly speaking, those white farmers who were regarded as ill-treating blacks were prime targets for the occupiers. A comment from a focus group discussion highlights this:

> There were other farmers, Mr Peters and Kevin Walters, who were responsible for the deaths of a lot of comrades during the war of liberation. They once put poison in the comrades' [war veterans'] food, some died immediately after consuming the food while others died in the Mazowe area. Mr Peters and his friends went on to cut off the heads from the corpses only to hang these as trophies in their farm houses. Can you imagine, the skulls were only recovered at Mr Peters' farm when we took over the farms? (Mushambanyama Village)

David Hastings of Robin Hood farm (the first farm occupied) was seen as particularly cruel. It was believed that Hastings received his farms as a reward for his father fighting during the Second World War and this irked the war veterans because, while these white soldiers were rewarded with land, ex-guerrillas received nothing of significance after the war of liberation. Indeed, as discussed, although some ex-combatants were successfully reintegrated into urban or rural life after 1980, a substantial number slipped into deep destitution and social ostracism (Chitiyo 2000). Hastings' parents were also killed by guerrillas during the war of liberation and this made him particularly spiteful towards blacks. He served in the Rhodesian army during the war (as had many members of farmer families in Shamva). Further, it was perceived that Hastings or his wife served on the Rural District Council and that he was deeply involved in the oppositional politics of the MDC, which had made "tremendous inroads in the district" (Matondi 2001: 90) by early 2000. Because Hastings was seen to have significant influence over other local farmers, occupying his farms first was a way of neutralising this influence, while also sending a clear message to other white farmers that the war veterans were serious about land repossession.

There were other considerations which applied to specific farms, such as Lion's Den. It bordered communal areas and white farmers near communal areas were perceived as particularly hostile to communal area villagers. These farmers confiscated communal area cattle that strayed onto their farms and prohibited crop farming near their farms for fear that their crops would be contaminated. Thus, based on their encounters with particular farmers or farms, local people had deeply personal reasons for occupying a specific farm. For instance, one occupier indicated that he decided to occupy Lion's Den because the water canal there was dug by his father. His father had been taken as *chibaro* to dig the canal, so it only made sense to him (the son) that he would occupy Lion's Den and farm on land near the canal and use the water for his crops.

9.4.3 Organisation of the Farm Occupations

Farm occupations in Shamva were spearheaded and organised by local war veterans without ruling party directives, though later we address more explicitly the involvement of party and state personnel. This of course goes contrary to the views of local white farmers. For instance, Kevin Walters (a former commercial farmer from Shamva) indicated that "all levels" of government were involved in "a high[ly] organised campaign" and with "the full knowledge and sanction" of the ZANU-PF portfolio with the main purpose to break the back of the MDC (email dated to one of the authors, 18 June 2015). War veterans and ordinary occupiers in the district invariably deny this.

A central figure in the occupations in Shamva was undoubtedly the district war veteran commander, who had his office under a tree near the sub-office of the Rural District Council in Wadzanai township in Shamva town (which is located thirty kilometres from Bindura town). The commander was war veteran Manhambara, who initially was the base camp commander at Lion's Den. Manhambara describes himself as the Land Possession Commander. He was in fact the acting district war veteran chairperson as the chairperson (Ziteya) was a teacher at Chindunduma School and was away most of the time.

Manhambara, who was 44 years old in 2000, remembers a national war veterans' meeting in late 1999 (either in November or December) in Harare where they talked about the need to undo the ongoing land dispossession in the country. Other local war veterans also refer to this meeting, and note how it led to a cascading down of war veteran meetings at provincial and district levels at least within Mashonaland Central. At Shamva district level, "[w]e sat down as a district association to strategise and agree on where to start" (war veteran Muzavazi).

The command centre in Shamva was commonly referred to as the HQ (headquarters) and it is where all the decisions regarding which farms to occupy were made.

Additionally, the HQ was the receiving centre where people who were interested in occupying farms gathered before being deployed onto farms. To shed more light on this and the overall organisation of the occupations, we quote the following from a focus group discussion:

> At our HQ, this is where everyone came to register.... We had very strong and capable leadership at the HQ headed by war veterans. They are the ones who would dispatch people to different base stations scattered around the farms. As you moved down to the base stations [on each farm], we had structures made up of war veterans, women and youths. ... Those seeking land would first get to the HQ where if necessary they would spend the night there; they would then be referred to the base stations the following day. Once they got to the base stations, they would donate the food that they would have brought with them and contribute money towards the upkeep of those standing guard at the farms. Within the structures we had those responsible for stores [storing the food], cooking and security. We needed the security because we had misguided fellow blacks who in connivance with the MDC and white farmers would come to the base station to fight us. Their aim was to remove us from the farms. (Chiraramo Village)

As well, there were senior war veterans under the command of the district war veteran leader who moved between a few occupied farms each to check on progress and problems on these farms, and then to report back to the district commander. As the occupations unfolded in the district, the circulating war veteran leaders might request that some occupiers move to other farms to bolster the occupying presence there or, alternatively, people were simply requested to take part in a particular occupation to add to the size of the initial occupying force. Overall, then, there was a loose structure led by war veterans which commanded and coordinated the land occupations (including war veterans Kamoto, Makasha, Mukunga, Chando and Muhomba). The provincial war veteran commander (Muropa), who was a nurse in Madziwa communal area, at times moved around farms to check on how the occupations were going.

The process of occupying farms involved (though not always in this strict order) identifying the farm, approaching the white farmer, deploying people and setting up of bases (or base camps). War veterans identified the farms. When a farm was targeted, war veterans would go to the farm with a group of occupiers. As in the case of Bindura district, the farm owner was approached by the war veterans and asked to produce a map of the farm. This entailed some form of negotiation, whereby war veterans would seek to identify their share of the farm land. The white farmer would either refuse and resist, or surrender the map. In either case, occupiers were deployed immediately on the farm and a base camp established. The number of people deployed on a farm varied but at times reached between 150 and 200 occupiers. The deployment of people signalled that the farm was occupied. These dynamics are noted in the following:

> When we got onto farms as war veterans, we would ask for a map or other questions like how big the farm was. Our intention was not to remove the white farmers but to share the land.... So as the commander I asked the white farmers which part of the land they wanted to retain and which part they wanted to give us. When they showed us the land, we occupied the part that they wanted to retain instead of the part they wanted to give us. I also instructed

base commanders that the deployed people could use resources at the farm like water but they should remain camped outside farm houses. (war veteran Manhambara)

The presence of occupiers at the farm houses was a constant reminder to farmers and their families that the occupiers had invaded their personal and privatised space and it served to intimidate farmers (see Buckle 2001). Even if the farmer wanted to visit his work sheds or fields, let alone leave the farm to visit Shamva town for supplies, he would invariably have to pass by the occupying force.

Each and every occupied farm had a base camp with an authority and coordinating structure, which was led by a (war veteran) base camp commander or chairperson. Members of this would oversee certain tasks, such as food provision, transport and pegging of plots, as well as security and maintaining discipline. Pegging, involving the measuring and allocation of plots for the occupiers, was an important activity in laying claim to the farm and in providing occupiers a sense of permanency on the farm. But security was considered as especially crucial. War veterans knew that white farmers were armed and they were not sure how farmers would respond to constant intimidation and threats. Base commanders also stressed the need for discipline amongst occupiers, and this had two facets. Firstly, discipline entailed ensuring the ongoing commitment of occupiers to remain on the farm. Secondly, it sought to limit the possibility of occupiers engaging in violence and asset stripping of farm property, though this certainly did take place.

9.4.4   Anatomy of Occupiers

Examining the anatomy of occupiers, and their sheer diversity, brings to the fore the broad-based character of the occupations and their irreducibility to directives emanating from the party-state. In Shamva district, it is unquestionable that war veterans were the main agents in the land occupations. Most of the war veterans were from the district but others came from other districts in Mashonaland Central Province (such as Murewa and Mt Darwin) or neighbouring provinces (particularly Mashonaland East Province). One prominent war veteran, who was responsible for security in the local Shamva war veteran structure, came from Uzumba in Mashonaland East.

A number of the war veterans leading the occupations had common personal histories that dated back to the war of liberation. Although the circumstances leading them to join the war as guerrillas were different, many had met either in Zambia or Mozambique, had trained together and came to the front in Rhodesia to fight under the same unit or branch. After the ceasefire was announced, they had gone to the same Assembly Point called Dzapasi. Some were integrated into the newly formed Zimbabwe National Army, while others were demobilised and moved to different parts of the country. In later years, most came back to Shamva district to live and often in the communal areas of their birth. Therefore, it was not difficult to regroup in 2000 for purposes of occupying land in the district.

Although war veterans (mostly male) played a significant role, the occupations were not possible without other categories of occupiers whose motivations were often based seemingly on localised experiences and events. Amongst the occupiers were *mujibas* and *chimbwidos* (former war collaborators, male and female, respectively), and ordinary women, men and youth who came from diverse backgrounds. Many of these occupiers came from other parts of the province (such as Mt Darwin, Rushinga, Guruve and Uzumba-Mamba-Pfungwe), but the occupiers interviewed were mainly from communal areas in Shamva. Local war veterans regularly engaged with traditional leaders and headmen in the communal areas for purposes of mobilising villagers to occupy farms. As Chief Bushu highlighted:

War veterans approached us as community leaders and they explained why they needed to take land and why support was needed from communities. Then youths from villages were urged to join.

Many of the occupiers indicated that they learned about the occupations through these engagements.

The youth, locally referred to as *vana vedu* (our children), were certainly part of the occupiers deployed on the farms. Their roles, however, were described as supportive. These included singing, beating drums, helping with cooking and carrying items (such as pegs used to demarcate the occupiers' plots). There were sentiments that the youth did not participate in the occupations as much as they should have, likely owing to the formation and subsequent presence of MDC in the district. It is believed that the opposition party coerced young people and led them astray (and made them abandon the virtues of the land struggle as espoused by the ZANU-PF party). The very few respondents who admitted to violence perpetrated by the occupiers indicated that the youth, often carrying knobkerries and axes, were used as the "army" to intimidate the white farmers and, if there was resistance by farmers and farm workers, the youth would be centrally involved in countering this resistance.

While male war veterans commanded the occupations in Shamva, women war veterans (and women more broadly) were largely sidelined. As one male war veteran (Manhambara) clearly put it: "In terms of women, women were not really involved in leadership but were active in the occupations". Women moved around farms with war veterans and were part of the groups deployed on the farms in Shamva, and a number of women lived for extended periods on the occupied farms. Though some women, especially young unmarried women, did so in their own right, most married women did so on behalf of their spouses who were, for various reasons, not available at the time of occupations. Overall, women's work on the farms was limited and related to reproductive roles such as cooking and cleaning. They also sang and danced during the *pungwes*. Both men and women confirmed this. One women occupier said:

So as women, we were there in those groups that went around repossessing the farms. We would sing loud and clear and informed the white farmers that we were not going back in terms of taking over the farms. Our comrades [war veterans] would usually lead from the front. We would sing and dance in solidarity with them. (Focus Group Discussion, Chiraramo Village)

A male occupier (Gandidze) made the same point:

Women did what they always do and even did during the liberation struggle [in the 1970s]; they sang and cooked like they did for freedom fighters who came to the villages [during the war]. There were female war veterans but these did not get into the same positions as those male war veterans leading the occupations.

Undoubtedly, women played very specific and marginalised roles during the land occupations and—like both female guerrillas and female villagers during the war of liberation—they took no pronounced leadership role.

Three examples from Lion's Den farm speak to the experiences of women. Chihuni was twenty-five years old when the land occupations took place. She lived in faraway Kwekwe and followed her brother who had gone to Shamva to occupy a farm, ending up at Lion's Den farm with her brother. For her, it was as if someone had called "*handei kuminda*" ("let's go to the farms") and at once people left for the farms. She highlighted that women, including female war veterans, did the cooking and cleaning on the farms. A second woman (Munyoro) occupied a farm in Shamva in April 2000. She found her way to the Shamva town command centre and was deployed to Somer farm. She though asked the Somer base commander to be moved to Lion's Den nearer to Shamva town as she had a baby. In the case of the third woman (Karichi), her husband had gone to Lion's Den and had his name registered as an occupant by the farm commander. Karichi made her way to the farm only in September as she cared for her young baby, and her husband only stayed on weekends on the farm because he worked at Jiti School. There were many female occupants at Lion's Den and, contrary to the general trend, both the base commander (Musora) and deputy base commander (Chanaiwa) were women war veterans. The deputy commander spoke about shortage of land in the communal areas as the reason for joining the occupations, as she was "farming in the fields of my in-laws". Her main role at the farm was to maintain order, give instructions and work on logistical matters.

Though war veterans used various means (and notably nationalist discourses around land) to mobilise people to occupy farms, women (perhaps more than men) had their own (non-nationalist) personal reasons for occupying farms. Therefore, personal aspirations were articulated by women, even though tied indirectly to the broader historical grievance over land. To quote one woman:

We decided to join the war veterans in land occupations because my husband's father has a polygamous marriage so there is no land for farming. We have been farming on a very small piece of land. (woman occupier Karichi)

Another woman added that:

I came to the farm in Shamva in April 2000 with my two [communal area] neighbours. I came to take part in the land occupations because I was facing problems. My husband and I had no land of our own, as we were living with my parents. I did not feel okay staying on my parents' land while my husband was away working at the mine. (woman occupier Munyoro)

These points speak to the problem of landlessness in the communal areas of Shamva but also to the specific challenges which married women faced, as they do not have primary rights to land and access it through their husbands. As well, unmarried women had their own set of problems in this regard. In this regard, as James (2014) also notes for Shamva villagers, reasons and motivations for occupying land were multiple and overlapping, and perhaps best thought of as situational.

9.4.5 Occupiers and the Party-State

We have examined the motivations of Shamva occupiers based on historical and contemporary grievances, the overall organisational character of the occupations and the broad social base of the occupations. Based on this, there is no clear evidence which suggests that the Shamva occupations were orchestrated by the party-state, and not even in a mediated way (for instance, through the war veterans). The evidence presented about Shamva district (and Bindura district) shows a decentralised land occupation process in which people joined the occupations for both nationalist and non-nationalist reasons, and independently of any directive from the ruling party or state. In Shamva and Bindura, the involvement of party and state was of some significance but not in their institutionalised form. Rather, politicians and officials were acting in a personal capacity though perhaps drawing upon state resources at times in doing so. The occupiers were well aware that, by occupying land, they were acting against the rule of law. Hence, they did not necessarily expect support from the ruling party and state. From their perspective, they did not act at the behest of the ruling party or even on its behalf. In fact, war veterans claimed that party-state sought to distance itself from the occupations. We demonstrate these points in relation to the Shamva occupations.

Given the heightened mobilisation of war veterans and the intense pressure they had placed even on the office of the presidency in the late 1990s, the ZANU-PF government "knew it had reneged on the land issue" (war veteran Manhambara). However, the ruling party did not know what the war veterans intended to do about the unresolved land question. Hence, they watched the land occupations unfold from afar, until they realised that war veterans were seriously committed to recovering land. Most of the Shamva interviewees argued that the government later intervened to legitimise or formalise the process, at a time when white farmers had already been defeated and the war for land had been successfully concluded. The chief executive officer of the Rural District Council at the time of the research seemed to suggest this as well:

> During the actual occupations, the government remained aloof because it did not want to appear to be supporting invasions but when it saw people had occupied farms in large numbers it formalised the occupations. Government had a clever way of doing it – it said people have demonstrated and they cannot go back home just like that. That is when it brought the [fast track] land policy.

Nonetheless, war veterans received support from prominent local politicians, notably Border Gezi (the provincial governor of Mashonaland Central Province and the ruling party's national political commissar), and Nicholas Goche (a war veteran, as well as national minister of state security and local ZANU-PF Member of Parliament) who had owned a farm in Shamva district since the early 1980s. As well, some local state officials were involved (including from the Rural District Council).

The Shamva occupation command structure led by war veterans was assisted by ruling party politicians at times with food, vehicles, fuel (from Goche's farm) and fines (for those who were arrested on occasion by the police). On the issue of the police, a war veteran (Kajauta) argued that the Police Officer in Charge (Gopo) "actually followed people in farms and arrested them" and "so we had him transferred", which speaks to the possible influence of war veterans within the security establishment. High profile security personnel in central government also assisted with security and surveillance:

> Although they did not give us instructions, security officials – Chihuri [national police commissioner] and Shiri [air force national commander] – were also involved because it was not clear how the whites were organising their resistance. (war veteran Sori)

Further, some of the leading war veterans at district level in Shamva were local ZANU-PF elites. For instance, the vice-chairperson of Shamva's district war veterans' association (Kanengoni) was at the same time the Secretary for Security in the local ZANU-PF branch and he was responsible for transport during the occupations (in part because, by the time he had retired from the Zimbabwe National Army, he had acquired two trucks). Likewise, war veteran Kajauta was the Secretary for Administration in the local branch of the ZANU-PF party.

At a ZANU-PF rally in Shamva town in April (or May) at which Border Gezi spoke, it was reported that Gezi indicated that the farms of specific farmers in particular should be occupied and, because of this, Walwyn farm (leased by two local farmers, including a well-known MDC supporter) was subsequently occupied. At Glencairn Farm, the occupiers apparently informed the owner that Border Gezi told them that that no crops were to be grown on the farm during the coming agricultural season (*Zimbabwe Situation* August 8th 2000). The occupiers in Shamva, however, underlined that particular politicians and government officials supported them in their individual capacity, and not in any official capacity. This is attested to by local (state) agricultural extension officers who occupied farms and assisted occupiers in various ways. As one officer noted:

> The war veterans had no technical background and proper records or documentation, so they relied on people who worked in government departments and others who knew about land use to advise them on the types of farms that existed and what was being done in farms. These people helped war veterans in an independent capacity.

9.5 Conclusion

By looking "inside" the occupations in both Bindura and Shamva districts, we argue that everyday concerns, challenges and aspirations, founded in historical and contemporary experiences, intersected with a war veteran-led nationalist project around land,

and gelled into diverse localised mobilisations in occupying farms. The criticisms lodged against the ruling party and state by the war veterans in particular had been simmering for many years, and the failure to bring about meaningful land reform for two decades became interpreted as an act of betrayal. However, in activating the land occupations in the two districts (and in Zimbabwe more broadly), the war veterans did not adopt an in-principle anti-statist position. For this reason, war veteran commanders were prepared to engage in a tactical manner as the occupations progressed, including receiving support from state and ruling party personnel. The occupations had an institutional autonomy vis-à-vis the party-state, and they entailed a specific criticism of it pertaining to land dispossession and reform, but the occupations were not a direct and fundamental challenge to the state per se and its logic, modalities and imperatives. Because of this, as a general tendency, the occupations became subsumed with relative ease under the state's fast track land reform programme, though not without some contestation.

References

Buckle C (2001) African tears: the Zimbabwe land invasions. Covos Day, Johannesburg

Charumbira R (2015) Imagining a nation: history and memory in making Zimbabwe. University of Virginia Press, Charlottesville

Chitiyo T (2000) Land violence and compensation: reconceptualising Zimbabwe's land & war veterans debate. Track Two Occasional Paper, Centre for Conflict Resolution, University of Cape Town

James GD (2014) Zimbabwe's 'new' smallholders: who got land and where did they come from? Rev Afr Polit Econ 41(141):424–440

James GD (2015) Transforming rural livelihoods in Zimbabwe: experiences of fast track land reform, 2000–2012. Unpublished PhD thesis, University of Edinburgh, UK

Kufandirori JT (2015) Fast track land reform in Matepatepa commercial area, Bindura district: effects on farm workers 2000–2010. Unpublished Master's Thesis, University of Free State, South Africa

Marongwe N (2002) Conflict over land and other natural resources in Zimbabwe. ZERO Regional Environment Organisation, Harare

Masuko L (2013) Nyabira-Mazowe war veterans' association: a microcosm of the national level occupation movement. In: Moyo S, Chambati W (eds) Land and agrarian reform in Zimbabwe: beyond white-settler capitalism. African Institute for Agrarian Studies, Harare, pp 123–155

Matondi P (2001) The struggle for access to land and water resources in Zimbabwe: the case of Shamva district. Unpublished PhD thesis, Swedish University of Agricultural Sciences, Sweden

Moyo S (2000) Land reform under structural adjustment in Zimbabwe: land use change in the Mashonaland provinces. Nordiska Afrika Institutet, Uppsala

Moyo S, Yeros P (2005) Land occupations and land reform in Zimbabwe: towards the national democratic revolution. In: Moyo S, Yeros P (eds) Reclaiming the land: the resurgence of rural movements in Africa, Asia and Latin America. Zed Books, London, pp 165–205

Mukonyora I (2011) Religion, politics, and gender in Zimbabwe: The Masowe Apostles and Chimurenga religion. In: Smith J, Hackett R (eds) Displacing the state: religion and conflict in neoliberal Africa, University of Notre Dame Press, Notre Dame, IN, pp 136–150

Sadomba WZ (2008) War veterans in Zimbabwe's land occupations: complexities of a liberation movement in an African post-colonial settler society. Unpublished PhD thesis, Wageningen University, Netherlands

234 Local Fast Track Occupations: The Cases of Bindura and Shamva

Sadomba WZ (2013) A decade of Zimbabwe's land revolution: the politics of the war veteran Vanguard. In: Moyo S, Chambati W (eds) Land and agrarian reform in Zimbabwe: beyond white-settler capitalism. African Institute for Agrarian Studies, Harare, pp 79–122

Selby A (2005) Losing the plot: the strategic dismantling of white farming in Zimbabwe 2000–2005. QEH Working Paper Series—Working Paper Number 143

Sukume C (n.d.). Agrarian transformation and emerging constraints: a case study of Shamva district. Study Commissioned by Shamva District Development Association

Yeros P (2002) The political economy of civilization: peasant-workers in Zimbabwe and the neo-colonial world. Unpublished PhD thesis, University of London, UK

Chapter 10
Post-Third *Chimurenga* Land Politics and *Zvimurenga* Analysis

Abstract Overall, while fast track land reform may have "resolved" certain matters, it also left other matters unattended or even facilitated the emergence of new dilemmas (and new sites of struggle) which continue to be played out, even within post-coup Zimbabwe under the presidency of Emmerson Mnangagwa. This chapter makes no attempt to provide a comprehensive overview of the post-2000 period in Zimbabwe and the fast track programme more specifically, as there is abundant literature on this. Rather, the chapter points to certain post-2000 issues arising from the analysis of the three *zvimurenga* and in particular of the third *chimurenga*. In other words, the chapter uses the *zvimurenga* examination as a lens through which to speak about post-2000 Zimbabwe. In particular, analytically, the chapter considers the ways in which the third *chimurenga* occupations were subdued and institutionalised by way of a state-driven restructuring of land and agrarian spaces (namely, through fast track), and the possibilities and existence of further land contestations—specifically in the light of a neoliberal capital-driven process under the presidency of Mnangagwa. It does so with reference to what we label as an autonomist commoning perspective.

Keywords Fast track · Mnangagwa · Land struggles · Land politics · War veterans · Autonomist commoning

10.1 Introduction

The third *chimurenga*, understood not simply as the nation-wide land occupations but the state's fast track land reform programme as well, led to significant restructuring—including de-racialisation—of Zimbabwe's agrarian economy, without addressing the ongoing agricultural and livelihood challenges which continue to mark the communal areas. Because of the introduction of small-scale A1 farms and commercially based A2 farms under fast track, with the latter in the main allocated to political and economic elites aligned to (or within) ZANU-PF and the state, fast track also resulted in new forms of class differentiation in the countryside. Meanwhile, the patriarchal configuration of land remains largely unaddressed, despite fast track's land quotas for women. Overall, while fast track may have "resolved" certain matters, it left other nagging matters untouched and even facilitated the emergence

© The Author(s), under exclusive license to Springer Nature Switzerland AG 2021
K. Helliker et al., *Fast Track Land Occupations in Zimbabwe*,
https://doi.org/10.1007/978-3-030-66348-3_10

of new dilemmas (and new sites of struggle) which continue to play out, even within post-coup Zimbabwe under the presidency of Emmerson Mnangagwa.

In this chapter, we make no attempt to provide a comprehensive overview of the post-2000 period in Zimbabwe and the fast track programme more specifically, as there is abundant literature on this (for instance, Scoones et al. 2010; Hanlon et al. 2013). Rather, we point to certain post-2000 developments which arise from our examination of the three *zvimurenga* and the third *chimurenga* more specifically. In other words, we use our *zvimurenga* analysis as a lens through which to see and speak about post-2000 Zimbabwe. In particular, we consider the ways in which the third *chimurenga* occupations have been subdued and institutionalised by way of a state-driven restructuring of land and agrarian spaces (namely, through fast track), and the possibilities and existence of further land contestations—specifically in the light of a neoliberal capital-driven process under the presidency of Mnangagwa. We do this with reference to what we label as an autonomist commoning perspective.

10.2   Ongoing Land Contestations in the Countryside

In mid-2001, the Zimbabwean parliament passed the Rural Land Occupiers Act which indicated that any occupations after 1 March 2001 were to be considered as illegal (while, simultaneously, protecting pre-March 1 occupiers). However, in certain cases even before March 2001, occupiers were subjected to forced removal by the state. At the Makande and Southdown Estates in Chipinge, for instance, war veterans early on had led wide-ranging occupations; but, the occupiers were driven out violently by anti-riot police in late 2000 (Zamchiya 2011). In this particular case, the evictions arose because the Zimbabwean state, under fast track, had made concessions to some sections of international and domestic capital, including the banana, coffee and tea estates in Chipinge. These evictions in fact occurred after Vice-President Joice Mujuru had visited Chipinge town earlier in 2000 and told local state officials that agro-estates should not be occupied in Chipinge because they provided decent employment, schools for local children and subsidised electricity costs for town residents.

In the period since March 2001, and for a number of years, occupations continued, including around the time of the 2002 presidential election. Despite the 2001 legislation, examples abound of the state turning a blind eye to many of these occupations, though significant levels of violence (including violent farm-house takeovers) took place. In most of these instances, high-ranking war veterans, ZANU-PF members or state officials were typically staking claim to the farms. Those occupiers who were evicted post-March 2001 tended to lack political connections. This led to a diverse range of localised contestations around land (Munyuki-Hungwe 2008) and entailed unresolved and new tensions between ordinary occupiers, war veterans, the party-state and others. At the same time, there arose ever-deepening land-use conflicts on fast track farms between A1 farmers and gold panners in some parts of the country.[1]

In the case of fast track farms, corruption in land allocations became a sore point, for war veterans, occupiers and A1 farmers. Sadomba (2008: 168) for instance reveals that war veterans presented a corruption document at a Mashonaland West Provincial Stakeholder Dialogue meeting in 2004, accusing ZANU-PF officials of "changing farms willy-nilly", "leasing farms to former white farmers" and "deliberately ignoring the mandatory twenty percent allocations for war veterans". In some cases, war veterans with land fell victim seemingly to circumstances beyond their control including competing land uses such as mining. An example is the attempted eviction in 2014 of 40 war veterans from Yotam Farm in Masvingo Province after the discovery of minerals at the farm. Such actions, when successful, led to the displacement of occupiers by elites often connected to the ruling party. Likewise, Cliffe et al. (2011: 915) note that politicians have used their political muscle to dispossess recipients of fast track land, especially A1 farmers. Indeed, these forced displacements became quite common, as early as the year 2002 (Marongwe 2011). Sadomba (2008) speaks of these displacements as entailing *jambanja* on *jambanja*, involving land-grabbing by ZANU-PF connected elites as they seek to transform A1 farms into A2 farms.

Overall, A2 farmers are the new landed elite who reproduce some form of "domestic government" on their farms, though there are instances (such as in Mazowe) where A2 farmers share their farms with A1 farmers. One well-known example of elite capture of land is the case of Manzou Estate in Mazowe which is made up a number of A1 farms (Spenenken, Celtic and Arnold farms). Former first lady Grace Mugabe is alleged to have "terrorised" and attempted to evict these A1 farmers.[2] At Little England Farm in 2003 in Mashonaland West, A1 farmers resisted being removed and they had to defend themselves because state agencies such as the police did not provide them with security in the face of possible eviction. In the presence of police, they protested and accused politicians, saying:

> You are corrupt and delivering a corrupt message. How did you come up with people (selected to remain at the farm)? We have suffered enough and we are ready to be killed by the evicting policemen and soldiers. The President must know that his ministers and officials in Mashonaland West are corrupt and receiving bribes from rich people. (ZWNEWS online, 19 August 2003)

Chiweshe (2017) also shows how internal contestations and conflicts within ZANU-PF, particularly post-2010, continue to reshape land dynamics in Zimbabwe. The party expulsion in 2014 of certain senior ZANU-PF leaders including then vice-president Joice Mujuru and Didymus Mutasa, popularly referred to collectively as Gamatox (which is a chemical used in storing maize and killing weevils) shows the uncertainty of land rights on fast track farms. With the loss of political power, some

---

[1]http://www.irinnews.org/report/90651/zimbabwe-fourth-chimurenga-gold.

[2]https://nehandaradio.com/2015/01/08/villagers-mazowe-evicted-pave-way-grace-mugabe-wildlife-park/.

of the politicians who were expelled from the ruling party are finding it difficult to maintain their land interests. For example, a report in the *Herald* of 4 May 2015 notes how then Manicaland Governor Mandiitawepi Chimene (aligned to Grace Mugabe) took over Mona Agro Farm in Rusape from Didymus Mutasa. Another case involved Themba Mliswa (former ZANU-PF provincial chair in Mashonaland West) who had to obtain an order from the High Court to evict youths and war veterans who had occupied his farm.

As well, communal villagers have continued to occupy white farms (and other landholdings) not listed under the fast track programme so that, even after fast track was apparently completed, renewed demands for land have led to "fresh" occupations (as we discussed earlier, in Chapter 8, with reference to the Lowveld in Masvingo Province). For example, 300 people occupied a farm in Chipinge owned by a South African farmer under the Bilateral Investment Protection and Promotion Agreement.[3] At times, though, the occupation of the remaining white farms has been engineered by political elites. Elites have identified lucrative or viable white-owned land-based projects and they have then mobilised and organised people to occupy the properties. One case is the unsuccessful attempt to take over Kuimba Shiri bird sanctuary.[4]

In other instances, political elites have arranged for the allocation of white-owned lands to themselves (or to family members or close acquaintances) through the Ministry of Lands and Resettlement. These "new beneficiaries" have then received offer letters for the land leading to the eviction of white farmers. Examples of this include the case of Lesbury Estate where Robert Smart was evicted to pave way for amongst others a prominent clergyman, Trevor Manhanga,[5] as well as the acquisition of 22 white-owned farms in 2015.[6] In 2016, then-ZANU-PF provincial chairperson Ezra Chadzamira was allocated Crest Ibeka Farm in Masvingo which then belonged to Yvonne Goddard.[7] There are other cases which involved significant levels of litigation over eviction orders given to some white farmers in places such as Chiredzi.[8]

As well, there are mobilisations of resistance against land-grabbing for large-scale biofuel projects which reflect localised concerns about the effects of displacement on livelihoods (Mutopo and Chiweshe 2014; Chiweshe and Mutopo 2014; Mujere and Dombo 2011; Makochekanwa 2012). Like the third *chimurenga* occupations, these efforts are not centrally organised but emerge at local levels. However, in these cases, communal area (and even resettled) communities are fighting displacement

---

[3]http://www.cfuzim.org/index.php/newspaper-articles-2/sa-citizens-in-zimbabwe/698-sa-farmer-arrested-and-evicted-from-chipinge-farm.

[4]https://www.thestandard.co.zw/2011/01/23/zanu-pf-supporters-invade-lodges-bird-sanctuary/.

[5]https://www.theindependent.co.zw/2018/06/15/manhangas-farm-seized/; It is however important to note that Mr Smart got his farm back after the fall of Robert Mugabe post November 2017.

[6]https://www.theindependent.co.zw/2015/08/31/fresh-land-grabs-govt-keeps-trampling-on-property-rights/.

[7]https://www.chronicle.co.zw/masvingo-farm-take-over-sparks-row/.

[8]https://bulawayo24.com/index-id-news-sc-national-byo-157665.html.

rather than seeking to access land. These include people from Nuanetsi and Chisumbanje taking a stand against land displacement meant to pave way for massive land projects. In Chisumbanje in particular, a diverse coalition has responded to land access challenges posed by large-scale deals, notably a community-based organisation called the Platform for Youth Development. In struggling against the establishment (and extension) of an ethanol project in Chisumbanje, villagers have engaged in protests, petitions, litigation, advocacy and various other activities such as sabotage, trespassing, stealing and destroying crops and machinery.

After the overthrow of Mugabe in November 2017, the displacement of A1 farmers has continued across the country with newspaper reports noting that in, January 2019 for example, 1000 villagers at Mzaro Farm in Masvingo had been given eviction orders.[9] What is of interest in these new evictions is the use of pre-2000 language of defining the villagers as "illegal settlers".[10] At Hope Farm in Kadoma, families were evicted recently by armed police to pave way for a new settler.[11] Similar to when Mujuru was removed from ZANU-PF, the G40 faction to which the Mugabes were linked (and which lost out after the change of power in November 2017) has increasingly faced serious challenges in maintaining their land claims. For example, the former chairperson of the ZANU-PF Youth League, Kudzanayi Chipanga, was reportedly set to lose Wakefield farm to the previous white owner. His former deputy was also said to be fighting off eviction from a farm in Chipinge which has been earmarked for one of the aides of the new president (Emmerson Mnangagwa).[12] As well, Sports Minister Kirsty Coventry has been allocated a 232-hectare farm in Mashonaland West repossessed from Patrick Zhuwawo, Robert Mugabe's nephew.

Beyond this continuation of evictions and elite capture, the post-2017 period has provided fresh impetus for new demands by war veterans on Mnangagwa. Some of these demands are acknowledged by the ruling party because war veterans were instrumental in the demise of Mugabe and the rise of Mnangagwa. For instance, at the December 2018 ZANU-PF conference, the party passed the following resolution related to war veterans:

> All war veterans who have not yet benefited from land allocation should be given land. The war veterans should be given their quota promised by government as this phenomenon has not been followed. The liberation war fighters want to be exempted from land tax on the land they liberated.[13]

A few years earlier, war veterans had been marginalised from the constitution-making process which led to the new constitution in 2013, though they had shaped the conditions under which the process occurred (through for instance public protests

---

[9] https://www.newsday.co.zw/2019/02/govt-slammed-over-fresh-displacements/.

[10] https://www.herald.co.zw/mutirikwi-illegal-settlers-face-eviction/.

[11] https://nehandaradio.com/2019/03/15/zimbabwe-farm-invasion-haunts-natives-the-cruel-eviction-at-hope-farm-in-pictures/.

[12] https://www.newsday.co.zw/2018/06/chipanga-to-lose-farm-to-previous-owner/.

[13] https://www.dailynews.co.zw/articles/2018/12/20/war-vets-demand-13-5bn.

and threats).[14] War veterans and others drove *Operation Vhara Muromo* (Operation Shut Your Mouth) in which violence and coercion were used to suppress oppositional views in the constitutional outreach programmes (Dzanetsa 2012). Though, like ZANU-PF, war veterans wanted to ensure that the new constitution defended and validated the gains of the fast track programme, they also had other interests focused on their own recognition and benefits. These were in part realised by way of Section 84 of the new constitution, which refers to access to pensions and health care for veterans.[15] Post November 2017, war veterans (at least initially) became more assertive in seeking to redress their unrealised claims. Some war veterans marched to the offices of the president making demands around their welfare and upkeep.[16] In a meeting with Mnangagwa in May 2018, war veterans made numerous demands including: 20% share in the state's command agriculture scheme, mining concessions, increases in their pensions and hunting concessions.[17]

By November 2017, there was no clear and consistent land policy to guide land administration and use. Large-scale land deals provide an example of what appears to be ad hoc land arrangements. There is certainly no clarity on how the two large investments (at Nuanetsi and Chisumbanje) fit into the wider context of fast track land reform and how the promotion of such foreign-funded commercial large-scale agricultural projects fit into the anti-colonial rhetoric of the 2000s. The deals signal a clear warning of fragile security of tenure for communal farmers, just as A1 farmers have been subject to displacement—given that, in both instances, the land is owned by the state. At the same time, the ruling party continues to pay lip-service to corruption in the land allocation processes. Thus, at the ZANU-PF conference in 2018, corruption in land allocation was once again cited as spoiling the success of the land reform programme.[18] The Land Commission in Zimbabwe is currently undertaking a land audit whose mandate includes dealing with such questions, but it remains to be seen how far it will be able to provide solutions to land corruption. Other audits including the Utete Commission in 2003 were unable to provide any resolution to this question.

10.2.1   Post-coup Zimbabwe: Land, Politics and the Economy

The events leading to the rise of Mnangagwa as party and state president are still shrouded in some mystery. Despite the initial significant celebrations by ordinary

---

[14]https://bulawayo24.com/index-id-news-sc-national-byo-11086-article-zimbabwe+war+vet erans++besieged+constitution+making+body+offices+(copac).html.

[15]http://www.justice.gov.zw/imt/wp-content/uploads/2017/10/WAR-VETERANS-ACT-REV IEW-IMT-DISCUSSION-PAPER-MINISTRY-OF-WELFARE-SERVICES-FOR-WAR-VET ERANS.pdf.

[16]https://www.newsday.co.zw/2018/12/war-vets-storm-mnangagwa-office/.

[17]http://www.sapst.org/outrage-over-war-vets-demands/.

[18]https://www.herald.co.zw/corruption-double-allocations-spoil-land-reform-zanu-pf/?fbclid   = IwAR3cJ3557rLUy9faOpWnXDf2Eo-XVUYY_AHikOvkvOxYPuFvXDB9Psx8u7g.

Zimbabweans of the fall of Robert Mugabe, most were not so naïve as to believe a bright future necessarily lay ahead. The rise of Mnangagwa and the fall of Mugabe, at one level, may have entailed a mere changing of the guard within ZANU-PF. Certainly, the new cabinet showed that the system of exclusion would continue under what is termed the "New Dispensation". Simultaneously, there are indications that certain worrying changes are afoot with regard to land, at least rhetorically.

There appears to be a shift to a more neoliberal macroeconomic approach. Indeed, such an approach inspired much of the Zimbabwe Transitional Stabilisation Programme launched in September 2018, which involves stabilising the national economy and financial sector, and introducing policy and institutional reforms which translate into a private sector-led economy. This is informed by Vision 2030 of a Prosperous and Empowered Upper Middle-Income Society. Monetary policy has also been promulgated. The cabinet approved the 2019 Pre-Budget Strategy Paper which speaks for example to fiscal consolidation and competitiveness of the local export sector. Tied to this restructuring (at least as set out officially) is a renewed concern over freehold tenure, agricultural productivity and land investments. In his inaugural speech on 24 November 2017, while re-emphasising the rhetoric around the inevitability of land reform, it was curious how Mnangagwa spoke the language of productivity and optimal land use. The speech noted the following:

> I exhort beneficiaries of the Land Reform Programme to show their deservedness by demon strating a commitment to the utilisation of the land now available to them for national food security and for the recovery of our economy. They must take advantage of programmes that my Government shall continue to avail to ensure that all land is utilised optimally.[19]

These statements mirror a neoliberal concern about land as a productive asset beyond any historical and social claims to land, as embodied in fast track initially. One of the objectives of the current land audit is to reduce the land sizes for farmers that are not fully utilising their land. This focus on underutilisation of land by the new black farmers masks the fact that this is a historical issue, with Scoones arguing that, in colonial Rhodesia and postcolonial Zimbabwe, the state has always noted with concern the underutilisation of commercial agricultural land (including by white farmers).[20] The way forward may be precarious for fast track farmers and it has implications on how people organise around land in Zimbabwe.

The obsession with productivity is linked to ideas around land as an economic resource and resonates with market-led land reform programmes supported by the World Bank over the years—these ideas focus on creating or restoring private property rights and functioning rural markets around land, credit and inputs (Wolford 2007). Such an approach is supported by many scholars who argue that the security of title is central to increasing efficiency and productivity (Zikhali 2008; Deininger and Feder 1998; Van Zyl et al. 1996). Under Mnangagwa, questions around this approach of efficiency and market orientation is being openly debated and promoted.

---

[19]https://www.chronicle.co.zw/president-mnangagwas-inauguration-speech-in-full-2/.

[20]Read an excellent blog piece on this by Scoones at https://zimbabweland.wordpress.com/2017/03/06/underutilised-land-in-zimbabwe-not-a-new-problem/.

This discussion is largely made without nuance around the various types of settlement and tenure under fast track; as Scoones[21] argues, it is "particularly in the A2 areas" that "there needs to be a step-change in investment and production". The move towards increasing access to 99-year leases for A2 farmers is a way of securing tenure and allowing for the land to be a bankable asset.

As it stands, though, the question of land tenure for the black farmers now on the land remains unresolved. The constitution in Section 291 ensures that the fast track programme will not be reversed, but there is near silence on the question of security of tenure for the new landholders. Section 292 of the constitution does implore the state to take appropriate measures, including legislative measures, to give security of tenure to every person lawfully owning or occupying agricultural land. It is, however, still contested as to what the appropriate measures should include, particularly for A1 farmers.

Related to this is the leasing of land by both A1 and A2 farmers. The government's public position is that land leasing is illegal, but this contradicts the statutory instruments on land tenure. Land leasing is in fact *stricto senso* legal, provided that the lessor obtains official permission from the Minister of Agriculture, Lands and Rural Resettlement.[22] What is clear is that informal land leasing is taking place and seems to be on the rise. This includes leasing by white farmers of A2 farms (in Goromonzi for instance) and the leasing out of land by A1 farmers in areas such as Marondera. Earlier studies (such as Moyo 1995) noted that the state's criticism of land leasing was based on the claim that markets should not be used to facilitate access to communal land. However, post-fast track, land leasing has been constructed by the ruling party as counter-revolutionary, that is, selling out and indirectly giving back land to white farmers (though this has continued, with senior government officials being implicated).[23]

Under the mantra that Zimbabwe is "open for business" through re-engaging with international capital, one key feature is embracing the return of white farmers. Speaking to white people at a rally in Harare in July 2018, Mnangagwa promised a non-racial land policy which protects the land rights of everyone including white farmers.[24] The commitment also includes encouraging displaced white farmers to enter into partnerships with black farmers. In many ways, the government's heightened concern with the agricultural productivity of fast track farms has promoted a return of white farmers. Chiweshe (2017) thus highlights a court case in which a judge ruled that a new black farmer would have his offer letter for a farm withdrawn by the government for non-utilisation. In this case, the white farmer returned

---

[21] https://zimbabweland.wordpress.com/.

[22] See Land Commission Act of 2017, section 28, part a and b.

[23] https://www.newsday.co.zw/2015/01/zimbabwe-land-policy-u-turn/.

[24] https://www.iol.co.za/news/africa/mnangagwa-assures-zimbabwes-white-farmers-their-land-is-safe-16168717.

to the land. This ruling and other cases of white farmers gaining back land[25] raises important questions on how land tensions will evolve under the "new dispensation".

Related to this is the issue of compensating white farmers. The importance of this issue was also underlined in Mnangagwa's inaugural speech:

> My Government is committed to compensating those farmers from whom land was taken, in terms of the laws of the land. As we go into the future, complex issues of land tenure will have to be addressed both urgently and definitely, in order to ensure finality and closure to the ownership and management of this key resource which is central to national stability and to sustained economic recovery. We dare not prevaricate on this key issue.[26]

Negotiations in the Constitution Parliamentary Committee (COPAC)[27] had led to a compromise on land, which was that compensation for land acquired before the constitution came into effect would only be for indigenous Zimbabweans and farms owned under a government-to-government agreement. Those of foreign descent (mainly, whites) would only be compensated for improvements made to the land. Further, all land acquired post the new constitution would only be compensated for improvements (Vollan 2013). In the 2019 budget proposal, the Government of Zimbabwe set aside US$53 million for compensation of white farmers but the Commercial Farmers Union was concerned that this was not sufficient, arguing that the farmers are owed over nine billion dollars.[28] In a parliamentary debate, the Minister of Lands and Agriculture noted that the onus of paying the compensation should be, in the first instance, on the individual resettled farmers who have benefited from the land arguing that: "It makes common sense that instead of labouring the taxpayer, the person who is directly benefiting from those improvements contributes towards the compensation of the former farmers".[29]

As part of re-engagement with capital under the "new dispensation", 99-year leases were extended to white commercial farmers. White commercial farmer Rob Smart was given back his Lesbury Estates farm in Headlands in December 2017. In July 2019, Mnangagwa's government stopped the eviction of a Chipinge commercial farmer, Richard Le Vieux (an exporter of coffee and avocados to Europe) by Manicaland's provincial minister, Ellen Gwaradzimba, who wanted to give the farm to her son.[30] But the double standards are very poignant. In July 2020, Marondera white farmer David Tippett was given an eviction order to vacate 337-hectare Extent

---

[25] See the Robert Smart story here: https://www.independent.co.uk/news/world/africa/zimbabwe-white-farmer-land-back-emmerson-mnangagwa-lesbury-robert-mugabe-zanu-pf-a8124786.html.

[26] https://www.chronicle.co.zw/president-mnangagwas-inauguration-speech-in-full-2/.

[27] This was the Constitution Select Committee of the Parliament mandated with drawing up a new constitution for Zimbabwe by the Government of National Unity between 2009 and 2013. It was established in April 2009 and headed by representatives from the three major political parties including ZANU PF, MDC T and MDC N.

[28] https://www.newzimbabwe.com/white-ex-farmers-unimpressed-with-53-million-compensation-say-owed-9-billion/.

[29] https://www.fin24.com/Economy/zim-land-beneficiaries-to-compensate-white-farmers-201 81004.

[30] https://foreignpolicy.com/2019/07/31/zimbabwes-new-land-reforms-dont-go-far-enough-mug abe-mnangagwa-white-farmers/.

of Magar Farm after the (now late) agricultural Minister Perence Shiri did not act on his offer letter, in order to pave way for Chenjerai Hunzvi's sons[31] (Hunzvi was the war veterans' leader at the time of 2000 land occupations). Others facing threats in the same region are Phil Frost of Lot 1 of Subdivision 7 of Winimbi Estates and Dave Palmer's Journey's End Farm. The military has also been implicated in a wave of land grabs from both white and black farmers. In July 2018, Zimbabwe National Army's Headquarters 5 Brigade took over Victory farm owned by Reverend Moyo in Kwekwe who had been allocated the farm under fast track land reform.[32]

Then, on 29 July 2020, the Government of Zimbabwe signed a Global Compensation Deed with representatives of former white farmers—Commercial Farmers Union (CFU) of Zimbabwe, the Southern African Commercial Farmers Alliance (SACFA)—Zimbabwe and Valuation Consortium (Private) Limited (Valcon). The agreed-upon negotiated figure for compensation for the value of improvements, biological assets and land clearing costs, is US$3.5 billion. This is to be paid in three instalments over five years (a 50% deposit payable 12 months after signature of the agreement and one quarter of the balance in each subsequent year). A critical issue within the agreement is the reaffirmation by the government that the agreement does not have an obligation for compensation for the land or create any liability whatsoever. Some of the farmers like Miki Marffy who has now resettled in Zambia remained sceptical of the agreement, arguing: "If we got something, that would be wonderful … And if we don't, or they try to do a currency-conversion trick, then we will just continue to hang on to our title deeds: nothing ventured, nothing gained".[33] Financing of this agreement is based on government borrowing via the issuing of a long-term debt instrument of 30 years maturity in international capital markets, as well as seeking willing international funders. Through this agreement, the Zimbabwean state seeks to fundamentally realign land reform processes within a neoliberal context in which "rule of law" and "respect for property rights" is re-established as a way to access international debt.

In the meantime, under the neoliberal promises, local communities continue to face evictions to make way for large-scale investments. For instance, Shangaan people in the area of the Chiredzi Rural District Council are facing eviction from 600 hectares of land to allow for the establishment of a Lucerne grass project by Dendairy (a private local milk processor based in Kwekwe) (Masvingo Centre for Advocacy and Development 2020). Many Shangaan people would be facing a second eviction, having been displaced in the 1960s to facilitate the formation of Gonarezhou National Park. Further, under Statutory 188 (of 2020), villagers of Batoka in Hwange (Matabeleland North Province) were given four months to vacate their communal lands for the building of a new township to service the Batoka Hydro-power plant.

In the case of women and the new constitution, they have emerged with specific guarantees around access to agricultural land. Indeed, various constitutional clauses

---

[31] https://allafrica.com/stories/202007110134.html.

[32] https://www.theindependent.co.zw/2018/07/13/military-evicts-new-farmer-at-gunpoint/.

[33] https://www.economist.com/middle-east-and-africa/2020/08/07/zimbabwes-white-farmers-are-promised-a-speck-of-compensation.

on women's rights to property and land provide the opportunity to advocate further for gender-sensitive reforms on land and other matters, since the scope of fundamental rights to which all citizens are entitled has been broadened; at the same time, a range of women's rights are given greater prominence (Moyo 2015). For example, Section 17 1(c) of the 2013 Constitution states that "the state and all institutions and agencies of Government at every level must take practical measures to ensure that women have access to resources, including land and on the basis of equality with men". This section ensures that, in terms of land distribution going forward, women and men should have equal access to land (Women and Law in Southern Africa 2017). For instance, women's land rights cannot be arbitrarily abrogated, as the constitution declares that:

> No person may be compulsorily deprived of their property except where, (a) the deprivation is in terms of a law of general application and (b) the deprivation is necessary in the interests of defence, public safety, public order, public morality, public health or town and country planning or in order to develop or use that or any other property for a purpose beneficial to the community. (Section 71 [3a and b])

Statutory Instrument 53 of 2014 was promulgated in order to give women rights to resettled land at the death of their spouse or due to divorce. Despite this, few studies have emerged to interrogate women's access and possession of land post the 2013 Constitution. The Ministry of Lands, Agricultural and Rural Resettlement at both provincial and national levels highlights that land allocation is generally guided by the "One family, One farm" policy. Yet, in a patriarchal society like Zimbabwe, a household implies male headship, and thus male control of land. Officials within the Ministry argue that their responsibilities end with the formal allocation and registration of permits; issues beyond this are private matters aligned to personal politics within households, and thus outside of the reach of the state. Despite the constitutional provisions, there are numerous questions around the operational modalities required to ensure equal access to land between men and women in Zimbabwe. A baseline research study conducted by Women and Law in Southern Africa (2017) indicates that the implementation of the various provisions is marred by complexities and challenges.

In this book, we have sought to pay due attention to the question of land and gender, during the three *zvimurenga* and in the intervening periods. The problems facing women in relation to rights of authority over—and access to—land became particularly pronounced during the colonial period and they have been in large part left unresolved in the postcolonial period. Until now, despite official pronouncements and statutory and institutional arrangements in place, land challenges for women remain intact and they will continue to exist until a broader patriarchal restructuring of Zimbabwean society takes place. Insofar as a further (or fourth) *chimurenga* is necessary, it appears that a "shemurenga" (Essof 2013) should receive priority. Perhaps this, more so than the second and third *zvimurenga*, would capture the spirit of Mbuya Nehanda.

10.3   Autonomist Commoning Perspective of the Third *Chimurenga*

As the occupations proceeded beyond the year 2000, state personnel and the state institutionally became increasingly involved in configuring what was taking place on occupied and fast track farms. This is a marked empirical tendency which is not possible to deny, and which we recognise. Recognising this tendency, however, does not undermine the central argument contained in this book about the third *chimurenga* and the two earlier *zvimurenga* as well. More specifically, our analysis of the third *chimurenga* occupations, and of the two earlier *zvimurenga*, highlights their localised and decentralised character. In the case of the third *chimurenga*, we argue that the very emergence of the occupations is simply not reducible to the machinations of the party-state, and that local initiatives and dynamics as well as localised political contestations continued to animate the occupations even while state practices became increasingly embedded in them.

In making this claim, and thereby arguing against the dominant scholarly interpretation of the third *chimurenga* occupations, we strongly suggest that this dominant interpretation has taken Zimbabwean studies one step backward by falling within a nationalist historiography—one similar to the state's own nationalist historiography, albeit with different moral connotations. In refraining from engaging in sustained empirical research, this scholarly approach reproduces the analytical rhetoric of the ruling party (namely, that the occupations were state-led) and merely gives state action a negative moral twist. In fact, because the ruling party has at times highlighted (via its discourse) the self-initiative of occupiers in reclaiming land, its depiction of the occupations has more empirical validity than the hegemonic scholarly perspective. This is a step backward because, when considering the scholarly literature on the first and second *zvimurenga*, it is clear that there has been a discernable and important shift away from a nationalist historiography to a historiography of nationalism. It is the latter perspective which informs our understanding and interpretation of the third *chimurenga* occupations.

At the same time, to examine the occupations from the perspective of a historiography of nationalism is not to adopt any particular moral stance towards them, or to leave them bereft of any critical analysis. The same could be said of the first and second *zvimurenga*. In fact, we argue that the one scholar (Sam Moyo) who—at least implicitly—has sought to foreground a historiography of nationalism tended to overstate the progressive impulses embodied in the occupations and thereby to romanticise them. For Moyo, the occupations seem to entail a proto-revolutionary moment, which became subdued and institutionalised by the Zimbabwean state. In this sense, in terms of any regressive political tendencies swirling around the occupations, the problem is the state, or at least the manner in which the state intervened and reconfigured the occupations as a political process.

In this respect, despite deeply contrasting claims about the effects of fast track on Zimbabwe's agrarian and national economies, and about the agricultural productivity and potential of A1 farmers, both Moyo and his (nationalist historiography) critics are—for different reasons and to different degrees—critical of the state when it comes to the occupations. From Moyo's position, the state intervened in an existing occupation process. This (fast track) intrusion legitimised a progressive decentralised occupation movement, while also—simultaneously and regressively—closing down and ending the occupations. The nationalist historiography position argues differently, in that the state constituted the occupations from the very beginning, for purposes of ruling party survival, and the occupations are criticised on this very basis. In both cases, the state is the main if not the only "cause" of regressive tendencies intrinsic to the third *chimurenga* occupations.

We argue to the contrary, and highlight the regressive tendencies of the occupations independent of any state intrusions. We do this from what we call an autonomist commoning perspective. For us, the third *chimurenga* occupations entailed, at first and in part, both autonomist and commoning tendencies. However, certainly in the post-2000 period, these commoning and autonomist impulses became subordinated to, and disciplined by, the imperatives of the state. This has not involved an undifferentiated totalising subordination by the state, as it has been characterised by unevenness and incompleteness as well as—as indicated—ongoing land contestations across both space and time. However, states are threatened by any autonomist and commoning impulses emerging "from below", and act accordingly.

Autonomist theory is heterogeneous and difficult to circumscribe (Eden 2008; Marks 2012; Weeks 2011).[34] Broadly speaking, however, autonomism problematises the pursuit of centralised control and direction (via the state) as a suitable foundation for reconstructing society, and therefore it tends to question state-focused strategies of political change. The autonomism of John Holloway (Holloway 2002, 2010) in particular is an optimistic argument about the power of ordinary people—through pre-figurative politics—to transform the conditions of society away from hierarchy and power towards new horizons of dignity and egalitarianism. Drawing significantly on his views of the Zapatista movement in Chiapas (Mexico), Holloway (2002) privileges the construction of (and experimentation with) social relations which stand asymmetrical to the status quo in the arena of everyday social life. This is part of a broader analytical argument about anti-power politics or anti-politics (Cuninghame 2010; Franks 2008; Katsiaficas 2006; Newman 2010; Binford 2005).

This version of autonomism is not unlike the commoning perspective of such theorists as Linebaugh (2008) and Federici (2012). This perspective also questions the possibility and validity of state-centric transformation and stresses the importance of building alternative communities from the ground-up in a way which is not subordinate to the logics and imperatives of the state (and capital). Commoning involves reasserting and reinvigorating locally based autonomous processes and practices, and typically in an emergent manner rather than a predesigned fashion. In delineating the political progressiveness of "the commons", a distinction is made between "the commons' on the one hand, and the public and private spheres under capitalism on

---

[34]Much of the following discussion about autonomist commoning is taken from Alexander and Helliker (2016).

the other. Hence, a politics of commoning is "not mediated by the State or capital" (Thorburn 2012: 254). For instance, the public space, which is different from the (autonomous) commons, is "given from a certain [state] authority to the public under specific conditions that ultimately affirm the authority's legitimacy" (Stavrides 2012: 587). To reclaim the public, while of political relevance in an age of neoliberalism specifically, may simply entail legitimising the state and thereby reasserting its authority. Because of this, radically different forms of authority and sociality would be associated with the commons, out of which threats to state authority arise.

In this context, Linebaugh (2010: 2) refers to "the actuality of communing" (or commoning) and "communing practices", while Federici (2011: 2) speaks about the "movement for the commons". Commoning involves communities in processes-of-becoming so that "the community is developed through communing, through acts of organisation oriented toward the production of the common" (Stavrides 2012: 587). The manner in which these appear in practice would embody different forms, but they would include "self-established rules, self-determination, self-organisation and self-regulating practices particularly vis-à-vis the state and capitalist social, economic and cultural relations" (Bohm et al. 2008: 6).

This does not mean that commons-as-process exists outside, certainly in any complete sense, the presence and reach of state and capital, as commoning simultaneously entails processes and tendencies within, against and beyond the logic and hierarchies of state and capital. As Stavrides (2012: 594) claims, for example, "the realm of the common emerges in a constant confrontation with state-controlled 'authorised' public space". Likewise, Bohm et al. (2008: 10) highlight that autonomous practices intrinsic to commoning are in a "permanent and ongoing struggle" with existing structures, practices and discourses of hegemonic power. Indeed, "the commons" can be (and often is) domesticated, institutionalised, incorporated and subdued by hegemonic regimes of (state) authority, and repressed if the necessity arises. Hence, it is important not to romanticise "the commons", as "discrepancies, ambiguities, and contradictions are ... ingredients of a potential community in action" (Stavrides 2012: 591), including "politics of separation and division" (Jeffrey et al. 2012: 1254) intrinsic to commoning processes.

Central to processes of enclosure (or anti-commoning practices) under capitalism historically has been the enclosure of women under conditions of patriarchy, and specifically the almost natural connection drawn between women and social reproduction. Because of this, de Angelis (2014: 304) speaks of a "patriarchal form of commons" which simply reproduces prevailing patriarchal structures and practices. Reference is thus made to social reproduction commons in which, by trying to disentangle social life from the rhythms of state and capital, commoning must pursue "autonomous spaces from which to reclaim control over the conditions of our reproduction" (Caffentzis and Federici 2014: 101). Women and patriarchy are critical in this regard, because of the historical centrality of women in social reproduction at the household level (Mies and Bennholdt-Thomsen 1999; Dalla Costa 2008; Linebaugh 2008). Likewise, Linebaugh (2008: 141) speaks of the "feminist strengths of the commons", and Federici (2012) brings to the fore the significance of recollectivising social reproduction. A "commons" is not a "commons" if it continues to subject women to forms of enclosure (Federici 2011).

In this context, land dispossession and enclosures are not past processes rooted exclusively in the phase of so-called primitive accumulation under capitalism, and anti-enclosure (or commoning) practices and movements do not belong simply to "a distant irreversible past" (Linebaugh 2008: 131). Certainly, in the case of colonial Zimbabwe, land dispossession was an ongoing process of colonising space, land and people, rather than restricted to the early phase of massive land expropriation. Hence, it is important to highlight that "[t]hose who had been expropriated had not only a grievance but a living memory" (Linebaugh and Rediker 2000: 22) of past commoning (or at least of a pre-expropriation period of time). Both enclosure and anti-enclosure movements exist today in a multiplicity of forms, and the third *chimurenga* occupations are worthy of critical investigation in terms of an autonomous commoning moment in Zimbabwean history.

Before detailing the autonomous commoning impulses in the third *chimurenga* occupations, and the countervailing (regressive) tendencies intrinsic to the occupations, we first outline what the state did and did not do through its fast track intervention, so that it becomes clear that not all countervailing tendencies were "caused" by the state. Without doubt, the Zimbabwean state sought, successfully, to assert its hegemony over the occupied farms by transforming private freehold space into public space, thereby undermining the emerging commoning space. With these rural spaces no longer under the effective ownership and control of white farmers (or capital), the state felt compelled to restore order.

In Moyo's argument, as indicated, the situation in the year 2000 was marked by a near-revolutionary situation, with war veterans establishing embryonic centres of power on occupied farms. These local centres of authority (or base camps) established on farms, and led by war veterans, were akin to the base camps established by guerrillas in Tribal Trust Lands during the second *chimurenga*. In certain cases, occupiers vigorously defended their occupations not only against white commercial farmers but against state and ruling party intrusions as well, including years after fast track first began. Additionally, with the state (at an institutional level) initially remaining aloof as the occupations reverberated around the countryside, an autonomous land occupation "movement" gathered steam during the year 2000. As a result: "[F]ormal democratic norms and procedures were partially suspended, agrarian property rights were abrogated in a *fundamentally progressive way*, and bureaucratic hierarchy itself was suspended in the countryside" (Moyo and Yeros 2007: 105 their emphasis).

Through the fast track programme, the Zimbabwean state—according to Moyo—legitimised and protected the occupations and occupiers. But it also effectively subdued the occupations and thereby stabilised and brought order to the countryside, leading to an "interrupted revolution" (Moyo and Yeros 2007: 103). In this way, the interventions by the party-state were a contradictory two-edged sword (legitimation as progressive and co-option as regressive): the land occupation movement was both "adopted *and* co-opted" by the ruling party (Moyo and Yeros 2007: 106 their emphasis). Ultimately, and contrary to the autonomist perspective, Moyo and Yeros (2011: 93) argue for the potential of state-centric change: "[T]he war veterans did not go far enough ... within the state, to guarantee the momentum and working class character of the revolutionary situation [under fast track]". This stabilisation in the countryside did not appear evenly and immediately but, by 2003, "the institutions responsible for implementation [of fast track] appeared to have stabilised" (Moyo and Yeros 2007: 109). As Cliffe et al. (2011: 914) argue:

> Whatever had been the role of veterans or even spontaneous movements onto farms, once state control of past occupations and future [fast track land] designations was established, power over the process of redistribution of the land acquired was spread to include a mélange of other actors: party and government leaders (local and national), traditional authorities as well as veterans (sometimes operating through district land committees, sometimes self-appointed).

Moyo and Yeros (2007) note that a central feature of the apparent near-revolutionary situation was a temporary suspension of the state bureaucracy, notably through the rapid establishment of at first ad hoc land committees at district and provincial level, including war veterans, local ministry officials and chiefly authorities. They argue that, up until 2002, "the authority of local state structures tended to be subordinated to the war veterans' structures" (Moyo and Yeros 2007: 108). But, within one year—that is, by 2003—it seems a "co-optation of war veterans was more or less complete in the countryside" (Moyo and Yeros 2007: 109, 114).

In general, during the early years of fast track, there was "an ongoing attempt by the leadership of the ruling party to reign in radical elements within its ranks, and especially among lower-echelon war veterans" (Moyo and Yeros 2007: 114). The war veteran-established base camps on occupied farms were incorporated into, and subordinated to, the local structures of power on the new fast track farms, in the form of the "committees of seven". These new committees "managed, monitored, and regulated the land occupations" as fast track proceeded apace, but the power of the veterans was whittling away. Sadomba and Helliker (2010: 214) also note this:

> [A] high-level meeting on 18 December 2000 attended by Ministers, senior civil servants, provincial governors and district administrators on the one hand and representatives of war veterans on the other. The meeting was clearly held to allow the government to regain its lost authority and to subdue war veterans specifically and the land occupation movement as a whole. Government structures were strengthened and given powers to control the land allocation processes. The period that followed (2001–2005) saw an empowered elite forcing its way into the [land] movement.

On occupied farms where fast track was implemented, state intervention thus undoubtedly dissipated the energy of the occupations. Functions of land allocation and administration became centralised and placed under the purview of increasingly formalised structures (such as district and provincial land committees) on which war veterans were subjected to marginalisation. As Chaumba et al. (2003: 545) argue, "[t]he violent political demonstration element of the farm invasions during the time of *jambanja* of 2000 was to be replaced with the imposition of a particular type of order and planning, and a shift in register [by the state] from the political to the technical". On fast track farms, government disregarded and often dismantled any farm-based forms of authority initiated and set up by war veterans, and state agricultural planners and technocrats in large part undercut the power of veterans and took over the restructuring of these agrarian spaces as per fast track specifications. War veterans felt left out of the fast track process because of the intervention of technocrats and bureaucrats (Sadomba 2008).

But there were many instances in which war veterans sought to prevent any simple take over by state agencies, and at times they took the lead in demarcating plot size, location, boundaries and allocations, or at least collaborated with state technocrats in doing so (Chaumba et al. 2003). However, they were acting in the main within the confines of the state's land-use planning and reconfiguration of fast track farms. At the same time, as A1 farmers were allocated land and started to pursue farming, the occupation-based social arrangements morphed into post-settlement institutional formations focusing on agricultural production. These new formations sought to respond to the new challenges of farm settlement such as access to productive assets, water and credit (Chiweshe 2011; Murisa 2011; Mapuva 2011). Hence, forms of associational life on the farms, notably farmer groups, arose around mutual goals (Murisa 2011, 2018). Like the authority amongst occupiers during the period of *jambanja*, these new modes of civil organisation soon became marked by forms of exclusion—including along class and gender lines (Chiweshe 2011).

Added to this has been the expansion of chiefly power, as chiefs with jurisdiction in communal areas began to assert their authority (including at times ancestral authority) over the A1 farms in particular (Sinclair-Bright 2016). Chiefly control is enhanced by the reinvention of traditional authorities on the fast track farms through a form of traditionalism imposed by government to manufacture leadership structures. This new kind of traditional leader is the *sabhuku*, who is chosen by the pertinent traditional chief for every A1 scheme. This title is not hereditary, and the chief can replace or remove the *sabhuku*. It is thus open to patronage, as chiefs normally appoint people they have known for some time. Though they claim traditional authority and impose traditional rules and norms, *sabhuku* represent a new manufactured form of traditionalism imposed upon A1 farmers to facilitate the influence of traditional chiefs and by extension the central state (Chiweshe 2011). For these (and other) reasons, the authority and influence of war veterans waned overtime as new centres of authority and power emerged on the A1 farms. In this way, the imposition of the Zimbabwean state's authority in and through fast track dissipated any commoning and autonomist moments within the third *chimurenga* occupations. As Moyo and Yeros (2011: 93) argue, war veterans did not "prepare the ground for a sustained struggle" in defending their decentralised occupations against state intrusions.

While the state's fast track intervention undercut autonomist and commoning tendencies in the long term, countervailing tendencies were built into the occupations from the start. There were both autonomous commoning impulses and countervailing impulses embedded in the third *chimurenga* occupations. The latter impulses did not only emanate from outside (the state) but were inherent within the very constitution and character of the occupations. In seeking to rein in the autonomist and commoning impulses, in order to protect and shore up its public authority, the Zimbabwean state reinforced the regressive tendencies.

The occupations entailed, explicitly, a rejection of market-led land reform (which prevailed until the year 2000) and, additionally, they contributed to undermining the racially based private property regime dominating the agrarian landscape in Zimbabwe. This undermining of colonial land enclosures signifies a commoning moment in the occupations. As Linebaugh (2008: 45) argues, "[c]ommoners think first not of title deeds, but of human deeds". Further, once on the farms, the decentralised local forms of authority led by war veterans arose amongst the occupiers. Before the state became involved in the occupations in a direct and forceful manner (via fast track), the situation on the ground (on the occupied farms) had a significant autonomous character in terms of authority structures.

However, the occupations, as a "commoner"-driven land redistribution "programme", simply restructured the agrarian economy on a racial basis. Most importantly, in terms of the type of sociality and social relations constructed on the occupied farms, this anti-enclosure movement in no sense had an anti-patriarchal moment. Patriarchal relations and practices were maintained on the occupied farms, as specifically evident in the sphere of social reproduction where the supposedly naturalised link between women and social reproduction was never challenged. There were also pronounced forms of exclusion (or enclosure-like measures), with reference in particular to the tens of thousands of farm labourers who worked and lived on the occupied commercial farms and were never properly incorporated into the occupations. Hence, while women were subordinated, farm labourers were excluded.

From our perspective, then, the third *chimurenga* occupations were marked by significant internal tensions and ambiguities, including regressive tendencies irreducible to the state's interventions. On the one hand, the occupations involved an anti-enclosure movement in taking over farms characterised by the private property regime, as well as an anti-state moment in that the occupiers—in seizing land— acted against a postcolonial state which had failed to address the nagging colonial land question. On the other hand, occupiers—mainly from communal areas—failed in the most part to incorporate farm labourers into the occupation, while women were incorporated into the occupations in a subordinate and gendered way. Despite some cases to the contrary, exclusion and subordination became ingrained in the occupations and on-farm practices.

In the end, the anti-state stance of the occupiers was a conjunctural moment arising from the Zimbabwean state's land reform failures and the tension existing in the late 1990s between the party-state and war veterans. Thus, it was not an (in-principle) anti-statist stance, which meant that the war veteran leaders were prepared to engage tactically with the state during the occupations, including by way of police protection and material resources from state and ruling party personnel. If only for this reason, the occupations became easily subsumed under the state's fast track land reform programme. Because of this, the autonomist character of the occupations is questionable.

The occupations may have had an institutional autonomy vis-à-vis the state but they did not go against the fundamental manner in which the state institutes and reproduces modalities of hierarchy, exclusion and subordination. While local forms of authority and solidarity existed at the base camps on the occupied farms, there was no real attempt to bring about a new kind of sociality in terms of everyday practices, and certainly less so than what happened in many liberated zones during the second *chimurenga*. Of course, even if the occupiers were inclined to undo existing practices and initiate new ones, it is likely that there was simply no time to do so, given the speed at which the state moved in through fast track. But the absence of any visible signs of reconfiguring everyday practices questions likewise the commoning character of the occupations.

10.4 Conclusion

It is now twenty years since the third *chimurenga* occupations began, yet in many ways our understanding of the occupations remains incomplete. The volume of scholarly literature about the second *chimurenga*, by the year 2000, far exceeds the current literature on the third *chimurenga* occupations. In fact, since 2000, more has been written about the war of liberation in the 1970s than about the occupations. However, it is clear that the effects of the occupations were substantial in terms of restructuring the political economy of Zimbabwe. Even the war of liberation, in large part because of the form of political settlement at independence and the post-independence trajectory of the Zimbabwe state, was unable to undo what the third *chimurenga* occupations accomplished—the end of the hegemony of white commercial farmers and farming. In this context, we encourage other scholars (including young scholars) to consider the importance of fieldwork-based research on the occupations, as memories of the occupations continue to fade or become blurred.

One of the central components of at least some of the second *chimurenga* literature, in particular those studies involving localised studies, is the richness of their agrarian histories. In other words, understanding the convoluted complexities and dynamics of locally based guerrilla war in the 1970s entailed historical contextualisation. Any localised studies of specific third *chimurenga* occupations should do likewise, but normally historical depth is missing because the central focus is fast track farming and not fast track occupations. In this particular study of the third *chimurenga* occupations, we sought to highlight the spatial diversity of the occupations, without necessarily offering the necessary historical context to these occupations. In lieu of this, we chose to adopt a broad historical-comparative approach to the third *chimurenga* occupations by way of locating them historically in relation to the earlier *zvimurenga* and intervening periods of agrarian history. In doing so, we hope that we have given the reader a sense of the specificities of the third *chimurenga* occupations while also setting out important themes and debates pertaining to the earlier *zvimurenga* which resonate with the fast track occupations.

References

Alexander T, Helliker K (2016) A feminist perspective on autonomism and commoning, with reference to Zimbabwe. J Contemp Afr Stud 34(3):404–418

Binford L (2005) Holloway's Marxism. Hist Mater 14(3):251–263

Bohm S, Dinerstein A, Spicer A (2008) (Im)possibilities of autonomy: social movements in and beyond capital, the state and development. Working Paper No. WP 08/13, School of Accounting, Finance and Management, University of Essex, November 2008

Caffentzis G, Federici S (2014) Commons against and beyond capitalism. Community Dev J 49(s1):i92–i105

Chaumba J, Scoones I, Wolmer W (2003) From jambanja to planning: the reassertion of technocracy in land reform in south-eastern Zimbabwe? J Mod Afr Stud 41(4):533–554

Chiweshe MK (2011) Farm level institutions in emergent communities in post-fast track Zimbabwe: case of Mazowe district. PhD thesis, Rhodes University, South Africa

Chiweshe MK (2017) Zimbabwe's land question in the context of large-scale land based investments. Geogr Res Forum 37:13–36

Chiweshe MK, Mutopo P (2014) National and international actors in the orchestration of land deals in Zimbabwe: what is in it for smallholder farmers? In: Mihyo P (ed) International land deals in east and southern Africa. OSSREA, Addis Ababa, pp 243–274

Cliffe L, Alexander J, Cousins B, Gaidzanwa R (2011) An overview of fast track land reform in Zimbabwe: editorial introduction. J Peasant Stud 38(5):907–938

Cuninghame P (2010) Autonomism as global social movement. Work USA: J Labor Soc 13(4):451–464

Dalla Costa GF (2008) The work of love: unpaid housework, poverty and sexual violence at the dawn of the 21st century. Autonomedia, New York

De Angelis M (2014) Social revolution and the commons. South Atl Q 113(2):299–311

Deininger K, Feder G (1998) Land institutions and land markets. Policy Research Working Paper No. 2014, The World Bank Development Research Group

Dzanetsa GA (2012) Zimbabwe's constitutional reform process: challenges and prospects. Institute for Justice and Reconciliation in Africa. http://ijr.org.za/home/wp-content/uploads/2017/05/IJR-Zimbabwe-Constitutional-Reform-OP-WEB.pdf

Eden D (2008) Against, outside & beyond: the perspective of autonomy in the 21st century. Unpublished PhD thesis, The Australian National University, Australia

Essof S (2013) Harare Shemurenga: The Zimbabwe women's movement, 1995–2000. Weaver Press, Harare

Federici S (2011) Feminism, finance and the future of #occupy (Interview with Max Haiven). Communications. www.libcom.org/library/feminism-finance-future-occupy-interview-silvia-federici

Federici S (2012) Revolution at point zero. PM Press, Oakland

Franks B (2008) Postanarchism and Meta-ethics. Anarch Stud 16(2):135–153

Hanlon J, Manjengwa J, Smart T (2013) Zimbabwe takes back its land. Kumarian Press, Sterling

Holloway J (2002) Change the world without taking power. Pluto Press, London

Holloway J (2010) Crack capitalism. Pluto Press, London

Jeffrey A, McFarlane C, Vasudevan A (2012) Rethinking enclosure: space, subjectivity and the commons. Antipode 44(4):1247–1267

Katsiaficas G (2006) Subversion of politics: European autonomous social movements and the decolonization of everyday life. AK Press, London

Linebaugh P (2008) The Magna Carta Manifesto. University of California Press, London

Linebaugh P (2010) Meandering on the semantical-historical paths of communism and commons. The Commoner:1–17

Linebaugh P, Rediker M (2000) The many-headed hydra. Beacon Press, Boston

Makochekanwa A (2012) Foreign land acquisitions in Africa—an analysis of the impacts of individual land deals on local communities. Presented at the World Bank conference on land and Poverty, 23–26 April 2012

Mapuva J (2011) Enhancing local governance through local initiatives: residents' associations in Zimbabwe. Afr J Hist Cult 3(1):1–12

Marks B (2012) Autonomist Marxist theory and practice in the current crisis. ACME: Int J CritAl Geogr 11(3):467–491

Marongwe N (2011) Who was allocated Fast Track land, and what did they do with it? selection of A2 farmers in Goromonzi district, Zimbabwe and its impacts on agricultural production. J Peasant Stud 38(5):1069–1092

Masvingo Centre for Advocacy and Development (2020) Enquiry into evictions of Shangaan people in Chiredzi Rural District Council. MACRAD, Masvingo

Mies M, Bennholdt-Thomsen V (1999) The subsistence perspective: beyond the globalised economy. Zed Books, London and New York

Moyo H (2015) Pastoral care in the healing of moral injury: a case of the Zimbabwe National Liberation War Veterans. HTS Teol Stud/Theol Stud 71(2):1–11

Moyo S (1995) The land question in Zimbabwe. Sapes Trust, Harare

Moyo S, Yeros P (2007) The radicalised state: Zimbabwe's interrupted revolution. Rev Afr Polit Econ 34(111):103–121

Moyo S, Yeros P (2011) After Zimbabwe: state, nation and region in Africa. In: Moyo S, Yeros P (eds) Reclaiming the nation: the return of the national question in Africa, Asia and Latin America. Pluto Press, London, pp 78–102

Mujere J, Dombo S (2011) Large scale investment projects and land grabs in Zimbabwe: the case of Nuanetsi Ranch Bio-Diesel Project. Paper presented at the international conference on global land grabbing, University of Sussex, Brighton, United Kingdom, 6–8 April 2011

Munyuki-Hungwe MN (2008) Challenges in constructing social space in newly resettled areas in Mazowe: empirical evidence from Mazowe. Land and Livelihoods Programme Working Paper, Centre for Rural Development

Murisa T (2011) Local farmer groups and collective action within fast track land reform in Zimbabwe. J Peasant Stud 38(5):1145–1166

Murisa T (2018) Land, populism and rural politics in Zimbabwe. ERPI 2018 International Conference Authoritarian Populism and the Rural World Conference Paper No. 51

Mutopo P, Chiweshe MK (2014) Water resources and biofuel production after the fast-track land reform in Zimbabwe. Afr Identities 12(1):124–138

Newman S (2010) The politics of postanarchism. Edinburgh University Press, Edinburgh

Sadomba WZ (2008) War veterans in Zimbabwe's land occupations: complexities of a liberation movement in an African post-colonial settler society. Unpublished PhD thesis, Wageningen University, Netherlands

Sadomba W, Helliker K (2010) Transcending objectifications and dualisms: farm workers and civil society in contemporary Zimbabwe. J Asian Afr Stud 45(2):209–225

Scoones I, Marongwe N, Mavedzenge B, Mahenehene J, Murimbarimba F, Sukume C (2010) Zimbabwe's land reform: myths and realities. Weaver Press, Harare

Sinclair-Bright L (2016) This land: politics, authority and morality after land reform in Zimbabwe. Unpublished PhD thesis, University of Edinburgh, UK

Stavrides S (2012) Squares in movement. South Atl Q 111(3):585–596

Thorburn E (2012) A common assembly: multitude, assemblies, and a new politics of the common. Interface: J Soc MovS 4(2):254–279

Van Zyl J, Kirsten J, Binswanger H (eds) (1996) Agricultural land reform in South Africa: policies, markets, and mechanisms. Oxford University Press, Cape Town

Vollan K (2013) The constitutional history and the 2013 referendum of Zimbabwe: Nordem special report. Norwegian Centre for Human Rights, Oslo

Weeks K (2011) The problem with work: feminism, marxism, antiwork politics and postwork imaginaries. Duke University Pres, Durham and London

Wolford W (2007) Land reform in the time of neoliberalism: a many-splendored thing. Antipode 39(3):550–570

Women and Law in Southern Africa (2017) Women and access to land post-2013 constitution in Zimbabwe. WLSA, Harare

Zamchiya P (2011) A synopsis of land and agrarian change in Chipinge district, Zimbabwe. J Peasant Stud 38(5):1093–1122

Zikhali P (2008) Fast track land reform and agricultural productivity in Zimbabwe. Working Papers in Economics 322, University of Gothenburg, Department of Economics

Glossary

A1: a farm-plot held under a permit allocated according to the Model A1 scheme of the fast track land reform programme. An A1 farm consists of a number of A1 farm-plots, with at least three hectares of cropping per homestead and shared grazing land. Al farm-plots are occupied by small-scale farmers. Most of the A1 farmers originated from communal areas. The main purpose of the A1 scheme has been to decrease land pressure in the communal areas as well as to provide land based assets to the rural poor.

A2: an entire farm or a part of a farm held under a ninety-nine-year lease allocated according to the Model A2 scheme of the fast track land reform programme. An A2 farm consists of individual plots of land that are classified as small-, medium- and large-scale commercial schemes. Critics of fast track highlight that most of these farms were allocated to ruling party elites and high-ranking members of the security forces.

BIPPA farms: farms owned under Bilateral Investment Protection and Promotion Agreements (BIPPAs) between the Zimbabwean government and foreign governments. Under the fast track land reform programme, some of these farms were acquired for resettlement. In terms of Section 295 of the Constitution of Zimbabwe, farmers with BIPPAs, as well as indigenous farmers, must be compensated for land taken over for resettlement.

British South Africa Company: mercantile company incorporated on 29 October 1889 by a royal charter given by Lord Salisbury, the British prime minister, to Cecil Rhodes with a mandate to annex and then administer territory in south-central Africa, to act as a police force, and to develop settlements for European settlers. It governed and administered colonial Rhodesia until 1923, when white settlers received limited self-government under direct British authority.

Chimurenga: Shona word literally meaning revolutionary struggle or uprising. *Zvimurenga* is the plural of *chimurenga*.

Commercial Farmers Union: white-dominated farmers' organisation that represents and advances the interests of mainly white commercial farmers in

© The Editor(s) (if applicable) and The Author(s), under exclusive license to Springer Nature Switzerland AG 2021

K. Helliker et al., *Fast Track Land Occupations in Zimbabwe*, https://doi.org/10.1007/978-3-030-66348-3

Zimbabwe. It was founded in 1942 (originally as the Rhodesian National Farmers Union). It continues to exist, although in a much-depleted capacity.

Communal areas: areas governed under a so-called communal land tenure system. Under colonialism, these areas were first known as Native Reserves and then Tribal Trust Lands. Communal areas cover 42% of Zimbabwe's land area, where approximately 66% of the country's population resides. According to the Communal Lands Act, all communal land is vested in the state president who has powers to permit its occupation and utilisation. Traditional authorities (namely, chiefs), however, have a role in administering communal land. Officially, communal area land is not subject to market transactions.

Fast Track Land Reform Programme: announced on 15 July 2000 by then Vice President Joseph Msika, the Zimbabwean government sought to legitimise and bring order to the nation-wide land occupation movement through this resettlement programme. Formal state structures were set up to allocate farms and farm-plots to settlers under the A1 and A2 models. In its policy documents, the Zimbabwean government defined "fast track" as an accelerated programme in that adequate farm infrastructure and public services would be provided by government only subsequent to resettlement.

Freehold tenure: all land held by or under the authority of a title including a private individual or institution, or at times a state entity. This tenure system permits the almost unrestricted use and exchange of land. White commercial farms were held under freehold tenure.

Gukurahundi: a series of massacres of Ndebele civilians carried out by the Zimbabwe National Army from early 1983 to late 1987. It derives from a Shona language term which loosely translates as "the early rain which washes away the chaff before the spring rains". It resulted in the death of an estimated 20,000 Ndebele people.

Jambanja: a Shona term that literally means violence. It is used frequently to describe the period of supposed chaos and disorder that characterised the nation-wide land occupations led by war veterans. As well, it has been used to refer to politically-instigated violence, or the "taking of force" of land or property.

Land occupations: a contested term used to describe sporadic movements onto white farms from 1980 to the year 1999 by communal area villagers and then for the massive movement onto farms from the year 2000, as led by war veterans. Other scholars have preferred to use terms such as land invasions, land squatting or land grabbing.

Movement for Democratic Change: a political party formed in September 1999 and led originally by trade unionist Morgan Tsvangirai. It began as a coalition of workers, urban civic groups, youth and other urban discontents, and has contested both parliamentary and presidential elections since the year 2000. Over the years it has experienced many splits. It claims that most of the elections have been stolen by the ruling party.

Old resettlement areas: farms resettled in the early post-independence land reforms in Zimbabwe, based on different settlement and agricultural models. About 76,000 people benefited from resettlement on about 3.6 million hectares of land. Compared to the fast track programme, this early programme was insignificant.

Traditional leaders: persons (typically, male) recognised by the state as having some level of ancestral and customary claim to authority with reference to a particular communal area, though these claims are often contested by others. First existing in pre-colonial times, the colonial state reconstructed the chieftainship system for purposes of colonial rule. After some initial hesitancy, the Zimbabwean state now recognises the legitimacy of the chieftainship system and likewise tries to subordinate chiefs to its rule.

Spirit medium: known in Shona as *svikiro*. Spirit mediums are vessels through which ancestral spirits speak to and guide the living. Some of the most influential mediums in Zimbabwe include Nehanda and Kaguvi, who played an important role in the first *chimurenga*, while mediums were also central to the second *chimurenga*. Like chiefs, spirit mediums are associated with the communal areas, or former Native Reserves/Tribal Trust Lands. Unlike chiefs, they are not part officially of the Zimbabwean state's governance of communal areas.

War Veterans: ex-combatants (or ex-guerrillas) who fought in the war of liberation (second *chimurenga*) during the 1970s in Zimbabwe. They played a major role in the nation-wide land occupations from the year 2000.

ZANU-PF: The Zimbabwe African National Union—Patriotic Front (ZANU-PF) is the ruling party of Zimbabwe since independence in 1980. The party was led for many years by Robert Mugabe, first as Prime Minister and then as President from 1987 after the merger with the Zimbabwe African People's Union (ZAPU) and the establishment of an executive president.

Printed in the United States
by Baker & Taylor Publisher Services

Printed in the United States
by Baker & Taylor Publisher Services